Knowledge-Intensive Entrepreneurship and Innovation Systems

This book examines entrepreneurship from three interrelated perspectives. First, it links entrepreneurship to *innovation* and to the generation, transformation and use of *knowledge*. Second, it inserts entrepreneurship in *innovation systems* of various types – national, sectoral and local. Third, it views entrepreneurship not as a single event but as a *process* that evolves in time, from the pre-entry experience, to the entrepreneurial act, to the evolution of the entrepreneur and the new company.

Knowledge-Intensive Entrepreneurship and Innovation Systems is divided into four key sections. Part I presents the main themes and results of the book. Part II covers the main theoretical and conceptual foundations of knowledge-intensive entrepreneurship in innovation systems. In Part III knowledge-intensive entrepreneurship and innovation systems are examined empirically in Western Europe and in Central and Eastern Europe. In the final part of the book some policy implications on knowledge-intensive entrepreneurship in innovation systems are derived.

With chapters from a range of international contributors, the book answers questions such as; what are the main dimensions of knowledge-intensive entrepreneurship? What are the factors affecting its emergence, evolution and performance? How important is knowledge-intensive entrepreneurship for European growth and competitiveness? Is the situation of Central and Eastern Europe, engaged in a process of major economic and institutional transformation, similar or different from the one of Western Europe?

The book will be of particular interest to postgraduate students and researchers in entrepreneurship, innovation and industrial dynamics, as well as to public policy officials and entrepreneurs.

Franco Malerba (PhD Yale) is Professor of Industrial Economics at Bocconi University and Director of KITeS-Cespri, at the same university.

Routledge studies in global competition

Edited by John Cantwell

University of Reading, UK

and

David Mowery

University of California, Berkeley, USA

1 **Japanese Firms in Europe**
Edited by Frédérique Sachwald

2 **Technological Innovation, Multinational Corporations and New International Competitiveness**
The case of intermediate countries
Edited by José Molero

3 **Global Competition and the Labour Market**
Nigel Driffield

4 **The Source of Capital Goods Innovation**
The role of user firms in Japan and Korea
Kong-Rae Lee

5 **Climates of Global Competition**
Maria Bengtsson

6 **Multinational Enterprises and Technological Spillovers**
Tommaso Perez

7 **Governance of International Strategic Alliances**
Technology and transaction costs
Joanne E. Oxley

8 **Strategy in Emerging Markets**
Telecommunications establishments in Europe
Anders Pehrsson

9 **Going Multinational**
The Korean experience of direct investment
Edited by Frédérique Sachwald

10 **Multinational Firms and Impacts on Employment, Trade and Technology**
New perspectives for a new century
Edited by Robert E. Lipsey and Jean-Louis Mucchielli

11 **Multinational Firms**
The global–local dilemma
Edited by John H. Dunning and Jean-Louis Mucchielli

12 **MIT and the Rise of Entrepreneurial Science**
Henry Etzkowitz

13 **Technological Resources and the Logic of Corporate Diversification**
Brian Silverman

14 **The Economics of Innovation, New Technologies and Structural Change**
Cristiano Antonelli

15 **European Union Direct Investment in China**
Characteristics, challenges and perspectives
Daniel Van Den Bulcke, Haiyan Zhang and Maria do Céu Esteves

16 **Biotechnology in Comparative Perspective**
Edited by Gerhard Fuchs

17 **Technological Change and Economic Performance**
Albert L. Link and Donald S. Siegel

18 **Multinational Corporations and European Regional Systems of Innovation**
John Cantwell and Simona Iammarino

19 **Knowledge and Innovation in Regional Industry**
An entrepreneurial coalition
Roel Rutten

20 **Local Industrial Clusters**
Existence, emergence and evolution
Thomas Brenner

21 **The Emerging Industrial Structure of the Wider Europe**
Edited by Francis McGowen, Slavo Radosevic and Nick Von Tunzelmann

22 **Entrepreneurship**
A new perspective
Thomas Grebel

23 **Evaluating Public Research Institutions**
The U.S. Advanced Technology Program's Intramural Research Initiative
Albert N. Link and John T. Scott

24 **Location and Competition**
Edited by Steven Brakman and Harry Garretsen

25 **Entrepreneurship and Dynamics in the Knowledge Economy**
Edited by Charlie Karlsson, Börje Johansson and Roger R. Stough

26 **Evolution and Design of Institutions**
Edited by Christian Schubert and Georg von Wangenheim

27 **The Changing Economic Geography of Globalization**
Reinventing space
Edited by Giovanna Vertova

28 **Economics of the Firm**
Analysis, evolution and history
Edited by Michael Dietrich

29 **Innovation, Technology and Hypercompetition**
Hans Gottinger

30 **Mergers and Acquisitions in Asia**
A global perspective
Roger Y.W. Tang and Ali M. Metwalli

31 Competitiveness of New Industries
Institutional framework and learning in information technology in Japan, the U.S. and Germany
Edited Cornelia Storz and Andreas Moerke

32 Entry and Post-Entry Performance of Newborn Firms
Marco Vivarelli

33 Changes in Regional Firm Founding Activities
A theoretical explanation and empirical evidence
Dirk Fornahl

34 Risk Appraisal and Venture Capital in High Technology New Ventures
Gavin C. Reid and Julia A. Smith

35 Competing for Knowledge
Creating, connecting and growing
Robert Huggins and Hiro Izushi

36 Corporate Governance, Finance and the Technological Advantage of Nations
Andrew Tylecote and Francesca Visintin

37 Dynamic Capabilities Between Firm Organisation and Local Systems of Production
Edited by Riccardo Leoncini and Sandro Montresor

38 Localised Technological Change
Towards the economics of complexity
Cristiano Antonelli

39 Knowledge Economies
Innovation, organization and location
Wilfred Dolfsma

40 Governance and Innovation
Maria Brouwer

41 Public Policy for Regional Development
Edited by Jorge Martinez-Vazquez and François Vaillancourt

42 Evolutionary Economic Geography
Location of production and the European Union
Miroslav Jovanovic

43 Broadband Economics
Lessons from Japan
Takanori Ida

44 Targeting Regional Economic Development
Edited by Stephan J. Goetz, Steven C. Deller and Thomas R. Harris

45 Innovation, Knowledge and Power in Organizations
Theodora Asimakou

46 Creativity, Innovation and the Cultural Economy
Edited by Andy C. Pratt and Paul Jeffcutt

47 Co-opetition Strategy
Giovanni Battista Dagnino and Elena Rocco

48 Knowledge-Intensive Entrepreneurship and Innovation Systems
Evidence from Europe
Edited by Franco Malerba

Knowledge-Intensive Entrepreneurship and Innovation Systems

Evidence from Europe

Edited by Franco Malerba

Routledge
Taylor & Francis Group
LONDON AND NEW YORK

First published 2010
by Routledge

4 Park Square, Milton Park, Abingdon, Oxon OX14 4RN
605 Third Avenue, New York, NY 10017

Routledge is an imprint of the Taylor & Francis Group, an informa business

Typeset in Times by Wearset Ltd, Boldon, Tyne and Wear

British Library Cataloguing in Publication Data
A catalogue record for this book is available from the British Library

Library of Congress Cataloging in Publication Data
A catalog record for this book has been requested

ISBN13: 978-0-415-55791-7 (hbk)
ISBN13: 978-0-203-85740-3 (ebk)

In memory of Angelo Pagani,
my uncle and my first academic mentor on entrepreneurship
FM

Contents

Figures

Tables

Contributors

David B. Audretsch is one of the Directors of the Max Planck Institute of Economics in Jena, Germany.

Kate Bishop is a research officer at NIESR. Her research interests span applied labour economics, innovation, and business strategy within the knowledge-based economy. Currently, Kate is engaged on an ESRC-funded project examining skills gaps and training provision across five sectors within the UK. Previously, Kate worked on the *Innovation and Productivity Grand Challenge* at Imperial College, London. This work included research on the determinants and outcomes of university–industry collaboration. Kate's other academic projects have focused on knowledge-based entrepreneurship and networks in the UK, the Czech Republic and Hungary.

Stefano Breschi is Associate Professor of Industrial Economics and Director of the Master of Science Programme in Economics and Management of Innovation and Technology at Università Commerciale L. Bocconi, Milan.

Johan Brink is Researcher, Institute of Innovation and Entrepreneurship, University of Gothenburg, Gothenburg.

Guido Buenstorf is a research group leader at the Max Planck Institute of Economics in Jena, Germany. He studied Economics and Political Science at the University of Freiburg and the University of Massachusetts, and has a PhD in Economics from the Friedrich Schiller University in Jena. His main fields of research are industry evolution, entrepreneurship and the economics of science and technology.

Roberto Camerani is a DPhil Candidate and an Associate Tutor at SPRU, Science and Technology Policy Research, University of Sussex.

Patrick Cohendet is Professor of Economics at the University of Strasbourg, France. He is presently Visiting Professor at HEC Montréal Business School in Canada. His research interests include economics of innovation, technology management, knowledge management, theory of the firm and economics of creativity.

Chiara Franzoni is Research Associate at the Politecnico di Torino and research fellow at BRICK (Bureau of Research on Innovation, Complexity and Knowledge), Collegio Carlo Alberto. Prior to joining the Politecnico di Torino, she was post-doc at CNR (Consiglio Nazionale delle Ricerche) and visiting post-doc at Andrew Young School of Policy Studies. Franzoni's research expertise is in the area of knowledge diffusion and transfer mechanisms, financing of science and academia, and patents and Intellectual Property Rights policies, with a strong focus on academic patenting. Her work has appeared in *Research Policy*, *Industrial and Corporate Change*, *IEEE Transactions on Engineering Management* and *Scientometrics*.

Manuel Mira Godinho is Professor of Economics at ISEG, the Economics and Management School of the Technical University of Lisbon and is member of UECE/ISEG, a research unit on complexity and economics. He is the coordinator of a Master's in Economics and Management of S&T and Innovation and of a Postgraduate Programme in Economics and Management of Industrial Property. His current research interests include entrepreneurship and industry dynamics, intellectual property, catching up and technological change, and the evaluation of innovation systems.

Max Keilbach is Research Scholar at Max Planck Institute of Economics Development and Entrepreneurship, Growth and Public Policy.

Camilla Lenzi is Assistant Professor of Applied Economics at Politecnico of Milan. She joined KITeS as a post-doctoral research fellow in 2005. She received a PhD in Economics from the University of Pavia in 2006 and MSc in Industry and Innovation Analysis from SPRU, University of Sussex. Her main research interests are in the field of entrepreneurship, skilled labour mobility, geography of innovation and social network analysis.

Francesco Lissoni is Associate Professor of Applied Economics at the University of Brescia and Senior Fellow of KITeS, the centre for Knowledge, Internationalization and Technology Studies of Bocconi University, Milan. He is a member of the DRUID scientific committee and the former director of ESSID, the European Summer School of Industrial Dynamics. Most recently, he has been nominated chairman of the research programme on *Academic Patenting in Europe*, funded by the European Science Foundation.

Patrick Llerena has been Professor in Economics since 1988; CEO of the Foundation University of Strasbourg, former director of the Bureau d'Economie Théorique et Appliquée (CNRS UMR no. 7522), University of Strasbourg, France. He has published in the following fields: innovation economics; economics of science; theories of the firm and of organization; scientific and innovation policies, and decision theory under uncertainty. He is also coordinator of a Network of Excellence funded by the EU: DIME 'Dynamics of Institutions and Markets in Europe'; and of the LEA (Laboratoire Européen Associé) 'Knowledge, Science and Innovation', a joint research programme between ISI, Karlsruhe; MPE, Iéna and BETA, Strasbourg.

Maureen McKelvey is Professor at University of Gothenburg School of Business, Law and Economics Department of Business Administration.

Franco Malerba is Professor of Industrial Economics at Bocconi University and Director of KITeS-Cespri, at the same university. He holds a PhD in Economics from Yale University. He has previously been President of the International Schumpeter Society and of EARIE (European Association of Research in Industrial Economics). He is editor of the journal *Industrial and Corporate Change*, advisory editor of the journals *Research Policy* and *Journal of Evolutionary Economics* and member of the Advisory Board of the International *Journal of Industrial Organization.*

Ricardo Mamede is Assistant Professor at ISCTE – Lisbon University Institute, and is presently heading the research department of the Portuguese National Strategic Reference Framework Observatory. His main research interests are in the fields of structural change, industry dynamics and public policy.

Maria Luisa Mancusi is Assistant Professor of Applied Economics at Bocconi University, Milan, and research fellow at KITeS-Cespri (Bocconi University, Milan).

Daniel Mota graduated in Applied Mathematics and works presently as research consultant. He worked at the Portuguese National Statistics Institute for ten years, where he developed his interests on applied econometrics. His research interests are focused on microeconometrics, especially in the field of the labour market. Recently, he has been working on stated choice and revealed preference models.

Slavo Radosevic is Professor of Industry and Innovation Studies at School of Slavonic and East European Studies, University College London. His research interests are in the area of science, technology, industry, innovation and growth in the countries of Central and Eastern Europe and he continues to be involved in international projects in this area. He is acting as an expert for the various DGs of the European Commission and as consultant to the UN Economic Commission for Europe, World Bank, Asian Development Bank and UNESCO.

Maja Savic is a National Endowment for Science Technology and Arts (NESTA) PhD/CASE Fellow working on a three-year project that looks at location, knowledge sourcing and internationalization practices of Knowledge Intensive Business Service SMEs in the UK. Maja has an MSc in European Political Economy-Transition from the London School of Economics and a BA in Management from Birkbeck College, University of London.

Nicholas Vonortas is Professor of Economics and International Affairs at The George Washington University, Director, Center for International Science and Technology Policy, Director, International Science & Technology Policy Program.

Richard Woodward has been a researcher for the Warsaw-based think tank CASE – the Center for Social and Economic Research – since 1994 and Lecturer in International Business at the University of Edinburgh Business School since 2005. He studied Economics at the Pennsylvania State University in the United States and the University of Łódź in Poland and has authored numerous publications on the socio-economic transformation of the post-communist countries of Central and Eastern Europe. Most of his research has dealt with privatization and the relationships between enterprise networks and innovation.

Preface

Franco Malerba

This book examines entrepreneurship from three interrelated perspectives. First, it links entrepreneurship to *innovation* and to the generation, transformation and use of *knowledge*. Second, it inserts entrepreneurship in *innovation systems* of various types – national, sectoral and local. Third, it views entrepreneurship not as a single event but as a *process* that evolves in time, from the pre-entry experience, to the entrepreneurial act, to the evolution of the entrepreneur and the new company.

The book aims to answer the following questions. What are the main dimensions of knowledge-intensive entrepreneurship? In which ways is entrepreneurship related to innovation systems? What are the factors affecting its emergence, evolution and performance? How important is knowledge-intensive entrepreneurship for European growth and competitiveness? Is the situation of Central and Eastern Europe, engaged in a process of major economic and institutional transformation, similar or different from the one of Western Europe?

The book is organized in the following way. Part I presents the main themes and results. Chapter 1 by Franco Malerba lays down the major themes of the book, discusses knowledge-intensive entrepreneurship very broadly and illustrates the main contributions. This first chapter draws not only from the chapters in this book, but also from the results coming from the research project KEINS on entrepreneurship supported by the European Union.

Part II presents the main theoretical and conceptual foundations of knowledge-intensive entrepreneurship in innovation systems. The link between entrepreneurship and knowledge is discussed by Patrick Cohendet and Patrick Llerena in Chapter 2; the relationship between knowledge-intensive entrepreneurship and innovation systems is examined by Slavo Radosevic in Chapter 3 and the new role of science and the emerging scientific entrepreneurship is discussed by Chiara Franzoni and Francesco Lissoni in Chapter 4.

In Part III of the book, knowledge-intensive entrepreneurship and innovation systems are examined empirically in Western Europe and in Central and Eastern Europe. Here the book takes into account several types of innovation systems: the national, the technological and the sectoral, and their relationships. In Chapter 5 Roberto Camerani and Franco Malerba provide an introductory broad picture of the patterns of technological entry and new innovators in different technological fields and in different countries. Then two chapters go in depth into the data on new innovators. In Chapter 6 Stefano Breschi *et al.* analyse the survival of innovative entrants in some knowledge-based sectors, and in Chapter 7 Francesco Lissoni analyses the specific role of academic inventors as knowledge brokers. Then two chapters present two new in-depth field surveys on knowledge-intensive entrepreneurship in Western Europe and in Central and Eastern Europe. In Chapter 8 Camilla Lenzi *et al.* examine the main characteristics of knowledge-intensive entrepreneurship in several European countries (Germany, France, the UK, Italy, Sweden and Portugal) and in three broad sectoral systems (biotechnology, ICT and medical devices), and in Chapter 9 Slavo Radosevic *et al.* move the analysis to Central and Eastern Europe by examining Hungary, Lithuania, the Czech Republic, Croatia, Poland and Romania. Finally, the remaining chapters of Part III examine knowledge-intensive entrepreneurship in specific sectors. In Chapter 10 Johan Brink and Maureen McKelvey analyse entrepreneurship in biotechnology and the effects of technological and knowledge opportunities on firm strategies; in Chapter 11 Guido Buenstorf investigates the micro-dynamics of knowledge-intensive entrepreneurship in the German laser industry; in Chapter 12 Ricardo Mamede *et al.* examine the relationship between knowledge-intensive entrepreneurship, firm survival and industrial dynamics in knowledge-intensive industries in Portugal, and in Chapter 13 David Audretsch and Max Keilbach illustrate the relevance of knowledge-intensive entrepreneurship in high-technology and knowledge-based industries for economic and regional growth in Germany.

In the final part of the book (Chapter 14) some policy implications on knowledge-intensive entrepreneurship in innovation systems derived from this book are discussed by Franco Malerba and Nicholas Vonortas.

The chapters of this book initially originated from the three-year-long STREP project KEINS (Knowledge based entrepreneurship: institutions, networks and systems, EU project no. CT2–CT–2004–506022) supported by European DG Research and carried out by seven European research centres. Cespri (now KITeS) (Bocconi University), Max Planck Institute Jena, Chalmers University Sweden, Beta – University Louis Pasteur, University College London, Cisep University of Lisbon and Case Research Centre Warsaw. The chapters in this book have been chosen among the many outputs of research that compose KEINS, because we think that they represent interesting novel contributions to the literature and to the conceptual and empirical understanding of knowledge-intensive entrepreneurship. The interested reader may go to www.kites.unibocconi.it, then to 'Research networks' and then to KEINS and to the various 'Deliverables' in order to download these papers. In the preparation of this book, the chapters went through major redrafting, refocusing and rewriting.

The book could not have been possible without the joint effort of the large number of researchers involved in the KEINS project. Their contribution has been invaluable. Steve Klepper, Wesley Cohen and Yannis Caloghirou have provided very valuable comments and have actively participated in the various stages of this intellectual and research enterprise. I would also like to thank the participants to the workshops and conferences in which drafts and papers have been presented and discussed (Milan, Jena, Gothenburg, Lisbon and the two final conferences at Bocconi University and at DG Research of the European Union in Brussels). Finally, Pierre Valette, Pia Laurila and Otto Prange of EU DG-Research have provided major advice, guidance and support in all the stages of the project: I thank them very much.

Part I

Introduction

1 Knowledge-intensive entrepreneurship and innovation systems in Europe

Franco Malerba

1 Introduction

Knowledge-intensive entrepreneurship plays a major role in innovation, transformation of the industrial system and economic growth. New ventures that use, absorb and generate new knowledge and introduce new products and processes into the economy are a vital part of the economic progress and one of the backbones of the international competitiveness of countries. In addition, knowledge-intensive entrepreneurship rejuvenates technologies and industrial structure and contributes to the creation of new jobs. Increasingly, policy-makers are targeting these types of ventures in the belief that they are essential elements for innovation and prosperity.

This book examines knowledge-intensive entrepreneurship according to three related perspectives.

- First, the book links entrepreneurship to innovation and knowledge. Therefore the generation and use of knowledge become an essential part of entrepreneurship. In this way entrepreneurship becomes a major element of the dynamics of the economic systems and of the change of knowledge. It represents a major engine of growth and transformation of technologies and industries.
- Second, the book moves away from a person-centric view of entrepreneurship and links entrepreneurship to innovation systems. In this perspective entrepreneurship is dependent on, and related to, the presence and working of various types of actors (such as financial organizations, suppliers, users, public agencies, the local and the central government), networks of collaborations of various types and the institutional setting (in terms of regulation, standards, public policies and so on). In a sense knowledge-intensive entrepreneurship is examined as context dependent, specific to the systems in which it operates – national, sectoral or local.
- Third, the book adopts a process view of entrepreneurship. Entrepreneurship becomes characterized by various stages – from the pre-entry experience of the entrepreneur, to the start-up phase, to survival and to growth. Therefore knowledge-intensive entrepreneurship is intrinsically dynamic. However,

because it is inserted into innovation systems and has transformational properties, knowledge-intensive entrepreneurship is not just affected by innovation systems, but also contributes to change them.

Along these three perspectives, this book tries to answer the following questions. What are the conceptual building blocks of knowledge-intensive entrepreneurship? What are the factors affecting it? How relevant is it across technologies? What are its main characteristics in advanced European countries? And in transition economies? What are its main features across industries and sectoral systems?

This chapter aims to set the broad framework that has shaped this book and at the same time intends to start answering these questions. It refers to the various chapters of this book that represent extensive revisions of the contributions developed and written for the European research project KEINS supported by EU DG Research (Knowledge base entrepreneurship: institutions, networks and systems, EU project no. CT2–CT–2004–506022). However this chapter refers also to other contributions that have been part of the KEINS project and that touch relevant issues related to knowledge-intensive entrepreneurship but that are not in this book for reasons of space. In the pages that follow there are papers that are referred to as 'KEINS'. These other contributions can be found at www.kites.unibocconi.it (then go to 'Research networks' and then to 'KEINS' and to 'Deliverables' in order to download these papers).

This chapter is organized in the following way. In Section 2 the means and measures of knowledge-intensive entrepreneurship are discussed. Then in Sections 3, 4 and 5 the main building blocks are examined: innovation (Section 3), knowledge (Section 4) and innovation systems (Section 5). Section 6 focuses on the European case and provides a quantitative view of new innovators and of academic inventors. In Section 7 the analysis then shifts to two surveys on knowledge-intensive entrepreneurship, one done in Western Europe and the other in Central and Eastern Europe. Finally, in Section 8, the main features of knowledge-intensive entrepreneurship in specific industries and sectoral systems are presented. Some policy implications close the chapter.

2 A broad view of knowledge-intensive entrepreneurship

This book adopts a broad definition of knowledge-intensive entrepreneurship: knowledge-intensive entrepreneurship concerns new ventures that introduce innovations in the economic systems and that intensively use knowledge.

From this broad definition, it follows that knowledge-intensive entrepreneurship may take place in various ways: through the foundation of new firms or through the display of entrepreneurial spirit within existing firms or through the action of single individuals within non-profit organizations such as universities or public laboratories. Some of these are start-ups in terms of new and small firms, involved in the commercialization of new knowledge. Other times knowledge-based entrepreneurship involves the generation of new activities within large corporations (corporate entrepreneurship). Here technological diversification is relevant. In other instances actors such as researchers and scientists within universities play a major role.

The main part of the book presents a variety of quantitative analyses, which are highly complementary. These analyses range from those based on existing micro data – such as new firms formation, spin-offs, companies that patent for the first time, inventors, academic entrepreneurs, firms with high skill intensity – to ad hoc surveys explicitly carried out in order to shed new light on knowledge-intensive entrepreneurship. Finally, the book contains detailed case studies of knowledge-intensive entrepreneurship.

In this book the quantitative analyses have started from the consideration that knowledge-intensive entrepreneurship may concern those companies that have *a high level of skills or high expenditures in R&D.* As a consequence, knowledge-intensive entrepreneurship does not characterize only high-tech sectors, but can also be present in traditional and medium- or low-tech sectors. More specifically, in this book three complementary definitions of knowledge-intensive entrepreneurship have been used for quantitative analyses using large sets of micro data.

1 The first one defines knowledge-intensive entrepreneurship as *those new firms in sectors that are highly knowledge-intensive.* This means that knowledge-intensive entrepreneurship is represented by new firms in high technology sectors, in ICT or in all those sectors with a high content of human capital and skills.
2 The second definition considers knowledge-intensive entrepreneurship *the new innovators* in a technology/sector. New innovators in a technology or sector can be either *de novo* entrants or established firms active in a process of technological diversification. In this book new innovators in terms of *de novo* entry or technological diversification are measured in terms of those firms that patent for the first time in a technology/sector.
3 The third definition considers *academic inventors*, and refers to academic start-ups as well as to academic patenting in various technological classes.

3 The first building block of knowledge-intensive entrepreneurship: innovation

Entrepreneurship is a multidimensional and multidisciplinary concept. Analyses of entrepreneurship have focused on various dimensions and elements: the individual, the group, organizations (firms, universities etc.), industries and countries. The disciplines that have studied it have ranged from economics, to management, organization, sociology, cultural studies and so on. So different fields of studies have tackled the issue. Surveys of entrepreneurship have been done by, amongst others, Busenitz *et al.* 2003; Gartner 1990; Acs and Audretsch 2003; Casson 1982 and 2003; Glancey and McQuaid 2000; Shane 2000; Storey 2000; Swedberg 2000; Westhead and Wright 2000. Only in economics we have at least three traditions dealing with entrepreneurship (Audretsch and Keilbach,

KEINS,[1] 2006; and Garavaglia and Grieco, KEINS, 2005): the Schumpeterian one focused on innovation, the Chicago one related to Knight's approach and the Austrian one related to Von Mises, Kirzner and Shackle.

In this book entrepreneurship is examined along three well-defined perspectives. Innovation is the first one. This follows a Schumpeterian approach, as Schumpeter's 1934 statement claims: 'The carrying out of new combination we call "enterprise" ... the individuals whose function is to carry them out we call "entrepreneurs".' In this perspective one we may want to define entrepreneurship as the setting up and the owning of a new business. However it often happens that, even if a new firm is per se 'something new', it has no characteristics of novelty or change (just think about the simple break-up of a firm in two subsidiaries). And also some studies emphasize the 'opportunistic and innovative behaviour and therefore include as entrepreneurs those who transform an existing company or organization rather than setting it up [of a new business]' (Schumpeter 1934). As a consequence, in this book entrepreneurship is considered an activity that faces uncertainty and has the goals of creating something new: a technology, a product, an organization, a market. In this respect one may advance the point that the field of entrepreneurship is 'the study of sources of opportunities, the processes of discovery, evaluation, and exploitation of opportunities; and the set of individuals who discover, evaluate and exploit them' (Shane and Venkataraman 2000: 218). In this perspective, Audretsch *et al.* (2002) propose to distinguish between demand factors (in terms of market opportunities) and supply factors (in terms of resources and talent/abilities/traits) and to examine how they shape the risk and reward profiles, the generation of new knowledge, the development of innovation, and the entry and exit in an industry.

4 The second building block of knowledge-intensive entrepreneurship: knowledge

A second building block of knowledge-intensive entrepreneurship is knowledge. Scientific and technological knowledge have become of paramount importance for innovation (Foray 2002). As a consequence, the share of the scientifically educated workforce has risen continuously over the past three decades in almost all the European countries, generating an additional supply of highly qualified human capital. So, in addition to the classical qualifications in applied sciences in the various engineering disciplines, the last decades have also witnessed rapidly expanding opportunities for applied scientific research in new sub-disciplines in physics and chemistry (e.g. material sciences), biology (e.g. gene technology), pharmacology, medicine and, last but not least, information sciences. Thus many highly specialized applications firms need to be capable of absorbing new technological opportunities into their commercial R&D outcomes and to integrate it with the complementary knowledge about markets and applications.

More generally, this second perspective considers entrepreneurs as knowledge operators, dedicated to the utilization of existing knowledge, the integration and coordination of different knowledge assets, the creation of new knowledge, and engaged in the development of new products and technologies. In their activities, knowledge-intensive entrepreneurship has to create, use and manage scientific and technological knowledge in various ways, and has to match it with knowledge of markets and applications. This leads to learning and the accumulation of competences in specific scientific, technological and market domains. In the realm of knowledge usually a division of knowledge precedes a division of labour (see Chapter 2, this volume). In a sense firms first decide the domain of knowledge in which they will be active and then choose the activities in which they will be involved (Pavitt 1998; Brusoni *et al.* 2002). As Pavitt (1998) claims, technological opportunities generated by new knowledge create room for diversity and experimentation. The entrepreneur is a key player in connecting the generation of new knowledge or the use of existing knowledge with the generation of activities and products. In a sense, knowledge-intensive entrepreneurship represents a capability of integrating the developments in knowledge with the reconfiguration of resources, organizational skills and external links (see Chapter 2, this volume). However, in a company, there is a complex interaction between the domains of knowledge (which could be more or less broad or more or less specialized), the organizational domain and the artefact domain. Knowledge-intensive entrepreneurship has to coordinate these three aspects (Brusoni and Prencipe 2006) and decide about internalization and externalization. In this respect, knowledge operates as both an enabling force and an obstacle to entrepreneurship. What kind of knowledge needs to be generated and transferred explains to a large extent what kind of social context embodies the entrepreneurial action. One of the key functions played by entrepreneurs is what Witt and Zellner (KEINS, 2005) call the 'self-sorting process', i.e. the process which precedes the funding of a new firm or division. This process involves several steps: to identify a business opportunity generated by new scientific or technological knowledge; to attract and coordinate resources; and to integrate new knowledge into the established organization. The outcome of such a process is the decision about the organization within which to transfer newly generated scientific or technological knowledge: a new firm or an established organization?

The types of knowledge at the base of knowledge-intensive entrepreneurship may drastically differ. On one extreme there is scientific knowledge. As Witt and Zellner (KEINS, 2005) claim, scientific knowledge may take the form of propositional knowledge (which can be encoded and stored by some information medium, and then transmitted and used widely), and procedural knowledge (which encompasses substantial tacitness and command on methodology and procedures). The acquisition of this knowledge is based on practice and interaction. Thus for knowledge-intensive entrepreneurship based on scientific knowledge, the type of scientific knowledge implies different ways of knowledge use, communication and transfer. Knowledge-intensive entrepreneurship may coordinate and integrate external scientific knowledge with the internal one, within a division of labour perspective. Propositional knowledge may be transferred among organizations. Procedural knowledge is usually carried in embodied form. Therefore the realization of the commercial value may take the form of a new entrepreneurial start-up firm funded by the scientist or of her/his participation as an employee or collaborator to an established firm. On the other extreme, for knowledge needed for developing complex products, modularity in product design and openness in the interfaces are very important elements that shape entry and competition. As Brusoni and Fontana (KEINS, 2005) show, product architectures that are modular and are also open (in the sense that the proprietariness of the interfaces is low) allow high entry. The same holds for open but non-modular platforms, such as the case of open-source software. Here it is the openness of the (non-modular) platform that may induce entry. However system integration is usually conducted by incumbents, which coordinate the new entrants and the specialist suppliers. Different is the case of modular but close architecture (such as local area networks) in which several interfaces are proprietary and often restricted: here incumbents dominate the scene. The same holds for non-modular and closed platforms (such as minicomputers and the early mobile phone industries): here again incumbents dominate the scene and knowledge-intensive entrepreneurship is limited to niche strategies targeted to a specific set of users (Brusoni and Fontana, KEINS, 2005).

The focus on knowledge brings to the forefront the learning environment in which knowledge-intensive entrepreneurship acts. In fact, the learning environment in terms of technological regimes (defined in terms of scientific and technological opportunity, appropriability, cumulativeness and knowledge base) greatly affects the entry rates of new firms and the emergence of new innovators in a technology or in a sector (Malerba and Orsenigo 1996; Breschi *et al.* 2000). High scientific and technological opportunity conditions indeed affect positively the entry of new innovators, particularly when incumbents cannot fully protect their innovations (i.e. low appropriability conditions) and cumulativeness of advance is not so high (i.e. low cumulativeness conditions). This is confirmed by Dushnitsky and Lenox (2005, in Godinho and Mamede, KEINS, 2005) who show that firms' investments in new ventures increase when technological opportunities are higher and IP regimes are weaker.

5 The third building block of knowledge-intensive entrepreneurship: innovation systems

A final building block of the conceptual framework used in this book is innovation system. As the United States National Commission on Entrepreneurship (NCOE) (2001) and Blithe (2000) point out, systems and networks are key for entrepreneurship. Successful entrepreneurs are consummate networkers who thrive in communities. Networks are essential because they are links to potential sources of capital, new employees, strategic alliance partners, and service providers (lawyers, accountants, consultants etc.). Networks also allow entrepreneurs to share information and assessments on markets and technologies as well as lessons learned from their own entrepreneurial activities. Thriving regions generally boast a wide array of both informal and formal networking structures.

In an innovation system perspective, the features, behaviour and performance of knowledge-intensive entrepreneurship are the result of the interaction of several actors. In the next pages, as well as in the rest of the book, four of them will be discussed: established producers, users, universities/public research institutes and venture capitalists.

5.1 Established companies and spin-offs

Established firms affect knowledge-intensive entrepreneurship in various ways. One is the role of knowledge-intensive entrepreneurship within established firms in the form of new activities or diversification. As Godinho and Mamede (KEINS, 2005) discuss, corporate venturing in terms of investments by established companies in the promotion of new business ventures is quite pervasive, and is driven by strategic reasons related to the preservation of existing companies and the access to external knowledge. Problems may arise about conflicts in terms of the risk aversion between the parent firm and the individual entrepreneur. This may involve the trust among the parts, the relevance of the parent firm's assets for the new venture and so on. One aspect of corporate venturing may regard the purchase of new business ventures (Audretsch *et al.* 2007) in which specific characteristics of the targets seem more important in the purchasing decision than the presence of abnormal profits in the acquired firm. Audretsch *et al.* (2007) show that corporate venturing through takeovers is related to intangible assets and Intellectual Property Rights (IPRs). Looking at 273 German IPOs between 1997 and 2003 and 58 takeovers, they show that the probability of takeovers decreases with the number of patents owned by the CEO of the target company, because this means that intangible assets (which are complementary to the physical assets) cannot be owned together with the physical assets. On the contrary the presence of patents owned by the target company – which therefore can be owned together with the complementary assets – increases the probability of takeovers.

The other way established companies affect knowledge-intensive entrepreneurship is through the spin-off process. Spin-offs from established companies are quite common in industries with a high rate of change. Spin-offs inherit knowledge about technologies, products and markets from the companies they come from. Therefore they are affected by the parent company in their choice of technology, product and markets (Klepper and Sleeper 2005; Klepper and Thompson 2005).

5.2 Users and demand

Users are often a source of knowledge-intensive entrepreneurship. When they are a source of ideas and information for new products, users may end up launching new activities within a company or set up new firms, as in software or instrumentation (Von Hippel 2005). Knowledge-intensive entrepreneurship may also stem out from downstream activities more related to demand, such as distribution. Distributors may have a deep knowledge about applications and uses, as the case of some entrants in the laser industry show (Buenstorf 2007; Chapter 11, this volume).

5.3 Universities and public research organizations

A perspective on innovation systems stresses that also universities and research organizations are involved in knowledge-intensive entrepreneurship. In academic organizations knowledge from scientific research may generate new technologies that often reach commercial exploitation. This takes place through a person ('the academic entrepreneur') that embodies the necessary scientific and technological knowledge. More and more, universities and their researchers have been, at least in many plans for local or industrial development, assigned the daunting task of generating both new, valuable ideas useful for innovation and also of creating a breed of new entrepreneurs capable of exploiting them.

More specifically, academic entrepreneurship can be examined in two ways. One is the common view of academic start-up firms, in which university scientists set up a business company in order to commercialize the results of their research. The example of biotechnology firms is a case in point. The extent of academic spin-offs depends on the legal framework of academic careers (easiness of tenures, leaves of absence, positions available, pecuniary benefits from external companies), regulations regarding patent property and the sharing of licence earnings and the effectiveness of IP conditions (effective IP conditions facilitate the taking up of projects by external companies, while low appropriability favours academic spin-offs) (Godinho and Mamede, KEINS, 2005).

The second way is broader and regards all the activities (carried out within universities) in which academics develop and market their research ideas or research-based artefacts. This broader view of academic entrepreneurship includes a large variety of activities, from patenting, to the involvement in large industrial projects or in the commercialization process within companies, to consulting and so on. This is a more pervasive, but equally relevant, form of academic entrepreneurship which cannot be disregarded in the knowledge-based society.

Franzoni and Lissoni (Chapter 4, this volume) look at this second broader view of academic entrepreneurship. They claim first of all that academic entrepreneurship is closely related to the features of the national innovation system and in particular of the institutional characteristics of national academic systems. Franzoni and Lissoni look at the history of the US and French systems, the last one taken as the extreme example of the European case. They claim that academic entrepreneurship is much more affected by the long-standing institutional features of national university systems, than by any ad hoc legislation affecting IPRs in university. More generally, they discuss to what extent commercial activities may or may not help scientific entrepreneurs to progress in their careers. Among those institutional features, university autonomy, personnel mobility and the principal investigator principle stand out as the most important. Patent-based and spin-off based technology transfer is by and large the product of a specific institutional history, that of the US research universities, where these features have been prominent. Every introduction of those issues within the various European university systems should require first and foremost strong reflections and adjustments that take into account institutional, organizational and environmental characteristics of academic research at the national level. Franzoni and Lissoni (Chapter 4, this volume) show that academic entrepreneurs who are active in patenting and firm-founding (and more generally in technology transfer) come disproportionately from the ranks of scientific entrepreneurs with a brilliant scientific record, possibly oriented to fundamental research. These scientists' economic agenda is centred upon entrepreneurial efforts within the university, aimed at gaining reputation through discipline building, creation and management of laboratories and research teams, and the search for the economic resources necessary to pursue those goals. To those scientists, patent licensing and spin-off creation are appealing not just because of the expectation of profits, but also because they offer valuable opportunities to enlarge their sphere of influence, to empower their internal and external consensus, and increase the budgets available for their research. Hence, any wise policy of technology transfer in academia should move from a broad consideration of the overall personal incentives faced by scientists within the context of academic careers.

5.4 Finance

One key networking issue for knowledge-intensive entrepreneurship is finance. Ben-Ari and Vonortas (KEINS, 2005) claim that certain factors of knowledge-intensive entrepreneurship limit the possibility of obtaining financing – no collateral, extensive intangible assets, difficult evaluation of potential innovations. So, debt financing through direct loans is usually not available, while equity financing and government grants are the only ways to get resources. But the types of financial actors and networks change over the life of knowledge-intensive entrepreneurship. Family and friends networks and public funds are relevant in the pre-seed stage while business angels and universities' venture capital funds become relevant in the seed stages. Venture capital enters the scene in the start-up and early stages. The reason is that venture capital usually focuses on large and safer deals and needs to have effective exit mechanisms, such as regulations for IPOs and M&As. The public actor is quite relevant in the creation of a regulatory framework for risk financing, the support for venture capital funds, the provision of information and advice to knowledge-intensive entrepreneurship and the support of knowledge-intensive entrepreneurship in specific technologies or sectors that are of immediate interest to the private sectors but that can generate spillovers of several kinds.

5.5 Institutions

A system view considers institutions as a key factor shaping entrepreneurship. Institutions include norms, routines, common habits, established practices, rules, laws, standards and so on, that shape entrepreneurship cognition and action and affect the interactions with other agents. Institutions and the related organizations differ greatly in terms of types, ranging from the ones that bind or impose enforcements on agents to the ones that are created by the interaction among agents (such as contracts); from more binding to less binding; from formal to informal (such as patent laws or specific regulations vs. traditions and conventions). A lot of institutions are national (such as the patent system), while others are specific to sectoral systems, such as sectoral labour markets or sector-specific financial institutions.

5.6 Opportunities

The innovation system view also redefines the concept of opportunity for knowledge-intensive entrepreneurship. Because knowledge-intensive entrepreneurship is considered as an essential part of innovation systems, for Radosevic (Chapter 3, this volume) entrepreneurship is explained as a systemic phenomenon driven by opportunities. But what constitutes entrepreneurial opportunity? Radosevic considers three types of opportunities: technological opportunities, market opportunities and institutional opportunities. Entrepreneurship is driven by complementarities arising from the favourable interaction of all three types of opportunities. In the absence of one of these, entrepreneurial opportunities cannot be realized. This perspective is actually an integration of three views on entrepreneurship: the ones from Kirzner, Schumpeter and List. In each of these views, entrepreneurship is a function of different driving factors. For Kirzner entrepreneurship is a function of imbalances/distortions/asymmetries/disequilibria on the market. For Schumpeter entrepreneurship is a function of technological opportunities. For List entrepreneurship is related to national system of political economy/institutional complementarities or synergies. For Radosevic the final outcome in terms of entrepreneurship is the result of alignment between these three components. Exploitation of technological opportunities is dependent on market opportunities and on institutional opportunities. On the other hand, if technological opportunities are not present, it may not be possible to exploit market opportunities. Similarly, institutional opportunities may hinder the coupling between technological and market opportunities. In this context a mismatch among the three types of opportunities may be due to that fact that one of the components (markets, technology or institutions) is undeveloped; or that wrong orientation of one of the components (anti-entrepreneurial institutional system; technology lock-ins; closed markets) is present; or that mismatches between technological, market and institutional opportunities (which have developed in isolation and thus are not compatible) are present. The extent of mismatches of these three opportunities will be smaller in some contexts and larger in others. This is a systemic property of innovation systems, which cannot be expressed as scarce factor, and which is based on compatibility between different sub-systems of innovation systems.

This point is reinforced by Holmén *et al.* (KEINS, 2005), Foster and Metcalfe (2001) and Loasby (2001) who add one key element: the perception, identification and realization of these opportunities. Perception has to consider several elements that constitute the opportunity, including the economic value, the mobilization of resources and the regime of appropriability of innovations. And the level of uncertainty affects the perceived opportunities. In sum, bringing to market activities and products related to new scientific knowledge and advanced technologies means placing major relevance on the commercial organization of knowledge transfer. Organizations have to perceive the market and the productive opportunities of the new scientific knowledge and of advanced technologies. They have to develop them into workable business conceptions and designs, attract resources and coordinate the processes, interactions and integration with the firm's existing capabilities (Witt and Zellner, KEINS, 2005).

5.7 Sectoral systems, national systems and regional systems

In sum, knowledge-intensive entrepreneurs are active in quite different innovation systems. One could identify three of them: national innovation systems (Nelson 1993; Lundvall 1993); sectoral systems (Malerba 2002) and local systems (Cooke 2008), each of which affects entrepreneurship in different ways and through different mechanisms.

The literature on national and regional systems has been extremely convincing in illustrating the major differences in the features, architecture and dynamics of national and regional systems. Here there is no space to summarize this discussion. It is enough to say that national and regional systems affect knowledge-intensive entrepreneurship through their actors, relationships and institutions. However, up to now the literature on national systems of innovation has not been rich with comparative studies of knowledge-intensive entrepreneurship in various countries and of the role of the complementarities in affecting its dynamics. In this book an analysis of knowledge-intensive entrepreneurship in Western Europe and in Central and Eastern Europe is carried out in Chapters 8 and 9.

Even less examined have been the effects of sectoral systems on knowledge-intensive entrepreneurship. The work on sectoral systems of innovation (see Malerba 2002 and 2004) shows that the knowledge base, actors and networks differ from sector to sector. So within different sectoral systems, knowledge-intensive entrepreneurship is affected by specific knowledge and technological domains, faces heterogeneous agents, interacts with different networks, and is shaped by different institutions. Similarly, technological regimes differ across sectors and affect innovation and entry differently (Breschi *et al.* 2000). Major differences exist across sectoral systems in the type and in the factors affecting knowledge-intensive entrepreneurship. Think, for example, of new biotechnology companies – often spin-offs from universities with a great command of science and strong links with a network of large firms and research laboratories – and compare them with new innovators in machine tools – where the entrants are spin-offs from existing companies and knowledge is based on experience and publications – or with entrants in the software – in which open-source and user–producer relations are a key element of knowledge-intensive entrepreneurship. As we will see later on, sectoral specificities and characteristics will be examined in depth in Chapter 8 for biotechnology, ICT and medical devices, and in Chapters 10 and 11 for biotechnology and lasers. An additional discussion on the differences between sectoral systems for corporate venturing has been done by Parhankangas and Arenius (2003), mentioned and discussed in Godinho and Mamede (KEINS, 2005). Parhankangas and Arenius (2003) examine 50 technology-related firms operating in seven industrial sectors and show that the type of corporate venturing (i.e. spin-off supported by the parent company) – developing new technologies, serving new markets or restructuring old businesses – depends on the sector. The first type of firm – developing new technologies – is typical of biotechnology, optical technologies, electrical components and medical devices or solutions for intelligent networks. Corporate venturing serving new markets (but with the same technology of the parent company with respect to products) characterizes specialized industrial machinery and components, software services and integrated IT systems. Corporate venturing restructuring the old core business is present in low and medium technology – metals, standardized industrial components and rubber and plastics. Finally, spin-offs occur more frequently in sectors in which product proliferation and market segments are relevant, scale economies are limited and human capital is important.

6 Knowledge-intensive entrepreneurship in Europe: a first quantitative broad view

After this conceptual introduction related to the main characteristics of knowledge-intensive entrepreneurship (which constitute Part II of this book), the main part of the book (Part III) examines the evidence of knowledge-intensive entrepreneurship in Europe. It does that according to several complementary directions of analysis: a broad quantitative view, two field analyses of knowledge-intensive entrepreneurship in different countries (Western Europe and Central and Eastern Europe), as well as some industry and firm studies. In this section, the first quantitative picture is presented, while in Sections 7 and 8 the country studies and the industry studies will be discussed.

What do we know about the relevance of knowledge-intensive entrepreneurship in Europe measured in terms of relevance of new innovators in a technology or a sector? A first answer may come from the examination of those firms and other organizations that enter for the first time as innovators in a technology or a sector. This can be done by looking at firms patenting for the first time in that technology or sector. Camerani and Malerba (Chapter 5, this volume) have examined new innovators in terms of companies and other organizations that patent for the first time in 12 technologies with a high rate of change, ranging from ICT, to semiconductors, to pharmaceuticals and biotechnology, to machine tools. They have used patents applications at the EPO in the period 1990–2003 in Europe, the United States and Japan. New innovators in a technology field can be *de novo* technological entrants if they patent for the first time ever, or technological diversifiers if the entrant has already patented in another technology. Camerani and Malerba find that innovative entry is a very frequent phenomenon: on average more than 40 per cent of the patenting firms in every period did not patent before in a given technological class, and more than 25 per cent of new innovators never patented before in any class. The importance of new innovators differs across technologies: *de novo* innovators are particularly high in ICT, medical engineering and measurement instruments, and low in semiconductors and chemicals. However the relevance of new innovators in terms of total number of patents is much less relevant. This means that innovative entrants start innovating with very few patents and are smaller than incumbent innovators: this is similar across technologies and across countries. If technological entry is quite common, persistence is more difficult: around half of the firms have only one patent. Those firms that enter and become persistent innovators keep patenting in the same technological field in which they entered. Their initial patents are usually more cited than the ones of the occasional innovators. Also in these cases differences across countries and sectors are relevant.

From the inventors and patenting data collected in Chapter 5, Breschi *et al.* (Chapter 6, this volume) go in depth into the analysis of new innovators in two knowledge-based industries: switches and lasers. Here new innovators are linked to the date of firm establishment, in addition to the date of the first patent. Here new innovators are firms that apply for their first patent within a relatively short time after their establishment. Breschi *et al.* examine if the performance of new innovators in terms of survival and persistence of innovations is affected by the type of founder (the type of founder can be the inventor or not), the inventor's experience in patenting, the technological scope and the relevance of the first patent and the link of the patent to science.

Another quantitative picture of knowledge-intensive entrepreneurship concerns academic entrepreneurship. As mentioned previously, academic entrepreneurship can be examined either through academic start-ups, or through the contributions that academics make to innovative activity. In this second view, academic patenting is part of the larger phenomenon of university–industry technology transfer. Patents represent a tool for protecting innovation in a number of science-based industries (pharmaceuticals, biotechnology, chemicals and electronics) when academic scientists generate advancements that may have industrial applications. Using these perspectives, some very interesting results emerge from the KEINS studies, some of which, for reasons of space, have not been included in this book.

Lissoni *et al.* (KEINS, 2007) have examined academic patenting in France, Italy and Sweden. They show that in these countries academic scientists have signed many more patents than previously estimated in other studies. This re-evaluation of academic patenting in Europe comes by considering all patents signed by academic scientists active in 2004 (both those assigned to universities and those to business companies, governmental organizations and public laboratories). This patent reclassification shows that academic patenting is a growing phenomenon in Europe: from 2 per cent of domestic EPO patents in 1985 to 4 per cent in 2000. So, from these new KEINS data, universities' contribution to total domestic EPO patent applications in France, Sweden and Italy appears not to be much less intense than that of their US counterparts: 4 per cent in the EU and 6 per cent in the United States. Specific institutional features of the university and research systems in the three countries (IPR arrangements, the institutional profile of national academic systems and the research contracts) contribute to explain these different ownership patterns between the United States and Europe. Professors that are inventors are concentrated in chemicals, engineering (electronics), biology and medical sciences. In Europe between 60 and 80 per cent of academic patents were owned by business companies and 14 per cent by individuals. Only 10 per cent are held by universities in France and Italy. In the United States 69 per cent of academic patents are owned by universities. In the various sectors in Europe businesses own 57 per cent of academic patents in pharmaceuticals and 85 per cent in electronics. This difference depends on the funding of research and on the exploitation strategies of patents. However, university ownership of patents, although small, has increased in all the three European countries considered. From this analysis one may conclude that there is a similar propensity to patent in Europe and in the United States, although not necessarily a similar scientific or technological level.

A further and deeper analysis can be done by looking at the networks of academic inventors. By examining networks in France, Italy and Sweden, Lissoni and Sanditov (KEINS, 2007) show that academic inventors occupy central positions in the small world networks of inventors (i.e. where small-world properties come from social network analysis and mean high knowledge variety but also a high speed of information diffusion). Academic inventors stand in between otherwise unconnected inventors and teams of inventors so that they can effectively control the information between inventors and teams. However, only a few of these academic inventors occupy a position of 'brokers', i.e. they stand in between two industrial researchers.

An analysis of these brokers in Italy is carried out by Lissoni (Chapter 7, this volume). Lissoni combines a relational analysis on inventors' data with the results of a short questionnaire submitted to a sub-set of Italian academic inventors, and with data on their scientific publication record and CVs. The aim is to explore in greater depth the personal relationships that academic inventors entertain with co-inventors with a different background and/or professional status. Lissoni finds that the brokerage and gatekeeping positions are very few, and they are held by scientists with both a large number of patents and a strong publication record. These scientists are not better than colleagues at further cooperating on research with co-inventors from industry after the patent experience. However, they do better when it comes to keeping in contact with industrial researchers. While brokerage and gatekeeping positions are not correlated to academic inventors' propensity to entertain stable research collaborations with their co-inventors from business companies, they are correlated with their propensity to keep in touch after patents for information exchanges of all sorts. Network ties between academic and industrial researchers may be short-lived as far as knowledge exchanges are concerned, but may serve other purposes well. In particular, the joint reading of the quantitative evidence and the top brokers' biographical notes suggest that these agents actively manage their relationships outside university. Some of them, especially those who have signed patents only with one or two different assignees, are likely to keep in touch mainly for research or research funding purposes. Others, such as those academic inventors with many different assignees and/or assignees such as public consortia and the like, may nurture their personal links outside universities for more strategic purposes. The existing literature on university patenting has focused almost exclusively on academic inventors' monetary incentives. In Chapter 7 Lissoni finds that the social contacts gained through collaboration with industry may be part of the reward, as they help boosting the academic inventors' reputation and career both inside and outside the university.

7 Knowledge-intensive entrepreneurship in Western Europe and in Central and Eastern Europe

A second type of evidence of knowledge-intensive entrepreneurship comes from country analyses conducted with surveys at the firm level. Two specific firm surveys have been administered by the KEINS project to a large number of firms in Western Europe and in Central and Eastern Europe.

The first survey explores the characteristics of knowledge-intensive entrepreneurship in Germany, France, the UK, Italy, Sweden and Portugal in sectors such as biotechnology, electronics and medical devices (Chapter 8, this volume). The field survey selected a group of new firms that innovate within a very short time after their establishment and are both knowledge-based (i.e. active in science-based and science-driven sectors) and technology-based (i.e. patenting in sophisticated and dynamic technological contexts). The KEINS database has been collected through a long and complex process of sample identification, data collection, questionnaire elaboration, and survey administration. Special attention has been dedicated to the background of the founder/s and the relationships of the new firm with the parent organization. The database has been inspired by contributions on human capital theory and on the parent firm's knowledge and on technological inheritance. The former concentrates on founder's individual characteristics and intellectual capital as principal determinants of start-up success. The latter stresses how new entrants' performance is significantly influenced by founders' pre-entry background and experience as well as their technology and knowledge endowments. Ninety-nine companies have been examined. They have been grouped according to three broad regions: Germany, Northern Europe and Latin Countries (France and Italy). They have been also divided according to three sectors: biotechnology, electronics and medical devices. The major findings are several. Venture capital is very important in countries such as Germany and Northern European countries, and in biotechnology. Banks, on the contrary, are more relevant for medical devices, and local and regional authorities for biotechnology. In general, IPRs (licensing and patent) are important for founding a new firm. All firms stressed that specialized labour is key for their survival and success. A large number of firms consider networks to be important, but more than 40 per cent do not consider the links with universities to be important. Collaborations with previous employers are more frequent in Nordic countries, and in biotechnology. R&D is the main area of collaboration. For all companies the success is based on the uniqueness of their products. In addition, marketing skills and customer services are important in electronics and medical devices. Interestingly, from the previous employer biotechnology spin-offs have received knowledge about products, while electronics and medical devices spin-offs receive knowledge about customers. Most of the firms offer a product for a specialized market which are unrelated to the founder's previous employer. The level of human capital is confirmed as high: most founders have a PhD or a master's degree. The main motivation of the founder is independence and the commercial exploitation of her/his own idea. Fifty per cent have become serial entrepreneurs. Differences across countries (Germany, Nordic countries and France/Italy) are related to the degree of development of, and the ease of access to, financial markets, and the different functions and effectiveness of the university system. Financial markets affect the probability of entry into entrepreneurship. The university system affects the educational profile of founders, the frequency of collaborations with the private sector and the rate of university spin-offs. In sum, differences across sectors are related to the type of knowledge and competences necessary for market success, to the knowledge endowment, to the inheritance from, and the relationship with the previous employer, and to the types of customers. Moreover the interplay of country- and technology-specific variables strongly affects and shapes the entry decision of knowledge-intensive entrepreneurs as well as the early evolution of new innovative firms.

The second survey examines Central and Eastern Europe (CEE) (Chapter 9, this volume). The survey examines 304 firms in Hungary, Lithuania, the Czech Republic, Croatia, Poland and Romania. The chapter aims to identify the factors that affect the strengths and the weaknesses of knowledge-intensive entrepreneurship in CEE countries. These factors can be broadly divided into those referring to the entrepreneur, to the firm, and to the environment. The results of the survey show that knowledge-intensive entrepreneurs usually start their careers in the business sector rather than in the scientific sector, and found knowledge-intensive firms in order to take advantage of market (i.e. commercial and financial) opportunities. These entrepreneurs bring knowledge about products and technology from their previous employment and then with their new firms they develop new markets. In that respect, knowledge-intensive entrepreneurship in CEE countries can be considered a market repositioning activity. Technological opportunities are frequently mentioned as a key rationale for establishing companies only in the Hungarian sample, where more entrepreneurs come from the science sector. Knowledge-intensive entrepreneurship in CEE countries does not consist of 'gazelles' (i.e. fast-growing new-technology based firms which have the potential to reshape the industrial landscape). Instead, it consists of distinct types of companies, of which new-technology based firms are only one example. The key factors in knowledge-intensive entrepreneurs' growth are most often firm-specific capabilities which do not always involve R&D. Based on what the firms have identified as their success factors, firms have been grouped in three: new-technology based firms, 'networkers' and companies whose success is based on 'customer-oriented organizational capabilities'. The most common developmental barrier in domestic markets is the low level of demand. This is followed by high labour costs, increased competition and lack of public support. Firms fall into two groups with respect to the kinds of barriers that are most important for them. The first group is one where the main barriers concern skills shortages and high labour costs. For the second group, the major barriers are related to finance (lack of access to finance and of public support). Compared to standard companies in CEE countries, which tend to limit their strategic interactions for innovation to value chain partners such as buyers and suppliers, the networks of knowledge-intensive entrepreneurship are broader and more frequently involve innovation system actors (such as research institutes and universities), including professional networks (related to fairs and exhibitions). Indeed, here again Radosevic *et al.* (Chapter 9, this volume) identify a number of distinct types of firms with respect to the sources of knowledge that are most important for their innovation processes: those for whom value chain partners are the key source of knowledge for innovation; those for whom formalized R&D is the key source, and those for whom in-house or firm-specific innovation activities are key sources. Radosevic *et al.* also propose another grouping of companies with respect to the intensity of their links with external organizations. They distinguish four types of firms: network dependent, public research system oriented, foreign and domestic value chain dependent firms. They clearly illustrate that for different types of knowledge-intensive entrepreneurship different networks are important. In general, they are either vertical (foreign and domestic value chains) or horizontal (links with the domestic public research system).

In addition to the survey on knowledge-intensive entrepreneurship, some additional findings on the cases of CEE countries emerge from other works in the KEINS project. CEE countries are transition countries, in that they moved from central planning to a more market-oriented system. However, transition countries have very different initial conditions in terms of industrial structure, institutions, norms and practices, which greatly affects the type of entrepreneurship during transition (Aidis, KEINS, 2005). These conditions can explain the divergence between entrepreneurship in the Baltic countries (CEEB – Central and Eastern Europe Baltic countries), and those of the previous Commonwealth of Independent States (CIS). In the CEEB countries the economy is similar to that of the EU and the role of the state is not so pervasive. In the former CIS countries, on the other hand, the role of the state is still relevant, and networks of various types are still strong. In both these sets of countries, entrepreneurship is mainly in retail trade and low-tech sectors, and knowledge-intensive entrepreneurship is extremely limited. Radosevic (KEINS, 2008) identifies some key features of knowledge-intensive entrepreneurship in CEE countries. In CEE countries there is a limited domestic demand for knowledge-based products and activities, including public sector demand. This is even more so for knowledge-intensive services. Local networks do not play a major role. However customers have an increasing relevance, particularly in software. This is so because customization is quite important for the success and growth of firms. Knowledge is usually developed in-house, except in Internet-based business. Interestingly enough, in countries that were previously part of the CIS two features continue to be common: research institutes try to commercialize their activities and large firms act as a substitute mechanism for venture capital.

Finally, some specific country case studies conducted by KEINS researchers illustrate some of the previous points. Bishop (KEINS, 2006) examines the cases of knowledge-intensive entrepreneurship in the Czech Republic and Hungary. She brings evidence that access to finance remains key for firms' start-up and survival but also of the reluctance of entrepreneurs to seek and accept external finance. Instead, personal networks and family networks remain relevant for finance. Two other main problems are the training for potential entrepreneurs and the difficulties of entering international markets. However the impact of privatization and the removal of regulation are opportunities for entrepreneurship in these countries. In some cases the government provides the initial demand (as in the case of environmental services). In these cases users' networks are also important. Woodward (KEINS, 2006) examines knowledge-intensive entrepreneurship in Poland and Estonia and finds differences in national contexts. In Poland demand for knowledge products is weak, while in Estonia university and transfer offices are much more active and foreign demand is relevant for knowledge-intensive entrepreneurship. In general, however, there is limited relevance of networking in the generation of knowledge: knowledge resources are developed largely in-house. Instead networks play a role in the distribution phase and in the search for new customers. In both these contexts, the public sector is unable to create demand for high-tech products.

8 Knowledge-intensive entrepreneurship in industries and firms

A third type of evidence refers to knowledge-intensive entrepreneurship in specific industries and firms. Here the differences in the features and dynamics of knowledge-intensive entrepreneurship become quite evident, because of the specific features of technology and products, the type of knowledge, the characteristics of production, the structure of competition, the conditions of appropriability. In this book a variety of sectors are examined: biotechnology, lasers, ICT, in addition to other industries analysed in other KEINS papers.

For biotechnology, Brink and McKelvey (Chapter 10, this volume) develop a taxonomy related to the actors' ability to develop and use technological opportunities for products and services. Four case studies explore the relationship between the theoretically based taxonomy and specific biotechnology companies. This perspective draws upon two bodies of theories, one based around knowledge and one based around the differentiated nature of technological opportunities. The taxonomy develops the two main dimensions along which to analyse knowledge-intensive entrepreneurship: differences within the opportunities for continued knowledge development (such as R&D driven etc.) and differences within opportunities for the re-use of knowledge in a narrower or wider range of goods and services. Two aspects of the knowledge-intensive activities of firms have therefore been analysed: the extent to which innovative activities can follow a technological or scientific regime of knowledge development and the extent to which broad knowledge bases and integration (rather than focused knowledge bases and specialization) gradually open up new opportunities. Brink and McKelvey claim that the main value of entrepreneurial firms is their potential for technological development for products and services. This suggests that a key competence of the firm must be innovation and knowledge integration in order to deliver products and services – and not the technological competencies and scientific competencies per se. These cases show that firms which draw benefits from knowledge integration and local learning, rather than from radical innovations, can prosper and grow. They do so even without being radical innovators and without being highly science intensive. Scientific and technological knowledge are only strategic assets at particular times and for particular reasons. This suggests the need for understanding how the dynamics of technological opportunities affect specific firms. In certain industries, such as biotechnology, the knowledge-intensive and entrepreneurial firm must devote resources to the internal application of capabilities and the replication of knowledge, as well as to the new possibilities opened up by the development of science and technologies. The ability to reconfigure assets and competencies of the firms is therefore crucial. A main focus upon radical innovations and the protection offered by patenting, and publications, are not the whole story of the bioscience industry. To innovate per se is important. However, a more important function of these biotech companies lies in their ability to integrate knowledge into coherent understanding for further development of science and technology and its use in products and services.

Biotechnology is also examined by Bureth *et al.* (KEINS, 2006). They focus on biotechnology start-ups in the Rhine Biovalley and find academic spin-off to be of major importance. Here the main motivation is not just money. In this sector public research and science greatly affect start-ups. Links to public research increase firms' reputations and credibility and provide access to new, relevant scientific knowledge. In this respect not only networking, but also the scientific capability and personality of the founder, plays a key role in the start-up and the growth phases. However projects that are too far from the market do not survive long enough. Patents play a role in the creation of new firms in drug development, but less so in non-drug products and services. In general, entrepreneurship in biotech is a collective process involving many heterogeneous actors.

For lasers, Buenstorf (Chapter 11, this volume) examines knowledge-intensive entrepreneurship in the German industry. This industry is characterized by high heterogeneity in product sub-markets, limited economies of scope, a high rate of innovations and the continuous development of new laser applications. Buenstorf identifies four types of entrants: *de novo* entry based on academic research, diversifying entrants, spin-offs and entry from distribution activities. The similarities between the German case and the US case (examined by Klepper and Sleeper 2005) are striking. Spin-offs play a major role in industrial dynamics. The background of the founder affects spin-off performance, whereas the academic background had the least impressive performance of all. In terms of companies both laser spin-offs – which have technological experience – and integrating distributors – which have knowledge of applications – have the best performance.

Evidence of a specific relationship between knowledge-intensive entrepreneurship (in terms of new innovators) and R&D alliances comes from the analysis of semiconductors, telecommunications and pharmaceuticals. Malerba and Zirulia (KEINS, 2007) analyse the relationship between new innovators in terms of patents and R&D alliances in these industries. They find that new innovators may access networks of R&D alliances soon after entry, but they usually do not form another alliance after the first one. Established networks are beneficial to new innovators because they provide new innovators with complementary assets. Similarly, new innovators bring new capabilities into existing networks. Both features can be illustrated by the high technological distance (in terms of patents) between new innovators and the other partners in the network. However, initially new innovators are in a marginal position in established networks. Over time, however, as a result of the mutual learning process they become close to the other partners in terms of technological capabilities. Those new innovators that continue to form alliances become more central to the network. Differences across sectors emerge from the analysis: in semiconductors firms are more technologically similar since the beginning, and in telecoms those new innovators that continue to form alliances remain less central.

The characteristics of knowledge-intensive entrepreneurship in high-technology or high-skilled industries are examined by Mamede *et al.* (Chapter 12, this volume). They focus on new firms in Portugal between 1995 and 2000 using two classifications. One is the OECD classification that considers high- and medium-technology industries plus knowledge-based services. The other classification considers industries that are in the top 10 per cent in terms of average proportion of employees holding a university degree. Entrepreneurship in industries that are knowledge intensive has higher survival rates and higher growth rates than entrepreneurship in other industries. In addition, in these industries entry is less responsive to incentives (in terms of price cost margins or industry growth), and more to other factors such as the behaviour of incumbents, competition and limited economies of scale.

Finally, one last contribution shows that knowledge-intensive entrepreneurship – examined in some high-tech industries – affects economic growth and regional development and is characterized by specific industrial dynamics. For 440 counties in Germany in 2000, Audretsch and Keilbach (Chapter 13, this volume) show that high-tech entrepreneurship (entry in sectors that have a R&D intensity greater than 2.5 per cent) and ICT entrepreneurship (entry in ICT sectors) are higher in these sectors (than in the rest of the economy) and have a positive role in the economic performance of the German regions. High knowledge intensity (i.e. high-technology sectors or ICT) affects entrepreneurship by providing more entrepreneurial opportunities. In turn entrepreneurship affects economic performance through increase in competition and in diversity. For the 440 German counties examined high-tech entrepreneurship covers 15 per cent of all entrepreneurial activity. As Audretsch and Keilbach show, GDP growth and R&D intensity have positive effects on high-tech start-ups but only the latter has an effect on ICT start-ups (this may be due to the high growth of ICT start-ups in the 1990s). Similarly, population density (i.e. cities) is a major factor for high-tech and ICT start-ups. Unemployment rate only affects low-tech start-ups, while social diversity (Florida 2002) (measured in terms of entropy index of voting behaviour) affects high-tech start-ups. Industry diversity affects all types of entrepreneurship, implying that an environment open to the acceptance of new ideas is conducive to new firm formation. In addition, external effects of the Marshall–Arrow–Romer type related to industry concentration have a positive effect on all types of entrepreneurship.

9 By way of conclusion

This book goes beyond the individual-centric approach to emphasize the major role of knowledge, innovation systems, networks and institutions in fostering knowledge-intensive entrepreneurship. Indeed, a system view of knowledge-intensive entrepreneurship sheds new light on entrepreneurship in contexts characterized by a high rate of knowledge growth, multiple sources of change and major interdependencies among actors, technologies and demand. In the following chapters entrepreneurship is seen as an activity related to the generation, use, coordination and integration of knowledge. Firms are considered as knowledge processors, integrating scientific and technological knowledge with knowledge about applications and markets. They coordinate their domains of knowledge, their organizational structure, and their product range. The following chapters look at entrepreneurship in terms of start-up entrepreneurship, corporate entrepreneurship and academic entrepreneurship. Finally, knowledge-intensive entrepreneurship is seen as highly affected by the national as well as the sectoral context. That is to say, for instance, that the specific learning and institutional environment affects the type and intensity of knowledge-intensive entrepreneurship through the creation of specific opportunity conditions, the cumulativeness of the advancements, and the rules and regulations affecting innovations.

However, this book only scratches the surface of the phenomenon of knowledge-intensive entrepreneurship in innovation systems. The full understanding of the relationship between knowledge, entrepreneurship and innovation systems requires advancements at various levels.

- First, deeper analyses of the different dimensions of knowledge-intensive entrepreneurship and their links with innovation systems are required.
- Second, consistent appreciative theorizing of the dynamics of knowledge-intensive entrepreneurship is needed.
- Third, formal models, in particular agent-based models, of knowledge-intensive entrepreneurship, innovation and industrial dynamics have to be produced.
- Fourth, quantitative comparative analyses of the generation and evolution of knowledge-intensive entrepreneurship in different sectoral, regional and national systems are requested.

Some clear public policy implications can be derived from this book. Malerba and Vonortas (Chapter 14, this volume) group the policy recommendations into general and specific ones. General recommendations relate to the targets, levels of intervention and context conditions of policy. Specific policy recommendations address issues such as the creation of new knowledge, the establishment of appropriate knowledge infrastructure, the build-up of networks for distributed knowledge, public policy awareness of the role of the sectoral context and the identification of the differences (in knowledge-intensive entrepreneurship, in the institutional setting and in the factors influencing the link between knowledge-intensive entrepreneurship and economic growth).

In Chapter 14 Malerba and Vonortas claim that this book implicitly highlights important overall messages which should guide future analyses of knowledge-intensive entrepreneurship and the requisite policy. They include:

- A *systemic perspective* in that knowledge-intensive entrepreneurship implies an inherently systemic policy that must be tackled from various angles simultaneously to obtain long-term results. Single policy fixes will work for a short while but will not change attitudes for the long haul.
- A role for *socio-economic incentives* in that the simplistic view of the rugged individualist entrepreneur who takes uncalculated risks and single-handedly builds great companies is far from reality. Builders of knowledge-intensive enterprises respond to economic and social incentives that can be influenced to a significant extent by policy.
- *Competence building*, in that new activities and new structures invariably depend on achieving 'new combinations' of capability building.
- *Knowledge application*, in that knowledge-intensive entrepreneurship is about the application of knowledge to new activities. Thus policies need to focus also on applications and not just on the creation of new knowledge.

Note

1 As mentioned above, in this chapter, those papers written for the KEINS project, and not included in this book, are referred to as KEINS. The interested reader may go to www.kites.unibocconi.it then to the 'KEINS project' in order to download the paper. Also in the references at the end of the chapter, these papers are referred to as KEINS.

References

Acs, Z. and Audretsch, D.B (2003) *Handbook of Entrepreneurship Research: an Interdisciplinary Survey and Introduction*, Dordrecht: Kluwer Academic Publishers.

Aidis, R. (2005) 'Entrepreneurship in transition countries: a review', KEINS, Knowledge base entrepreneurship: institutions, networks and systems, EU project no. CT2–CT–2004–506022, deliverable 4b.

Audretsch, D.B. and Keilbach, M. (2006) 'Entrepreneurship – a survey of the literature', KEINS, Knowledge base entrepreneurship: institutions, networks and systems, EU project no. CT2–CT–2004–506022, deliverable 3.

Audretsch, D.B., Lehmann, E.E. and Plummer, L.A. (2007) 'Creating strategic advantage through entrepreneurial governance in new ventures', Jena Economic Research Papers in Economics 2007–086, Friedrich Schiller University Jena, Max Planck Institute of Economics, Thueringer Universitaets und Landesbibliothek.

Audretsch, D.B., Thurik, R., Verheul, I. and Wennekers, S. (eds) (2002) *Entrepreneurship: Determinants and Policy in a European–U.S. Comparison*, Dordrecht: Kluwer Academic Publishers.

Ben-Ari, G. and Vonortas, N. (2005) 'Risk financing for knowledge-based entreprises: mechanisms and policy options', KEINS, Knowledge base entrepreneurship: institutions, networks and systems, EU project no. CT2–CT–2004–506022, deliverable 0 bis.

Bishop, K. (2006) 'Knowledge based entrepreneurship in the Czech Republic and Hungary: results from 4 case studies', KEINS, Knowledge base entrepreneurship: institutions, networks and systems, EU project no. CT2–CT–2004–506022, deliverable 12.

Blithe, A. (2000) *The Origin and Evolution of New Businesses*, New York: Oxford University Press.

Breschi, S., Malerba, F. and Orsenigo, L. (2000) 'Technological regimes and sectoral patterns of innovation', *Economic Journal*, vol. 110 (463): 388–410.

Brusoni, S. and Fontana, R. (2005) 'Incumbents' strategies for platform competition shaping the boundaries of creative destruction', KEINS, Knowledge base entrepreneurship: institutions, networks and systems, EU project no. CT2–CT–2004–506022, deliverable 9.

Brusoni, S. and Prencipe, A. (2006) 'Making design rule: a multi-domain perspective', *Organization Science*, vol. 17 (2): 179–89.

Brusoni S., Prencipe, A. and Pavitt, K. (2002) 'Knowledge specialization, organization coupling, and the boundaries of the firm: why do firms know more than they make?', *Administrative Science Quarterly*, vol. 46 (4): 597–621.

Buenstorf, G. (2007) 'Evolution on the shoulders of giants: entrepreneurship and firm survival in the German laser industry', *Review of Industrial Organization*, vol. 30 (3): 179–202.

Bureth, A., Pénin, J. and Wolff, S. (2006) 'Entrepreneurship in biotechnology: the case of four start-ups in the Upper-Rhine Biovalley', KEINS, Knowledge base entrepreneurship: institutions, networks and systems, EU project no. CT2–CT–2004–506022, deliverable 17.

Busenitz, L.W., West III, G.P., Shepherd, D., Nelson, T., Chandler, G.N. and Zacharakis, A. (2003) 'Entrepreneurship research in emergence: past trends and future directions', *Journal of Management*, vol. 29 (3): 285–308.

Caloghirou, Y. and Vonortas, N.S. (2000) 'Science and technology policy towards research joint ventures', Final Report of the STEP–TO–RJVs Project (SOE1–CT97–1075), Targeted Socio-Economic Research Programme (TSER), European Commission, DG Research.

Casson, M. (1982) *The Entrepreneur. An Economic Theory*, Oxford: Martin Robertson.

Casson, M. (2003) 'Entrepreneurship', in Mokyr, J. (ed.) *Oxford Encyclopaedia of Economic History*, vol. 2, Oxford: Oxford University Press, pp. 210–15.

Cooke, P. (2008) 'Regional innovation systems: origin of the species', *International Journal of Technological Learning, Innovation and Development*, vol. 1 (3): 393–409.

Dushnitsky, G. and Lenox, M.J. (2005) 'When do firms undertake R&D by investing in new ventures?', *Strategic Management Journal*, vol. 26 (10): 947–65.

Florida, R. (2002) *The Rise of the Creative Class: and how it's Transforming Work, Leisure, Community and Everyday Life*, New York: Basic Books.

Foray, D. (2002) 'The knowledge economy and society', *International Social Science Journal*, vol. 54 (171): 1–169.

Foster, J. and Metcalfe, S. (eds) (2001) *Frontiers of Evolutionary Economics*, Cheltenham: Edward Elgar.

Garavaglia, C. and Grieco, D. (2005) 'Hand in hand with entrepreneurship. A critical overview from entrepreneurship to knowledge-based entrepreneurship', KEINS, Knowledge base entrepreneurship: institutions, networks and systems, EU project no. CT2–CT–2004–506022, deliverable 0.

Gartner, W.B. (1990) 'What are we talking about entrepreneurship?', *Journal of Business Venturing*, vol. 5 (1): 15–28.

Glancey, K.S. and McQuaid, R.W. (2000) *Entrepreneurial Economics*, Basingstoke: Macmillan.

Godinho, M.M. and Mamede, R. (2005) 'Creating knowledge-based firms out of existing organizations: a survey of the literature', KEINS, Knowledge base entrepreneurship: institutions, networks and systems, EU project no. CT2–CT–2004–506022, deliverable 2.

Holmén, M., Magnusson, M. and McKelvey, M. (2005) 'What are innovative opportunities?', KEINS, Knowledge base entrepreneurship: institutions, networks and systems, EU project no. CT2–CT–2004–506022, deliverable 7b.

Klepper, S. and Sleeper, S.D. (2005) 'Entry by spinoffs', *Management Science*, vol. 51 (8): 1291–306.

Klepper, S. and Thompson, P. (2005) 'Spin-off entry in high-tech industries: motives and consequences', mimeo.

Lissoni, F. and Sanditov, B. (2007) 'Networks of inventors and academics in France, Italy and Sweden: evidence from the Keins database', KEINS, Knowledge base entrepreneurship: institutions, networks and systems, EU project no. CT2–CT–2004–506022, deliverable 28.

Lissoni, F., Llerena, P., McKelvey, M. and Sanditov, B. (2007) 'Academic patenting in Europe: new evidence from the Keins database', KEINS, Knowledge base entrepreneurship: institutions, networks and systems, EU project no. CT2–CT–2004–506022, deliverable 27.

Loasby, B.J. (2001) 'Time, knowledge and evolutionary dynamics: why connections matter', *Journal of Evolutionary Economics*, vol. 11 (1): 393–412.

Lundvall, B.Å. (ed.) (1993) *National Innovation Systems: Towards a Theory of Innovation and Interactive Learning*, London: Pinter Publishers.

Malerba, F. (2002) 'Sectoral systems of innovation and production', *Research Policy*, vol. 31 (2): 247–64.

Malerba, F. (ed.) (2004) *Sectoral Systems of Innovation*, Cambridge: Cambridge University Press.

Malerba, F. and Orsenigo, L. (1996) 'Schumpeterian patterns of innovation are technology specific', *Research Policy*, vol. 25 (3): 451–78.

Malerba, F. and Zirulia, L. (2007) 'An exploration of knowledge-based corporate entrepreneurship: the coevolution of networks and technological capabilities', KEINS, Knowledge base entrepreneurship: institutions, networks and systems, EU project no. CT2–CT–2004–506022, deliverable 26.

National Commission on Entrepreneurship (2001) *Five Myths about Entrepreneurs*, Washington, DC: NCOE.

Nelson, R.R. (1993) *National Innovation Systems: a Comparative Analysis*, New York: Oxford University Press.

Parhankangas, A. and Arenius, P. (2003) 'From a corporate venture to an independent company: a base for a typology for corporate spin-off firms', *Research Policy*, vol. 32 (3): 463–81.

Pavitt, K. (1998) 'Technologies, products and organization in the innovating firm: what Adam Smiths tells us and Joseph Schumpeter doesn't', *Industrial and Corporate Change*, vol. 7 (3): 433–52.

Radosevic, S. (2008) 'Knowledge based entrepreneurship in the accession countries of Central and Eastern Europe: policy issues', KEINS, Knowledge base entrepreneurship: institutions, networks and systems, EU project no. CT2–CT–2004–506022, deliverable 31.

Schumpeter, J.A. (1934) *The Theory of Economic Development*, New York: Oxford University Press.

Shane, S. (2000) 'Prior knowledge and the discovery of entrepreneurial opportunities', *Organization Science*, vol. 11 (4): 448–69.

Shane, S. and Venkataraman, S. (2000) 'The promise of entrepreneurship as a field of research', *Academy of Management Review*, vol. 25 (1): 217–26.

Swedberg, R. (ed.) (2000) *Entrepreneurship: the Social Science View*, Oxford: Oxford University Press.

Von Hippel, E. (2005) *Democratizing Innovation*, Boston: MIT Press.

Westhead, P. and Wright, M. (eds) (2000) *Advances in Entrepreneurship*, 3 vols, Cheltenham: Edward Elgar.

Witt, U. and Zellner, C. (2005) 'Knowledge-based entrepreneurship: the organizational side of technology commercialization', KEINS, Knowledge base entrepreneurship: institutions, networks and systems, EU project no. CT2–CT–2004–506022, deliverable 8.

Woodward, R. (2006) 'Knowledge-based entrepreneurship in two transition countries', KEINS, Knowledge base entrepreneurship: institutions, networks and systems, EU project no. CT2–CT–2004–506022, deliverable 14b.

Part II

Perspectives on knowledge-intensive entrepreneurship and innovation systems

2 The knowledge-based entrepreneur

The need for a relevant theory of the firm

Patrick Cohendet[1] *and Patrick Llerena*[2]

The purpose of the chapter is to define the main elements which are necessary to the existence of a theory of the firm leaving some 'room' for an 'entrepreneur'. Moreover we would like to define a very specific 'entrepreneur', i.e. a 'knowledge-based' entrepreneur. Our main argument will be to assume that such an entrepreneur can only have a role in a firm considered as a knowledge processor instead of an information processor. To consider a firm as a knowledge processor is in itself also a major challenge. The relevant analytical perspective changes: from a division of labour perspective to a division of knowledge one.

The chapter is organized as follows: we start with a focused interpretation of the role of 'entrepreneurs' as viewed by economists, who did integrate them as a specific agent into their theories of the firm or of the organization, mainly Knight, Schumpeter, Winter and more recently Loasby. In Section 2, we plead for a Babbagian perspective of the division of labour in order to be able to consider properly the entrepreneurial dimension of firms as knowledge processors, and finally in Section 3 we reconsider the relevant role of the entrepreneur as a 'knowledge-led entrepreneur'.

1 The role of 'entrepreneurs' as viewed by Knight, Schumpeter, Winter and Loasby

Paradoxically, the 'entrepreneur' has vanished from economic theory, but even more paradoxically, when considering the Schumpeterian heritage of the evolutionary approach, the entrepreneur is missing in the evolutionary literature on the firm.

In the words of the standard economics, the firm as a profit maximizer does not leave any room for an entrepreneur, even for a decision-maker who would do more than compute the optimal solutions once for all. The main element necessary to have some space available for an agent as an entrepreneur is the existence of uncertain opportunities, emerging from rather synthocratic expectations of the future by the entrepreneur himself.

Uncertainty is one of the main characteristics which justify the existence of a firm. The firm induces uncertainty by its tendency to introduce innovation, and benefits from it by its capacity to seize and generate opportunities.[3]

In Knight's conception of the entrepreneur (1921), the entrepreneur is inherently associated with the existence of uncertainty. For Knight, in a context of uncertainty, if the role of producers is to take responsibility for forecasting the consumer's wants:

> in a second place, the work of forecasting and at the same time a large part of the technological direction of control and production are still further concentrated upon a very narrow class of the producers, and will meet a new economic function, the entrepreneur.
>
> (Knight 1921: 268)

For Knight, entrepreneurship is intrinsically linked to uncertainty and out of equilibrium situations, where unexploited opportunities exist. For Schumpeter, the entrepreneur appears most of the time as a crucial agent, breaking the equilibrium of the economic circular flows.

It is common to distinguish between the Schumpeter I, the author of the *Theory of Economic Development* (1934) and the Schumpeter II, the author of *Capitalism, Socialism and Democracy* (1942), the former using the 'entrepreneur' as the innovative agent, breaking the equilibrium of the circular flows and the latter concerned by large firms where economic development gradually becomes depersonalized and automated. Consequently, 'innovation is being reduced to routine. Technological progress is increasingly becoming the business of trained specialists who turn out what is required and make it work in predictable ways' (Schumpeter 1942: 132, quoted in Hagedoorn 1996: 890). The entrepreneur as a change-agent is supposed to have disappeared and innovation to have become a routine activity of large firms. At the same time, the link between the way these innovation activities were organized in the firm and their relevance for the external environment seems to have been lost. As Langlois (2002) clearly showed, both the idea that there are two different positions of Schumpeter on the role of 'entrepreneurs' and the principle of an obsolescence of his role are wrong.[4] There is in fact a paradoxical situation: there is an evolutionary (also neo-Schumpeterian) approach to firms/organizations with no real theory of entrepreneurship/entrepreneur.

To this wisdom, we should add two remarks:

- First, in Schumpeter's mind the 'entrepreneur' can be a 'collective agent', it is not necessarily a stand-alone 'individual',[5] and it is not obvious that the later Schumpeter ever renounced the figure of an equilibrium-backing entrepreneur.[6]
- Second, Schumpeter (1934) made a strong distinction between a 'manager' and an 'entrepreneur'.

This distinction is crucial for our purpose. Winter (1967) already noted: 'In short, the manager of a firm, when the economic system is in an equilibrium circular flow, is just another guy who "knows his job" in a firm full of people who know their job' (Winter 1967: 16), whereas an entrepreneur is 'the leader who leads the firm to new techniques. In Schumpeter's terms, the "carrying out of new combinations"' (Winter 1967: 19). A possible way to define and specify the role of an entrepreneur is to contrast him/her with the manager.

Even if Schumpeter was right in criticizing Knight for not having distinguished the role of the entrepreneur from that of the resource owner of the firm, it seems that Knight's conception suggests some overlapping between the roles of entrepreneur and manager. The Knightian entrepreneur, who is focused on facing uncertainty, has certainly no vision and no sense of knowledge management; however he shares with the evolutionary entrepreneur the common characteristic of being an interface between the internal organization and the external environment. As Kirzner wrote:

> Now I choose ... to label that element of alertness to possible newly worthwhile goals and to possible available resources ... the entrepreneurial element in decision making. It is this entrepreneurial element that is responsible for our understanding of human action as active, creative, and human rather than as passive, automatic and mechanical.
>
> (Kirzner 1973: 35)

The pursuing of opportunities through alertness is a common trait that the evolutionary entrepreneur shares with the Kirznerian one. However, Kirzner is focused only on the imperfect distribution of information, while the evolutionary entrepreneur is mainly concerned with the distribution and creation of knowledge.

In fact, the entrepreneurial dimension includes the following elements: uncertainty, opportunities and creation of knowledge. However it would be insufficient: 'intelligence' is also needed. With reference to the Walrasian equilibrium Knight draws a conclusion that Schumpeter does not make explicit: 'it is doubtful whether intelligence itself would exist in such a situation' (Knight 1921: 268, quoted by Loasby 2004: 4). In fact,

> intelligence and entrepreneurship are both responses to uncertainty, to situations in which there is no correct procedure for deciding what to do. In such situations people must create their own structures for interpretation and decision, or find some ready-made structure that they are prepared to adapt; and these activities, rather than the rational choices that can be produced by automata, are the distinguishing characteristics of human intelligence. Thus rational choice theory excludes both entrepreneurship and intelligence: appraisal criteria are important.
>
> (Loasby 2004: 4)

Thus the integration of 'entrepreneurship' in the theory of the firm implies to consider a knowledge-based view of the firm (to allow for the detection and the understanding of opportunities, the invention of new combinations, and the implementation of new solutions).[7] But a theory of the firm as processor of knowledge means also reconsidering the foundation of the relation between division of knowledge and division of labour; which is intrinsic to all theories of organization and/or of the firm.

2 The Babbagian perspective of the division of labour: towards a knowledge view of the classical Smithian interpretation[8]

In the classical Smithian interpretation, the progressive specialization of work induces the creation of capabilities and a progressive increase of specialized knowledge (by mechanisms such as learning by doing) (Loasby 1999; Pagano 1985). The task one is assigned to carry out determines what one can learn by carrying out this task. For our argument, what is decisive about this idea is that even though specialization decides what capabilities are created, the criterion that decides how labour is divided is efficiency (i.e. thanks to higher dexterity of who is specialized). It is here that considering the 'second step' from a Babbagian perspective has something important to add.

Babbage (1832) analysed – as did Smith (1776) – England's transition from an agricultural to an industrial economy. Babbage's work is considered a crown jewel of management literature and arguably one of the first management texts ever written, and the beginning of what would later be called operations research. In his analysis of the transition to an industrial economy, Babbage applied Smith's idea of the division of labour not only to mechanical but also to mental operations (Stigler 1991). He paid attention to mechanization and the development of machine-based production in the factory, but at the same time considered the social relations in industry (Hyman 1982). The reason was that, according to Babbage, while the progress of technology played an important part in the emergence of the industrial economy, manufacturers had to pay attention to principles other than merely technological ones. Manufacturers also had to take care of what he called 'the domestic economy of the factory' – in modern terms, organization and management issues. Babbage's contribution that is most relevant in the present context, however, is the so-called 'Babbage principle' that he added onto the Smithian foundations of his analysis:

> That the master manufacturer, by dividing the work to be performed into different processes each requiring different degrees of skill and force, can purchase exactly that precise quantity necessary for each process; whereas, if the entire work is executed by one workman, that person must possess sufficient skill to perform the most difficult, and sufficient strength to carry out the most laborious of the operations into which the art is divided.
>
> (Babbage 1832: 137–8)

Babbage's contribution to Smith's analysis of the division of labour was to distinguish different dimensions of work tasks, such as skill and physical force (Stigler 1991).

What matters for our argument is that, although not inconsistent with Smith, Babbage puts the priority on skills in the sequence of steps in the cycle of division-of-labour–learning–division-of-labour based on the higher degree of specialization and higher knowledge (driven by increased complexity in product markets and technology). He also puts the priority on skills in terms of the criterion that should decide how labour is divided. In fact, one could say that with his argument he establishes existing knowledge and skills as the criterion that should decide how labour is divided.

Possibly, this observation was a result of Babbage's many factory visits in Britain and continental Europe. Having adopted a more fine-grained analytical lens, an additional problem in the Smithian endeavour of splitting up work into sub-tasks in order to maximize economies of scale and gains from specialization becomes visible. The problem is the following: if a work task has two dimensions, which of the two dimensions should be 'optimized', and thus used for the decomposition scheme? The solution Babbage proposed is to divide labour along its different dimensions: if you separate out those tasks that require a high level of physical force (and at the same time a low level of skill, which can be found in any average person), then you can 'optimize' the provision of physical force in the execution of the sub-task by allocating the task to the physically strongest person. Compared to the division of labour by a different logic, Babbage's system increases the economies of scale of specialization further. What is important for the present discussion is that Babbage thus presumed that the division of labour must itself be founded on differences of skills, a point also emphasized by Marshall (Hodgson 1993: 412). By granting logical primacy to skills (or more broadly speaking, characteristics of human resources), Babbage therefore turned the causal relation between the division of labour and of knowledge on its head: the division of labour is determined (caused) by the skill characteristics (the characteristics of the human resources), rather than the other way around.

From a classical Smithian perspective, the division of labour causes learning and skill development, and determines the division of knowledge by giving access to learning-by-doing. The implication of the development described by Martens (1999), however, is that considerations of knowledge have been sidelined in the literature that builds on Smith. Only one stream of literature along the line of Penrose–Arrow–Loasby pays attention to knowledge in considering the division of labour (Penrose 1959; Arrow 1962; Loasby 1999 and 2000). In particular, Loasby pointed out the consequences of a Smithian vision, as interpreted through a Penrosian lens: 'even within a single field of knowledge or activity, differences between individuals' environments, interacting with differences in their responses to these environments, lead to some differences between their knowledge and skills' (Loasby 2000: 7). We now turn to an analysis of the implications of the debate between the Babbagian vision and the classical Smithian interpretation of the division of labour in the theory of the firm.

2.1 The knowledge-based view of the firm

As emphasized above, the debate between the Smithian and the Babbagian views found strong echoes in the theory of the firm and consequently on the role of 'entrepreneurs'. Our opinion is that, in particular, the existing controversy between competences and transactions, which is the hottest issue right now in the literature on the firm, is inherently related to the question of the sequence between division of labour and division of knowledge. The literature dealing with transactions and competences does not in general recognize this underlying debate between division of labour and division of knowledge. Lacking awareness of this fact is, according to us, a source of misinterpretation and misunderstanding between the different theoretical approaches.

It is a rather relevant 'detour' to have a look at the approaches to strategy that are based on theories of the firm, in order to better understand the positioning of an 'entrepreneur' within an organization and/or a firm. It allows us at least to introduce some interesting dimension of the 'entrepreneur' tasks and missions in the design of strategies.

From the 1990s the debate on strategy has been dominated by the dispute between the 'classical' vision focused on activities promoted by Porter, and the challenging competence-based vision inspired by the influential article by Prahalad and Hamel published in the *Harvard Business Review* in 1990.

On the one hand, as is well known, the basic argument of the Porterian approach to strategy is that the superior performance of a firm is explained by its competitive advantage derived from its unique positioning in the industry, whether on costs or on differentiation. Such a positioning is based on activities performed by the firm and systematized in a value chain.[9] It is also based on efficiency as criterion, not on competence development.

On the other hand, the competence-based approach, which renewed the perspective of the resource-based approach due to Wernerfelt (1984), was clearly prescriptive as to the best way to set winning strategies for the firm, especially as to diversification and the excessive use of SBUs (Strategic Business Units) in highly decentralized profit centres. 'In the 1990s, [top executives will] be judged on their ability to identify, cultivate, and exploit the core competences that make growth possible' (Prahalad and Hamel 1990).

2.2 The firm as a 'processor of information', or why dynamic considerations should have priority over efficiency considerations

Let us consider Porter's strategic analysis, which relies on the traditional contractual approaches of the firm – transaction costs theory in particular – that are not designed to accommodate the fundamental characteristics of knowledge. As underlined by Fransman (1998), these traditional approaches consider the firm to be a 'processor of information', the behaviour of which can be understood as an optimal reaction to external signals and factors which are detected.[10] In the Porterian framework, the main result of the processing of information by firms is precisely those activities that emerge from the positioning of an end product within an industry structure. For Porter, once activities are shaped and defined by processing information, then these activities may drive competences (accumulation of collective knowledge and learning in the firm). Processing information is thus the core of the economic decisions of traditional firms, while creating and exchanging knowledge is just a by-product of the current activities.

There is, however, a condition for the above hypotheses (first focus on information processing, then consolidate knowledge by-products) to be valid: the economic environment is supposed to be stable (in particular, the industry structure is considered as given). In innovative environments, characterized by the development of hyper-competition, globalization and the knowledge-based economy, the hypothesis is severely questioned. As Bierly and Chrakrabati (1996) underlined:

> We believe the development of a dynamic knowledge strategy typology or taxonomy will offer more insight than the basic static strategy typologies developed to date. Other generic typologies such as Porter's (1980) rely on the basic assumptions that are not valid for many industries today. Specifically, they assume that a) the primacy focus of strategy is about the positioning of an end-product within an identifiable industry structure, and b) the industry structure is relatively stable and changes to the environment are mostly incremental, linear changes that do not redefine the product or industry. However, other researchers have observed the preponderance of boundary-less industry structures, hyper-competition, increasing globalization, an increasing rate of technological change and diffusion and a tremendous increase in access to information through the advance in computer and communications (Bettis and Hitt, 1995; D'Aveni, 1994; Hamel and Prahalad, 1994). The 'static' generic strategy typologies offer little practical value to top managers in determining how to develop a competitive advantage, which is the primary purpose of strategy.
>
> (Bierly and Chrakrabati 1996: 133)

This context is also the motive for the existence of an entrepreneurial behaviour.

For these reasons, following Fransman, we consider that the actual innovative economic environment invites a reconsideration of the foundations of the theory of the firm. In such environments, firms should be viewed *in primis* as processors of knowledge. As Nonaka and Takeuchi (1995) noted,

> when organizations innovate, they do not simply process information from outside in, in order to solve existing problems and adapt to a changing environment. They actually create new knowledge and information, from the inside out, in order to redefine both problems and solutions and, in the process, to re-create their environment.
>
> (Nonaka and Takeuchi 1995: 58)

2.3 The firm as a processor of knowledge: a programmatic way to implement the perspective we have argued for

As a processor of knowledge, the firm is conceived as a locus of setting up, construction, selection, usage and development of knowledge. The governance of the firm is not focused on the resolution of informational asymmetries, but on the coordination of distributed pieces of knowledge and distributed learning processes. Cognitive mechanisms are essential in this endeavour, and routines play a major role in keeping the internal coherence of the organization. The focus of the theory thus falls clearly on the process of creation of resources. This perspective on the firm has been taken forward, among others, by Cyert and March (1963); Cohen *et al.* (1972); Cohen (1991); Loasby (1976 and 1983); Dosi and Marengo (1994) and Marengo (1994 and 1996).

In a perspective focused on knowledge, the competence-based approach sees the firm *in primis* as a processor of knowledge, for which the focus of attention is the key limiting factor:[11] this theoretical approach to the firm supposes that a firm fixes its attention, first, on the core competence zone, which expresses the domain of knowledge for which the firm aims to be ahead of the competition (for example, Prahalad and Hamel pointed out that Sony had its core domain on 'miniaturization'). This domain of knowledge, with the highest focus of attention, encompasses different abilities to do things (the activities belonging to this domain), but also the capabilities to articulate them together. Moreover, in line with focusing attention on knowledge in this zone, activities belonging to the core domain can be ranked along a decreasing index of importance (from the most important activity to the less important one in this core domain).

Then, in a decreasing order of attention, the firm chooses its domain of competences, which encompasses what the firm 'knows well how to do' (but not necessarily better than the others). In the case of Sony, this could be the domain of 'mastering quality'. Here again, once this second domain of knowledge has been delineated, then activities that belong to this domain can be classified along a decreasing order of attention.

Finally, the remaining domain is the periphery of the firm. This domain encompasses activities that are necessary to support the functioning of the firm, but which are not under a high focus of attention, so that they do not require by definition a strong commitment in terms of knowledge management. The periphery is a domain of knowledge, where the firm has no ambition to do better than others, and would agree to buy from the outside, if necessary.[12] As for the other domains of knowledge, the activities that belong to the periphery (which generally correspond to the larger number of activities and employment positions in the firm) can still be ranked along a decreasing order of attention.

As a result, the 'division of knowledge' (as expressed by the delineation of the domain of core competences, the domain of competences and the periphery) 'precedes' the division of labour (expressed by activities that are ranked along a decreasing order from the core activity). In the co-evolution of the division of labour and the division of knowledge which is suggested above, it is clear that,

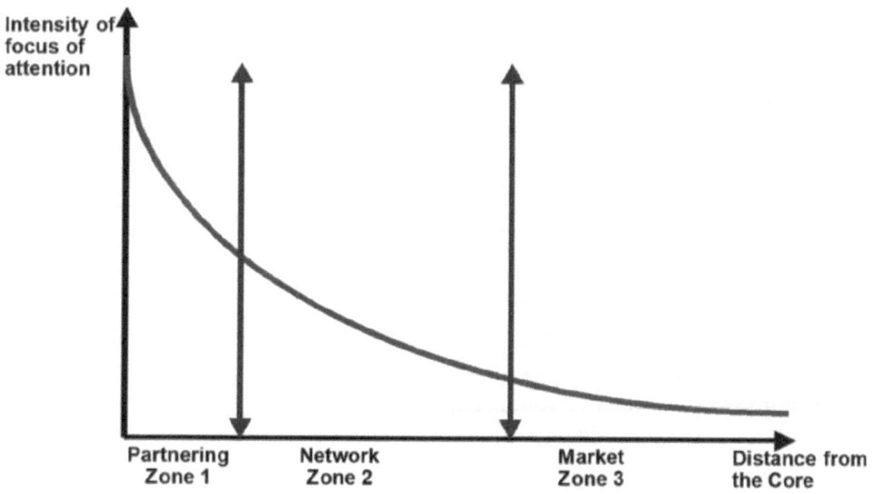

Figure 2.1 Ranking of activities of the firm (distance from the 'Core Competence').[a] (source: adapted from Amesse and Cohendet (2001).)

Note

a Illustrates the above ranking of knowledge activities within the firm. The first zone (Zone 1) is the core itself. The second zone (Zone 2) is one where the firm holds significant pieces of knowledge, but needs to access complementary forms of knowledge held by other firms to be able to develop and use the knowledge efficiently. This zone is characterized by 'networks'. The third zone (Zone 3) is the peripheral zone, where the firm does not hold any specific advantage in terms of knowledge.

to some extent, the choice of the different domains of knowledge is shaped by the pre-existing division of labour. The choice of competences is partly guided by the activities that the firm has experienced. However, the influence of the pre-existing division of labour is only partial. The cognitive deliberation that leads to the future division of labour could focus the attention on developing fields of knowledge of which the firm has no experience. Some of the activities required for the selected domains of knowledge could be acquired by the firm, and some are not even known in advance and could be progressively built.

Moreover, such an order implies that the firm manages competences and transactions simultaneously, but that it does so according to a specific lexicographic order of priorities.

First, within its sets of 'high focus of attention' (the 'core' competences) the firm functions as a knowledge processor giving full priority to the creation of resources. The activities belonging to this key domain of knowledge are not considered to be tradable on the market: they are 'disconnected' from the make-or-buy trade-off suggested by the transaction-cost theory. Second, once the set of activities that belong to the set of competences has been chosen, the other activities that do not belong to the core (the 'periphery' or 'non core activities') are then managed under traditional methods which may rely on the transaction-cost approach.[13]

A 'dual' structure of the firm emerges, as a consequence of a clarified view on the respective role of division of knowledge and division of labour.

3 The implication for a knowledge-based approach to the firm: more on the dual governance structure of the firm and on 'entrepreneurship'

3.1 More on the 'dual governance structure of the firm'[14]

If the common view nowadays is that transactions and competences are complements, the nature and intensity of the interaction between transactions and competences differ radically according to the theoretical approach of the firm. As we have seen, behind this debate, there is a very fundamental issue dealing with the relationships between the division of labour and the division of knowledge.

Keith Pavitt (1998 and 2002) emphasized already the importance of these distinctions in the growing knowledge-based economy. The starting point of Pavitt's analysis of the relationship between the division of labour and the division of knowledge is the large multi-divisional firm, with established R&D activities and 'a product range that has grown out of a common, but evolving technological competence' (Pavitt 1998). These firms are considered by Pavitt as the largest single source of the new technological knowledge on which innovation depends.

From these premises, following Nelson (1994), Pavitt distinguished two complementary elements in firm-specific knowledge.

- First, a 'body of understanding', based on competences in specific technological fields, 'and reflected in the qualification of corporate technical personnel and in the fields they patent and publish' (Pavitt 1998: 436).
- Second, a 'body of practice', related to the design, production and use of a specific product model, which is generally obtained through the combination of experience, experimentation and interactions between different parts of the organization. A body of practice 'consists largely of organizational knowledge that links a body of understanding with a commercial successful (or more broadly useful) artefact' (Pavitt 1998: 436).

The distinction between bodies of understanding (technologies) and bodies of practice (products) helps clarify the notion of 'diversity', which is confused in the evolutionary theory of the firm. As Pavitt emphasized, large firms are generally active in a range of technologies broader than the products they make. He showed that for the individual firm, technological diversity provides the basis to manufacture and improve its products. And that for the economy as a whole, more diversity amongst firms in their mixes of specialized technological knowledge enables them to explore and exploit a fuller range of product markets. The distinction between technological diversity and product diversity is well expressed in the following:

> At any given time, advances in some fields of technology open major opportunities for major performance in material, components, and subsystems (e.g. economies of scale in continuous processes, economies of miniaturization in information processing). The direction of these improvements is easily recognized, even if they require the commitment of substantial resources for their achievement, e.g. Moore's Law in the semiconductors. Thus experimentation and diversity do not take place between different technologies. On the contrary, rich and well-known directions of improvements in underlying technologies create opportunities for diversity and experimentation in product configurations. Technological opportunities create product diversity. There is no convincing evidence that technological diversity created product opportunities.
>
> (Pavitt 1998: 440)

Pavitt's developments on the notion of diversity bring forward some convincing arguments that help understand the limits of evolutionary analysts who put too much emphasis on firm-specific technological competences that lead to place, for instance, major technological improvements (technological discontinuities) at the heart of the theory of the innovative firm. He also adds very accurate remarks such as the fact that firms rarely fail because of an inability to master a new field of technology, but because they do not succeed in matching the firm's system of coordination and control against the nature of the available technological opportunities. For him, competition is not based on technological diversity, but on diversity and experimentation in products. The main problem is on coordination and control. However, in these developments, Pavitt implicitly assumed:

- First, that when we consider technological developments or product developments, the production of new knowledge is done 'in-house' by firms. When we consider improvements in a given technology it is done through activities performed within the firm. The division of knowledge is shaped by the division of work. Only those who do, have the possibility to know more.
- Second, that more or less, we start from a situation where firms have a given body of understanding (an existing level of technological diversity) and that their bodies of practice will adapt to find the 'right tuning' to the existing bodies of understanding.

We extend Pavitt's vision on the nature of the innovative firm, by questioning the idea that the division of labour 'comes first'. Pavitt himself, in a co-authored paper with Brusoni and Prencipe (2002) when asking 'why do firms know more than they do?' suggested such a perspective. But they considered mainly situations of large firms in cases of modular platforms. In a knowledge-based economy, we argue that the very essence of the meaning of knowledge-based economy is that the division of knowledge comes 'first'.

Our view is that when the economic environment is stable, the traditional view is acceptable: the division of labour can first operate with the determination of the activities to be carried out by each firm. Then from this division of labour firms can accumulate *ex post* some specialized knowledge. There is no room for any entrepreneurial behaviour. But as the environment gets more uncertain, as firms need to increase their ability to innovate in order to cope with changing environments, they have to create new knowledge 'from the inside out' to redefine both problems and solutions. They cannot rely on a pre-existing division of labour. The cognitive processes tend to come first, and drive the division of labour. Then 'entrepreneurs' come in as defined by Knight, Schumpeter, Winter and Loasby (see pp. 31–4).

3.2 Where 'entrepreneurship', and moreover 'knowledge-based entrepreneurship', has room in the framework[15]

Within this view of the theory of the firm, if we do not accept the idea that a specific agent is in charge of ensuring the coherence between the evolutionary forces and the cognitive evolution of the agents within the firm, we are condemned to face too many problems of indetermination. This situation will in turn prevent the theory from leading to relevant explanations of firm behaviours and in particular of entrepreneurial behaviour. The main types of indetermination that result from the evolutionary theory of the firm without an entrepreneur are the following ones.

1 There is indeterminacy in the degree of diversity that a firm can afford. Since diversity is not a free good, there should be an internal selection mechanism, in addition to the external ones, to shape the relevant routines. Let us imagine, for instance, that in a pure evolutionary perspective, all the routines of the organization happen to be equally successful, in such a way that the external environment does not reveal the hierarchy of choice between routines: without a specific agent within the firm in charge of selecting the efficient routines with relevance to limited resources consideration, the theory will lead to complete indeterminacy. Moreover, there is a need for delimiting those efficient routines that the selection mechanisms have detected, that is, those that will belong to the core competences of the organization. The external selection mechanisms cannot reveal what the core competences of an organization are. They certainly can select an efficient basket of routines, each one of them being a potential component of the core competences of the firm. But the domain of core competences has to be shaped according to a long-term strategy and a vision. The core domain is not just a juxtaposition of efficient routines. It is the result of a coherent and patient process of building specific knowledge, which requires some idiosyncratic connections between different routines and bodies of knowledge. And such a construction requires a specific agent in charge of orienting the process of creation and accumulation of resources of the firm.

2 There is a key distinction in the evolutionary theory of the firm between 'exploration' and 'exploitation' indeterminacy. As stated earlier, organizations always face the dilemma between concentrating their resources on the exploitation of the knowledge which is already available to them and the exploration of new possibilities. The theory calls for some internal agents in charge of tackling the problem of trade-off between exploration and exploitation, for there is no solution in the sole routine conception of the firm to solve the indetermination. Another way to put it is to say that the dynamic capabilities of the firm should not become critical rigidities as time elapses. But beyond the trade-off between exploitation and exploration, there is a specific need to 'allocate the intensity of learning' within the firm to those learning processes that appear strategic. As far as learning is concerned, as Cyert and March (1963) noted, attention is a scarce resource within the firm. The external selection mechanisms tell us nothing about where to focus the intensity of learning. There is critical room there for an agent in charge of orienting the process of resources through the allocation of the intensity of learning processes according to the 'vision' of the firm that is defined by this agent. Cyert and March indicated, for instance, the process of 'sequential attention to goals' as one of the mechanisms to drive the attention of participants of the firm in a problem-solving perspective. But there could be many others.

3 There could be some time discrepancies between, on the one hand, the speed of evolution of the external environment (and in particular the speed with which external selection mechanisms function) and, on the other hand, the speed of evolution of the cognitive processes at stake in the organization. It could take too much time for the external environment to reveal whether a given routine is to be successful, as it could require too short a time within a given firm to allocate the required resources and stimulate the learning processes that are needed to maintain a given efficient routine active: without a specific agent in charge of 'tuning' the balance between different speeds of evolution, there are strong risks of indetermination.

As a consequence of the above indeterminations, a specific agent in charge of the process of creation of resources has to be re-introduced to ensure the missing link between the internal and the external environments of the firm. The introduction of this specific agent is by no means an artificial one: it follows logically from the theoretical premises of the evolutionary firm. This agent, that we call the 'entrepreneur', has to be an active element of the interaction between both environments, designing the internal organization and being proactive towards the external environment. Of course, this agent may be a collective one: the 'entrepreneur' is in fact the repository of the 'entrepreneurial spirits' of the organization.

In the context of this evolutionary perspective, we can now try to define more precisely the content that should characterize this entrepreneurial function. First, the entrepreneurial function has to be characterized as a specific 'asset': it is a 'dynamic capability'; i.e. the ability to manage strategically the adaptation, the integration and the re-configuration of internal and external organizational skills, resources and functional competences towards a changing environment; where time to market and timing are critical, the pace of innovation is accelerating, and the nature of future competition and markets is difficult to determine (Teece and Pisano 1994: 538). This capability means that in particular the entrepreneur has a representation not only of the possible evolution of his external environment, but also of the corresponding internal configuration, that is relevant for meeting the requirements of the external environment. In other words, the key entrepreneurial function in an evolutionary approach is to organize the matching process between internal and external environments.

Second, to act as an active interface between the internal and external environments of the firm, the entrepreneur has to develop and diffuse a specific 'vision' of the firm's context and future. The vision of a firm is defined as the dominant set of beliefs in the firm regarding the firm's internal and external circumstances, the shape of things to come in the future and, in the light of these factors, the way the firm should 'play its cards'. Since vision depends on the particular construction of particular beliefs, vision is by definition always bounded. Bounded vision and the possibility of vision failure are, therefore, logical implications of the concept of vision (Fransman 1998). It is in accordance with this vision that the entrepreneur will position the firm in its environment, defining both its strategy and its internal structure.

It should be emphasized that by an active position towards the external environment, we mean that the entrepreneur is able to influence the competitive context in which he evolves. Innovations of different nature will allow him to determine at least partially the selection mechanisms at work outside the firm. In fact, he might endogenize the external environment, and build a flexibility of initiative (Amendola and Bruno 1990). This capacity of shaping the external environment is also a way to influence the nature of the industry evolution. Expectations and visions are of major importance for understanding not only the evolution of firms, but also the evolution of industries.

Third, the vision or the business conception, which is a primary entrepreneurial input (Witt 1998: 162), will impact the organization of the firm itself. In fact, we even strongly argue at this point that there is, in this respect, room in the evolutionary approach for a hierarchy and a managerial component. Some authors (Loasby 1991; Witt 1998) have already reappraised the role of entrepreneurs, and underlined the role of leadership as provision and enforcement frames in an evolutionary context.

> Cognitive commonalities, that is, socially shared tacit knowledge including knowledge about social models of behaviour, may emerge spontaneously from intense communication as an unintended collective outcome and may, as such, be difficult to influence.
>
> Sometimes, however, the institutional set-up of the interactions assigns certain individuals a position in which they get a chance to shape the com-

> munication processes and thus to exert an influence on the collective outcome. The firm organization is a case in point. Indeed, the social–cognitive implications of bounded rationality are the key to understanding why firms, as organizations, are able to achieve internal consistency and co-ordination of individual efforts.
>
> (Witt 1998: 166)

In the perspective of the firm as a processor of knowledge which is proposed by the evolutionary approach, the role of the entrepreneur and the role of the manager tend to overlap to a large extent. In the case of evolutionary entrepreneurship, the very reason for convergence is that managing knowledge cannot be distinguished from conducting the process of creation of resources. There are in fact two main specific contexts in which the overlapping is extreme.

When dealing with the core competences of the firm, the role of entrepreneur and the role of manager entirely overlap. The focus of the firm is completely on the process of management of knowledge. For the rest of the activities of the firm (those that do not belong to the core competences), they seem to require a classical type of management which is independent of any entrepreneurial consideration. However, it has been shown (Cohendet and Llerena 1999; Amin and Cohendet 2000) that, in this case, there is a specific problem of coherence of the firm between the management of activities that belong to the core competences and the other activities. This risk of incoherence supposes that the manager finds ways to diffuse his vision throughout the entire firm, so that brings again at least partially some overlapping with the entrepreneurial role.

When the environment is constantly changing, the process of allocation of resources cannot be separated from the process of creation of resources. Thus the role of the entrepreneur and the role of the manager also converge in such turbulent cases.

If we accept as a hypothesis that in the evolutionary approach, the roles of entrepreneur and of manager merge, we can call then the key actor of the evolutionary firm the 'entrepreneur/manager'. Beside the classical attributes of entrepreneur and manager in the traditional theories of the firm, the evolutionary approach allows this central actor to shape cognitive commonalities and socially shared interpretation patterns and frames. Through the introduction of the manager/entrepreneur as the key agent in charge of the process of creation of resources in the evolutionary firm, we are now in a position to rethink the two main mechanisms, the coordination mechanisms and the incentive mechanisms of the evolutionary theory of the firm.

As far as coordination mechanisms are concerned, the introduction of the entrepreneur/manager clarifies the ways the process of centralization/decentralization is taking place at the level of the firm. If he cannot by himself modify or change a given routine, the entrepreneur/manager may influence (directly or/and indirectly) the routines at all levels of the firm. More precisely, he may shape the learning processes at stake within a given routine, or between different routines by focusing the attention on certain characteristics of these processes (by rewarding, for instance, exploration instead of exploitation). It is by the development of common representation frames that he might obtain a certain degree of internal coherence, in particular of the sets of routines and rules governing the internal structure of the firm. The business conception plays in this case the role of a shared 'common knowledge', that Foss (1999) indicated as the founding element of leadership, i.e. defined as 'the ability to resolve social dilemmas by influencing beliefs' (Foss 1999: 1). The building of a state of common knowledge within the firm is one of the key tools that the entrepreneur/manager can control in a given firm. Not only can he define the intensity of the common knowledge in the firm, but he can also orient the form taken by the common knowledge in the enterprise (the common knowledge could be entirely identified with the 'culture of the firm', or rather through the type of structure of information that is implemented in the firm). However, as he contributes to shaping the common knowledge of the firm, he also implicitly defines *a contrario* what remains 'ambiguous' in the firm, in the sense used by March and Olsen (1976). In a dynamic perspective the degree of ambiguity of the organization is associated with the margin of interpretation of the members of the firm, which is to a certain extent associated, as it was said earlier, with the potential of creativity of the firm (provided that the common knowledge is strong enough). The efficiency of the entrepreneur/manager on the coordination mechanisms will thus result to a large extent in the fine-tuning between common knowledge and ambiguity.

With respect to the incentive mechanisms, the explicit introduction of a hierarchy in the evolutionary theory of the firm also contributes to clarifying the ways incentive schemes are set up. Moreover these additional characteristics of entrepreneurs/managers strongly plead for a richer design of incentives than in the pure transaction and incentive-oriented approaches. What becomes clear when taking into account the role of the entrepreneur/manager as it was detailed above is that this actor brings with him – beside the classical incentive mechanisms – another type of incentive mechanisms: those which have to do with the ways the entrepreneur/manager may diffuse his vision through the firm. The quality of the leadership (or 'business conception') of the entrepreneur/manager 'may induce the employees to believe in the success of the entire venture in which they are participating. It may also promise to be personally rewarding in terms of qualification enhancements, working conditions, career options and remuneration' (Witt 1998: 168).

At first sight, when considering the coherence between coordination and incentive mechanisms, the incentive-based and the transaction-costs approaches apparently reveal significant advantages over evolutionary ones, if we confront them to real-life situations such as the relationships between shareholders and managers on the one hand, and managers and employees on the other hand. It makes explicit who, in a given context, sets the incentive schemes and for which reasons, and to what extent the firm is centralized or decentralized. While in the evolutionary approach, it is often said that the role of hierarchy is generally hidden by the emphasis put on routines and the influence of shareholders is not even mentioned. This emphasis constitutes a priori a strong obstacle for the integration of incentive mechanisms in the framework of the evolutionary approach. We have tried to show in this contribution that this is not the case: the introduction of the evolutionary entrepreneur leads to an in-depth reconsideration of the main characteristics of the firm, including the key question of incentive mechanisms. It also leads to focusing the analytical attention on the role of entrepreneur and entrepreneurship as processes which create, develop and shape the knowledge base of firms, i.e. 'entrepreneurship' in an evolutionary approach is intrinsically a 'knowledge-based entrepreneurship'.

Notes

1 BETA–HEC, ULP–CNRS, 61 av. de la Forêt Noire, 67085 Strasbourg, France.
2 BETA, ULP–CNRS, 61 av. de la Forêt Noire, 67085 Strasbourg, France; pllerena@cournot.u-strasbg.fr.
3 Chamley wrote:

> One of the fundamental characteristics of the firm situation is the uncertainty which affects all its decision. The private firm creates or increases uncertainty through its creation initiatives and it derives its 'raison d'être' from the uncertain future of the economy. Economic activity can be conceived without uncertainty, but this is not the case for a private firm, neither is it for money. The role of the firm – like money but in the special field of production – is to standardize uncertainty, to integrate it systematically in economic activity.
>
> Schumpeter (1912) had situated the entrepreneur in his true perspective when he qualified him as the agent of the spontaneous evolution of the economy, as the disruptive factor of static equilibrium. If forty years later the private firm seems to him close to decay, one of the causes is that it is losing its 'raison d'être' in an economy without hazard and without risk (Schumpeter 1942).
>
> However to retain only the innovating role of the private firm and its instability would be to neglect one of its essential aspects. If the aim of the private firm is the creative initiative, the means consists mainly in organizing, planning and setting rules, laying down standards, establishing and maintaining some discipline: therefore the function of the private firm is also to create certainty.
>
> Actually the firm appears to be the point of equilibrium between uncertainty and security. Its existence – the existence of each particular firm as well as the existence of the private firm as an institution – depends on keeping this equilibrium.
>
> (Chamley 1956: 31)

Moreover, C. Chamley wrote:

> The firm, which was neglected by the classical theory as a distinct factor of production has been only superficially assimilated by the neo-classical theory. It could not be any other way. What prompts the search for maximum profit, without which the trend of the system towards the theoretical optimum disappears, can be effective only if the maximum profit is defined, i.e. if the economic world is transparent, which entails that the entrepreneur does not have anymore a specific role to play: [...] there is not even the residue left which can be considered as independent of any specific economic system. Facts contradict this interpretation of the economic world but they confirm indirectly its logic: the economy is not transparent. Therefore the search for the maximum profit is not what motivates the private firm.

> Actually the classical theory did not aim at providing a description of the mechanism of the firm but only at pointing the trends of liberal capitalism. The mistake of the neo-classical theory was its attempt at extending in a very rigorous way a determinism originally – and in a rather flexible way – conceived for the (economic) system as a whole to the internal functioning of the firm. Hence the paradox that following the end of the liberalism theory, the theory of the private firm remains to be written.
>
> (Chamley 1956: 36–7)

Historical note: Professor C. Chamley was professor at the University of Strasbourg, specialist of history of economic thought, and one of the main contributors to the creation of BETA as a research unit.

4 'The obsolescence thesis seeks to draw a distinction between early capitalism and later capitalism, perhaps, but not between an early and a later Schumpeter' (Langlois 2002: 2–3).

5 'Schumpeter (1949: 71) even mentioned the possible innovative role for organizations other than companies e.g. the role of state agencies, as collective entrepreneurial change-agents' (Hagedoorn 1996: 891).

6 See Becker *et al.* (2006), and Schumpeter (2005).

7 'Intelligent behaviour depends on making sense by making selective connections, thus creating new combinations; thus, of course, is Schumpeter's own summary of the entrepreneurial vision' (Loasby 2004: 5).

8 This section owes much to our common work with M. Becker (see Becker *et al.* 2007).

9 As Porter stressed in the new introduction to his 1985 book, 'the book's core is an activity-based theory of the firm' (Porter 1998: 15).

10 The behaviour of the firm as a rational information-processing machine means that the same signals will give rise through time to the same pattern of action, provided the technical conditions (as expressed by the production function) remain unchanged. The neo-classical theory of the firm, in particular principal/agent theory, has basically reduced the coordination principles to a bundle of bilateral contracts which are meant to achieve coordination by devising appropriate incentive schemes in order to align self-interested individual action with the common organizational goals. The transaction-cost approach, despite its different angle and its specific focus on the boundaries of the firm, comes to a similar fundamental conclusion: the firm could be seen as a 'nexus' of contracts. Its very reason of existence is to correct market failures, when the functioning of market mechanisms in terms of information processing is too costly. Transaction-cost theory agrees with the principal/agent vision that information is imperfect and that the existence of potential asymmetries of information authorizes unproductive rent-seeking behaviour. The firm is thus conceived as an institutional mechanism creating a governance structure to solve the problem of misaligned incentives attendant on imperfect information. The focus is thus on the process of allocation of resources needed to cope with such an adaptation.

11 Following Simon (1982), it is attention rather than information that is becoming a rare resource as screening and selection of information become important functions.

12 For further developments on the definition of the core domain and the periphery, see Amesse and Cohendet (2001), Cohendet and Llerena (2001, 2005 and 2009).

13 In other words, in the domain of competences, low-opportunism conditions prevail, and allow more valuable knowledge to be applied to the firm's activities. In the periphery domain, meanwhile, the risk of opportunism is high and leads to the situation described by the classical transactional approach. As Langlois and Foss (1999) claimed:

> as firms move increasingly from their core businesses, they confront increasing adverse selection and moral hazard, since management becomes increasingly

unable to efficiently monitor employees or to evaluate their human capital. Agency costs rise correspondingly, producing the net profitability disadvantage associated with further integration.

(Langlois and Foss 1999)

14 See also Cohendet and Llerena (2001, 2005 and 2009).
15 See also Cohendet *et al.* (2000) for related arguments.

References

Amendola, M. and Bruno, S. (1990) 'The behaviour of the innovative firms: relations to the environment', *Research Policy*, vol. 19 (5): 419–33.

Amesse, F. and Cohendet, P. (2001) 'Technology transfer revisited in the perspective of the knowledge based economy', *Research Policy*, vol. 30 (9): 1459–78.

Amin, A. and Cohendet, P. (2000) 'Organisational learning and governance through embedded practices', *Journal of Management and Governance*, vol. 4 (1/2): 93–116.

Arrow, K. (1962) 'The economic implications of learning by doing', *The Review of Economic Studies*, vol. 29 (3): 155–73.

Babbage, C. (1832/1989) 'The economy of machinery and manufacturers', in Campbell-Kelly, M. (ed.) *The Works of Charles Babbage*, vol. 8, London: William Pickering.

Becker, M.C., Cohendet, P. and Llerena, P. (2007) 'Division of labor and division of knowledge: why the nature of the causality really matters for the evolutionary theory of the firm?', in Cantner, U. and Malerba, F. (eds) *Innovation, Industrial Dynamics and Structural Transformation: Schumpeterian Legacies*, Berlin: Springer Verlag, pp. 49–65.

Becker, M.C., Knudsen, T. and March, J.G. (2006) 'Schumpeter, Winter and the sources of novelty', *Industrial and Corporate Change*, vol. 15 (2): 353–71.

Bettis, R.A. and Hitt, M.A. (1995) 'The new competitive landscape', *Strategic Management Journal*, vol. 16: 7–19.

Bierly, P.E. and Chakrabarti, A.K. (1996) 'Generic knowledge strategies in the US pharmaceutical industry', *Strategic Management Journal*, vol. 17: 123–35.

Brusoni, S., Prencipe, A. and Pavitt, K. (2002) 'Knowledge specialization, organization coupling, and the boundaries of the firm: why do firms know more than they make?', *Administrative Science Quarterly*, vol. 46 (4): 597–621.

Chamley, P. (1956) 'Théorie économique et principes de gestion des entreprises privées', *Revue Economique*, vol. 7 (1): 19–38.

Cohen, M. (1991) 'Individual learning and organizational routine: emerging connections', *Organization Science*, vol. 2 (1): 135–9.

Cohen, M.D., March, J. and Olsen, J.P. (1972) 'A garbage can model of organizational choice', *Administrative Science Quarterly*, vol. 17 (1): 1–25.

Cohendet, P. and Llerena, P. (1999) 'La conception de la firme comme processeur de connaissances', *Revue d'Economie Industrielle*, no. spécial 'Economie de la Connaissance', vol. 88, 2ème trim.: 211–36.

Cohendet, P. and Llerena, P. (2001) 'A dual theory of the firm between transactions and competences', paper presented at ETIC Conference, Strasbourg, October.

Cohendet, P. and Llerena, P. (2005) 'A dual theory of the firm between transactions and competences: conceptual analysis and empirical considerations', *Revue d'Economie Industrielle*, vol. 110 (110): 175–98.

Cohendet, P. and Llerena, P. (2009) 'Organisation of firms, knowing communities and limits of networks in a knowledge-intensive context', in Morroni, M. (ed.) *Corporate*

Governance, Organization and the Firm: Co-operation and Outsourcing in the Global Economy, London: Edward Elgar, pp. 104–20.

Cohendet, P., Llerena, P. and Marengo, L. (2000) 'Is there a pilot in the evolutionary firm?', in Foss, N. and Mahnke, V. (eds) *New Directions in Economic Strategy Research*, Oxford: Oxford University Press, pp. 95–115.

Cyert, R.M. and March, J.G. (1963, 1992) *A Behavioral Theory of the Firm*, 2nd edn, Oxford: Blackwell.

Dosi, G. and Marengo, L. (1994) 'Some elements of an evolutionary theory of organizational competences', in England, R. (ed.) *Evolutionary Concepts in Contemporary Economics*, Ann Arbor: University of Michigan Press, pp. 157–78.

Foss, N.J. (1999) 'Understanding leadership: a coordination theory', DRUID Working Paper no. 99–3, March.

Fransman, M. (1998) 'Information, knowledge, vision and theories of the firm', in Dosi, G., David, T. and Chytry, J. (eds) *Technology, Organization, and Competitiveness – Perspectives on Industrial and Corporate Change*, Oxford: Oxford University Press, pp. 147–92.

Hagedoorn, J. (1996) 'Innovation and entrepreneurship: Schumpeter revisited', *Industrial and Corporate Change*, vol. 5 (3): 883–96.

Hamel, G. and Prahalad, C.K. (1994) *Competing for the Future*, Boston: Harvard Business School Press.

Hodgson, G. (1993) *Economics and Evolution*, Cambridge: Polity Press.

Hyman, A. (1982) *Charles Babbage: Pioneer of the Computer*, Oxford: Oxford University Press.

Kirzner, I.M. (1973) *Competition and Entrepreneurship*, Chicago: University of Chicago Press.

Knight, F. (1921/1985) *Risk, Uncertainty and Profit*, Chicago: University of Chicago Press.

Langlois, R.N. (2002) 'Schumpeter and the obsolescence of the entrepreneur', University of Connecticut Working Paper no. 2002–19, August.

Langlois, R.N. and Foss, N. (1999) 'Capabilities and governance: the rebirth of production in the theory of economic organisation', *Kyklos*, vol. 52 (2): 201–18.

Loasby, B.J. (1976) *Choice, Complexity and Ignorance*, Cambridge: Cambridge University Press.

Loasby, B.J. (1983) 'Knowledge, learning and the enterprise', in Wiseman, J. (ed.) *Beyond Positive Economics?*, London: Macmillan, pp. 104–21.

Loasby, B.J. (1991) *Equilibrium and Evolution: an Exploration of Connecting Principles in Economics*, Manchester: Manchester University Press.

Loasby, B.J. (1999) *Knowledge, Institutions and Evolution in Economics*, London: Routledge.

Loasby, B.J. (2000) 'Organisations as interpretative systems', paper presented for a workshop in Amiens, 25–26 May.

Loasby, B.J. (2004) 'Entrepreneurship, evolution and the human mind', paper presented at the International J.A. Schumpeter Society conference in Milan, 9–12 June.

March, J. and Olsen, J. (1976) *Ambiguity and Choice in Organizations*, Oslo: Universitetsforlaget.

Marengo, L. (1994) 'Knowledge distribution and coordination in organizations: on some social aspects of the exploration vs. exploitation trade-off', *Revue Internationale de Systémique*, vol. 7: 553–71.

Marengo, L. (1996) 'Structure, competence and learning in an adaptive model of the firm', in Dosi, G. and Malerba, F. (eds) *Organization and Strategy in the Evolution of the Enterprise*, Basingstoke: Macmillan, pp. 124–54.

Martens, B. (1999) 'The evolution of the concept of division of labour in economics', Max Planck Institut für die Erforschung von Wirtschaftssystemen Discussion Paper no. 18–99.

Nelson, R.R. (1994) 'The co-evolution of technology, industrial structure and supporting institutions', *Industrial and Corporate Change*, vol. 3 (1): 47–63.

Nonaka, I.H. and Takeuchi, C. (1995) *The Knowledge-Creating Company*, New York: Oxford University Press.

Pagano, U. (1985) *Work and Welfare in Economic Theory*, Oxford: Basil Blackwell.

Pavitt, K. (1998) 'Technologies, products and organization in the innovating firm: what Adam Smiths tells us and Joseph Schumpeter doesn't', *Industrial and Corporate Change*, vol. 7 (3): 433–52.

Pavitt, K. (2002) 'Innovation routines in the business firm: what corporate tasks should they be accomplishing?', *Industrial and Corporate Change*, vol. 11 (1): 117–33.

Penrose, E. (1959, 1995) *The Theory of the Growth of the Firm*, Oxford: Oxford University Press.

Porter, M.E. (1980) *Competitive Strategy*, New York: The Free Press.

Porter, M.E. (1985) *Competitive Advantage*, New York: The Free Press (2nd edn 1998, with a new introduction).

Prahalad, C.K. and Hamel, G. (1990) 'The core competence of the corporation', *Harvard Business Review*, vol. 68 (3): 79–91.

Schumpeter, J.A. (1912, 1934 and 1980) *The Theory of Economic Development*, London: Oxford University Press.

Schumpeter, J.A. (1942, 1975) *Capitalism, Socialism and Democracy*, New York: Harper & Row.

Schumpeter, J.A. (1949) *Economic Theory and Entrepreneurial History – Change and the Entrepreneur; Postulates and Patterns for Entrepreneurial History*, Cambridge, MA: Harvard University Press.

Schumpeter, J.A. (2005) 'Development', *Journal of Economic Literature*, vol. 43 (1): 108–20 ('Entwicklung', translated by Becker, M.C. and Knudsen, T.).

Simon, H.A. (1982) 'The economics of information processing', in Simon, H.A. (ed.) (1982) *Models of Bounded Rationality, vol. 2: Behavioral Economics and Business Organization*, Cambridge, MA: MIT Press, pp. 71–8.

Smith, A. (1776) *The Wealth of Nations*, London: W. Strahan and T. Cadell.

Stigler, G. (1991) 'Charles Baggage (1791 + 200 = 1991)', *Journal of Economic Literature*, vol. 29: 1149–52.

Teece, D.J. (1998) 'Capturing value from knowledge assets: the new economy, markets for know-how, and intangible assets', *California Management Review*, vol. 40 (3): 55–79.

Teece, D.J. and Pisano, G. (1994) 'The dynamic capabilities of firms: an introduction', *Industrial and Corporate Change*, vol. 3 (3): 537–56.

Wernerfelt, B. (1984) 'A resource-based view of the firm', *Strategic Management Journal*, vol. 5 (2): 171–80.

Winter, S.G. (1967) 'Toward a neo-schumpeterian theory of the firm', mimeo, RAND Corporation.

Witt, U. (1998) 'Imagination and leadership: the neglected dimension of the evolutionary theory of the firm', *Journal of Economic Behaviour and Organization*, vol. 35 (2): 161–77.

3 What makes entrepreneurship systemic?

Slavo Radosevic

1 Introduction

The escalation in research on entrepreneurship has not qualitatively improved our understanding of how entrepreneurship contributes to growth. Due to its dominant focus on individual entrepreneurship, traditional research in this area has produced diminishing returns (Shane and Venkataraman 2000). According to Swedberg (2000: 32; see also Gartner 1988) attempts to single out 'one or several psychological traits as typical for the entrepreneurial personality is generally considered to have failed'. In contrast, Shane (2003: 61) argues that 'the entrepreneurship literature has shown that the people who engage in entrepreneurial activity are not randomly determined. Certain individual level characteristics are associated with the decisions to engage in entrepreneurship activity'. Although the role of personality traits remains unresolved it has created several myths on entrepreneurship that are all closely related to the idea of person-centric entrepreneurship (NCOE 2001 and 2002).

A thematic issue of *Research Policy* (see Shane and Venkataraman 2003) pointed to the limitations of this line of research and argued for a better understanding of entrepreneurship systems and networks. Scholars have begun to question the widely held view that entrepreneurs, as economic actors, are isolated and are depicting them as being tied through their social relationships to a broader network of actors. (For a review of this literature see Hoang and Antoncic 2003.)

The person-centric view of entrepreneurship has led to a very narrow policy focus on what it is and how it should be supported. By reducing the phenomenon to individual small firms and factors of their environment, a line of research has emerged in which the key parameters are the growth of new small firms and the obstacles they face. As this is a rather narrow view, focusing on only one dimension of entrepreneurship development, it is difficult to place the results of this work within a meaningful theoretical or conceptual framework (see, for example, GEM 2004).

Three issues motivate our exploration. First, a narrow person-centric view of entrepreneurship has become increasingly irrelevant as the focus is shifting towards systemic or network elements of firm growth and formation. Second, entrepreneurship is a vital property of systems of innovation (SI) and if we are to understand the dynamics of different innovation systems we must take explicit account of this dimension. Third, narrowly focused entrepreneurship policies risk irrelevance unless they encompass the systemic and network aspects of entrepreneurship.

In this chapter we develop analysis of entrepreneurship from a systems of innovation (SI) perspective in which entrepreneurship is a systemic (network) phenomenon, and emerges as an outcome of interaction (alignment) between technological, market and institutional opportunities. This framework builds on three research traditions in the entrepreneurship/systems of innovation literature (Schumpeterian, Kirznerian and Listian) which together form a multi-level and multidimensional framework for understanding entrepreneurship from a SI perspective.

The chapter is structured in the following manner. First, we discuss the notion of entrepreneurial opportunities presented in the literature. Second, we discuss each of the three types of opportunities that form our conceptual model. Third, we discuss why the simultaneous emergence of these three types of opportunities makes entrepreneurship a systemic phenomenon. Fourth, we analyse policies for entrepreneurship and finds that they are highly dependent on underlying and previously discussed conceptions of entrepreneurship.

2 What constitutes entrepreneurial opportunities?

Opportunities are at the core of entrepreneurship (Shane and Venkataraman 2000). As pointed out by Shane (2003: 10) 'the entrepreneurial process begins with the perception of the existence of opportunities, or situations in which resources can be recombined at a potential profit'. Wennekers and Thurik (1999: 46) define entrepreneurship as 'the manifest ability and willingness of individuals, on their own, in teams, within and outside organizations *to perceive and create new economic opportunities*' (emphasis added).

What constitutes entrepreneurial opportunity is generally seen as unproblematic. The dominant perspective is that entrepreneurship is a nexus of enterprising individuals and valuable opportunities (Shane 2003). Individual differences are seen as crucial in the discovery of entrepreneurial opportunities. Entrepreneurs discover opportunities that others do not recognize, because they have better access to information or because they are better at recognizing opportunities (Shane 2003).

The individual–opportunity nexus is taken as the basis for the entrepreneurial theory of firm developed by Dew *et al.* (2004), in which the emergence of the entrepreneurial firm is represented as the coming together of formerly dispersed knowledge about opportunity and resources, in a specific location at a point in time.

> The dispersion of economic knowledge over people, places, and over time leads to genuine uncertainty. A firm is created when an entrepreneur recognizes an opportunity in an environment of genuine uncertainty and is unable to sell this opportunity in the market.
>
> (Dew *et al.* 2004: 660)

Within this perspective

> the role of individuals is the primary issue that deserves explanation, since ... the origin of routines [which is presumably strongly influenced by the founder, my addition] must be explained in order for a theory of the firm based on routines to be robust.
>
> (Dew *et al.* 2004: 677)[1]

In their entrepreneurial theory of firm, Dew *et al.* (2004) see firm formation as the simple realization of an opportunity grasped by an individual. The process of firm formation and growth is perceived as unproblematic. This view of the entrepreneurial firm stands in contrast to Tsoukas (1996), which argues that a firm's knowledge is distributed, not only in the Hayekian sense (Hayek 1945) of dispersed knowledge on which Dew *et al.*'s (2004) theory is based, but also in the sense that 'it is inherently indeterminate; nobody knows in advance what that knowledge is or need be' (Dew *et al.* 2004: 22). Tsoukas argues that explicit and implicit knowledge, and individual and social knowledge, are mutually defined and cannot be separated from one another. If firms, argues Tsoukas (1996), lack the cognitive equivalent of a 'control room' then the role of individuals is much less important than is presumed by Dew *et al.* (2004).

Shane (2003: 18) defines an entrepreneurial opportunity 'as a situation in which a person can create a new means–ends framework for recombining resources that the entrepreneur believes will yield a profit'. For Casson (1982: 45), 'as a result, the process of opportunity discovery is cognitive and cannot be a collective act. Therefore, individuals, not groups or firms, discover entrepreneurial opportunities'. Indeed, this definition seems appropriate from the perspective of individual entrepreneurship. If we think of a new means–ends framework as acts of insight that are unique only to individuals, then this may be logical. However, if we take into account that entrepreneurship is very often cooperative, that it is also an act of skill that is often dispersed, and that due to uncertainty the organizational mechanisms of its reduction are essential to its exercise, then the logic of new means–ends frameworks becomes inadequate as a conceptual framework.

This framework may be sufficient in conditions of Kirznerian disequilibria where new ends–means frameworks can be operationalized as new strategies of arbitrage. However, when faced with pervasive uncertainties, institutional obstacles and the need to rely on a variety of supporting actors, it becomes rather inadequate to encapsulate opportunities. The creation of a new means–ends framework is better seen as a process whose result has been rationalized *ex post*, rather than as clear image of the future situation.

Rather than thinking about new means–ends frameworks for opportunities we prefer to think of opportunities as multidimensional and as 'objective' phenomena in the sense that these opportunities are common to all entrepreneurs. For example, the opportunities from ICT were obvious to entrepreneurs as they were inherent in the nature of the emerging technological trajectory. What we see as an outcome of successful entrepreneurship is much less a unique insight and much more an evolutionary process of the coming together and realization of different opportunities.

Indeed, variations in entrepreneurial opportunities account for a substantial part of patterns of entrepreneurial activity. From a system of innovation perspective it could be argued that this is the major part, as capable actors will emerge mainly depending on variations in opportunities. For example, if the accumulated technological and market opportunities which enabled the rise of Microsoft had not occurred, Bill Gates' capital, like that of as many others, would have been squandered. His entrepreneurial capabilities would have been secondary in a context of weak market and technological opportunities (see the discussion and related references in Holcombe 1999).

What constitutes entrepreneurial opportunity? There is probably not one general answer to this, as it depends on the level (firm, industry, national) as well as on the disciplinary scope of inquiry (business, economics, sociology). For example, Shane (2000) considers three major sources of opportunities: technological change, political/regulatory change and social/demographic change. From a systems of innovation perspective we consider:

- technological opportunities;
- market opportunities;
- institutional opportunities.

We argue that entrepreneurship is driven by complementarities arising from the favourable interaction of all three types of opportunities. In the absence of one of these, entrepreneurial opportunities cannot be realized. This perspective is actually an integration of three views on entrepreneurship: Kirznerian, Schumpeterian and Listian. In each of these views, entrepreneurship is a function of different driving factors:

- Kirzner: entrepreneurship = imbalances/distortions/asymmetries/disequilibria on the market.
- Schumpeter: entrepreneurship = technological opportunities.
- List: entrepreneurship = national system of political economy/institutional complementarities or synergies.

For Kirzner (1973), entrepreneurial opportunities are a function of imbalances, distortions, asymmetries and various disequilibria in the market. People use the information they possess to form a new means–ends framework, that guides their entrepreneurial action. For Schumpeter (1934), entrepreneurship is a function of innovation opportunities, which are a key precondition for the generation of entrepreneurial rents, and their erosion through subsequent imitation processes.[2] Generation of innovation, which is enabled by the state of inventions, is essential in explaining the existence of entrepreneurial opportunities. In our interpretation of List (1909), which here serves as an antecedent to the contemporary institutional economics and systems of innovation approaches, entrepreneurship is a function of the development of a national system of political economy and related institutional complementarities or synergies, which are conducive to entrepreneurship. From the institutional economics perspective, as defined by Schmid (2004: 1), institutions define the opportunity sets of interdependent transacting partners. Accordingly, institutional opportunities for one party are inevitably constraints for another, and vice versa. This is in contrast to new institutional economics, including the views of Douglas North, which define institutions primarily as constraints.[3]

Before we explain what makes this framework systemic, we elaborate these three views of entrepreneurship and entrepreneurial opportunities. The description of technological and market opportunities is brief as these dimensions are well covered in the literature.

2.1 Technological opportunity

Schumpeter does not, in fact, examine technological opportunities as he considers them unlimited. Technological opportunities come from inventions that are external to his scheme. He assumes that the supply of inventions, unlike the supply of innovation, is constant (Oakley 1990).[4] Innovative opportunities are realized at or near equilibrium because this provides the greatest opportunity for realization of entrepreneurial rents. So for Schumpeter, technology opportunities are exogenous to the economic system, while innovation opportunities are endogenous, i.e. their supply is influenced by market demand (Langlois 2002: 4). However, as inventions are economically irrelevant unless they are turned into innovations we can conditionally interpret that Schumpeter perceived entrepreneurship ultimately as a function of technological opportunities, which are latent and are exploited in near-equilibrium situations. For Schumpeter, the function of the entrepreneurship

> is to reform or revolutionize the pattern of production by exploiting an invention or, more generally, an untried technological possibility for producing a new commodity or producing an old one in a new way, by opening up a new source of supply of materials or a new outlet for production, by reorganizing an industry.
>
> (Schumpeter 1942: 84)

However, in his analyses he confined himself to technological innovations, both process and product.

Distribution of entrepreneurship opportunities over time is not even, but varies significantly in different historical periods. This proposition, which arises naturally from Schumpeter's scheme, conflicts with Baumol's (1968) argument that the supply of entrepreneurship is constant, but that the type of entrepreneurship varies. Schumpeter's bunching hypothesis has been elaborated through analysis of the long-term structural determinants of technological opportunities in Freeman and Perez (1988) and Perez (1983). Perez (2003) develops this further by bringing into the logic of long-term techno-economic structural change, the role of productive and financial capital. Production capital is fixed and knowledge-bounded, while financial capital is flexible and mobile. Their functional separation guarantees dynamism in the market system and produces dynamics related to the coupling and uncoupling of their relationship, which varies along different stages of the technology cycle.

Our understanding of technological opportunities has been further enriched through research on sector-specific technological regimes in which sector-specific differences in technological opportunities operate as one of the determinants of differences in technological regimes (Breschi *et al.* 2000). Shane (2003) reviews a large literature on entrepreneurship on the basis of which he concludes that 'industries differ in the entrepreneurial opportunities that they create, with some industries at some points in time being more fertile grounds for entrepreneurial activity than others' (Shane 2003: 19). Studies of business demographics (OECD 2003b: Chapter 4) have shown that the cross-industry variation for entry and exit rates is exceptionally high in young ICT-related services sectors, but a lot lower in more mature industries. Thus, some of Schumpeter's predictions are corroborated.

In summary, technological opportunities are essential to entrepreneurship as without them product and process innovations could not be realized. The question is whether these opportunities are permanent and unlimited or whether they are localized. Research since Schumpeter has shown that technological opportunities are localized and clustered in specific areas and bunched in specific periods.[5]

2.2 Market opportunity

The role of market opportunities in entrepreneurship is central to the views of the Austrian economists, and especially Kirzner. The point of departure for them is the uneven distribution of economic information across economic agents, which creates multiple arbitrage opportunities in which products and resources are shown to be incorrectly valued in their current uses (Hayek 1945). The entrepreneur exploits these disequilibria or distortions in the market, to produce a new equilibrium (Kirzner 1973). Uncertainty and asymmetric information underlie the market process that leads to different perceptions of market opportunities. As prices do not contain information about future goods and services, future technology and potential new entrant entrepreneurs are forced to make conjectures about the causes of price movements. If their conjectures are correct they have discovered new intertemporal and interspatial differences in demand and supply, which give rise to temporary entrepreneurial rents.

If market opportunities were the only determinant of entrepreneurship we would expect that entries would be driven by relatively high profits in a given industry, and exits would occur primarily in sectors with relatively low profits. Hence, there would be a negative cross-sectional correlation between entry and exit rates. However, a stylized fact in business demographics is that entry and exit rates are generally highly correlated across industries, in both the OECD and the developing countries (Bartelsman *et al.* 2005). The process of 'creative destruction' that occurs suggests that there are other factors, such as technological and institutional opportunities, that drive process of new firm formation and exits. Differences in entry/exit rates may also be interpreted as differences in institutional opportunities across different countries controlling for market and technological opportunities. This is not to deny the relevance of market opportunities in understanding the dynamics of entrepreneurship, but rather to point to the multidimensional nature of the opportunities.

The existence of market opportunities is not on its own a guarantee that technological opportunities will be realized. In fact, as Hirschman (1958) argues, excessive alertness to new opportunities may lead to the postponement of perfectly good projects on the grounds that far more profitable schemes are bound to exist. As

> the prospective profit rate on currently proposed projects does not come up to their (exaggerated) expectations ... [entrepreneurs] ... hold money or easily realized assets because they expect the profit rate on tomorrow's ventures to be higher than that of today's ventures, just as according to liquidity preference theory some investors are holding money because they anticipate higher interest rates.
>
> (Hirschman 1958: 20)

So, the market–technology opportunities relationship is not necessarily complementary; it is dependent upon type of market incentives, i.e. on institutional context or institutional opportunity. Also, for the discovery of opportunities the price system may not be all-important; however, on the experience of post-socialist economies it matters for the exploitation of opportunities (Franicevic 2001).

To sum up, the (non)existence and the type of market opportunities may greatly impact on the nature of entrepreneurship that emerges which in its turn may be greatly influenced by the role of the institutional system in conveying information and creating incentives among similar or identical technological opportunities. Kirzner's analysis is based on an environment in which entrepreneurial opportunities already exist; he does not take account of undeveloped markets. However, in transition economies and also in new technological areas, the market formation function is usually undeveloped. Market opportunities only exist where needs have been articulated. Yet, the process of articulation is deeply connected to the (non)existence of institutional opportunities.

2.3 Institutional opportunity

Market and technological opportunities have been accepted in the literature as important determinants of entrepreneurship. However, the role of institutional opportunities has not been explicitly taken into account in this context. Both Schumpeter and the Austrian economists abstracted from the institutional context of the market economy, or only briefly touched on this aspect,[6] and mainstream treatments of institutions as constraints rather than as opportunity sets of interdependent transacting partners (Schmid 2004) have played a role.

This is somewhat surprising given the current view that without the rule of law, including the property rights and the enforceability of contracts, capitalism might not have been possible (Baumol 2002; de Soto 2000).

In social terms, entrepreneurship involves connecting or bridging two separate social spheres (Barth 1978/2000). This anthropological counterpart to abstract Austrian theorizing on the matching of supply and demand via the equilibrating force of the entrepreneur has long been overlooked. In other words, historically entrepreneurship has been rooted in different economic organizational forms (individuals, banks, corporations, state and networks). These forms cannot be understood as isolated organizations unrelated to the political economy systems in which they operate. As Ebner (2000: 84) puts it: 'Schumpeterian entrepreneurship is not just simply a personal matter of embodied functions, but to a large extent also an institutional, structural and thus indeed a contextual matter, which is rooted in the historical diversity of various economic forms.'

So, any entrepreneurially driven change is deeply institutional in the sense that it requires and induces a series of institutional changes. As pointed out by Spinosa *et al.* (1997: 68) 'the first mass produced cars were not competing with other mass produced cars but with another way of life. The first personal computers were not competing with other personal computers but with our pastoral work habits'.

Entrepreneurship creates new institutional structures, but also becomes a part of them. For example, the Bill Gates phenomenon has impacted on the social environment of urban elites worldwide, and entered their economies independently of previous preferences and constraints. This induces a chain of institutional changes which eventually change the style of practices in at least some socio-economic domains.

In contrast, from the narrow Schumpeterian and Austrian perspectives, technological opportunities and market opportunities are institutionally free. As Harvey and Metcalfe (2005: 5–6) point out: 'Schumpeter spends little time on the instituted basis of the market economy, which is, as it were, taken for granted.'

Thus, 'Kirzner looks at the establishment of market order while Schumpeter looks at the transformation of market order' (Harvey and Metcalfe 2005: 9). The issue of market formation does not exist as it is solved by the entrepreneurs themselves who automatically will create markets as long as they are not blocked. As we have already pointed out Schumpeter does not explicitly take account of how differences in the economic environment affect entrepreneurship. Although in later work he states that 'every social environment has its own ways of filling the entrepreneurial function' (Schumpeter 1949/1989: 260, cited in Fagerberg 2003) he largely abstracts from the organizational dimension of entrepreneurship.

As Chris Freeman has pointed out on many occasions, the antecedent to the systems of innovation approach was Friedrich List (1909). In his book *The National System of Political Economy* he was the first to point to the importance of national technological capabilities, to the importance of what we would today call NIS, for individual businesses, and to the trade-off between static allocative and dynamic allocative efficiency.

List (1909) distinguished between the causes of the wealth and wealth itself. He finds the causes of wealth in *the power of producing wealth* (immaterial capital and productive power) which today we would describe as technological capability. In his arguments against Adam Smith he pointed out that mere accumulation is of minor importance compared with the organization of the productive forces of society. In 1909 he wrote that:

> The present state of the nation is the result of the accumulation of all discoveries, inventions, improvements, perfections, and exertions of all generations which have lived before us; they form the mental capital of the present human race, and every separate nation is productive only in the proportion in which it knows how to appropriate these attainments of former generations and increase them by its own acquirements.
>
> (List 1909: 113)

And that:

> The prosperity of a nation is not … greater in the proportion in which it has amassed more wealth i.e. values of exchange but in the proportion in which it has more *developed its power of production*. Although laws and public institutions do not produce immediate values, they nevertheless produce productive powers and Say is mistaken if he maintains that nations have been enabled to become wealthy under all forms of government, and that by means of laws no wealth can be created.
>
> (List 1909: 25, emphasis added)

List was aware of the trade-off between short-term and long-term efficiency, pointing out that 'from the national standpoint of productive power the cheapness of the moment might be far more than counterbalanced by the losses of the future measured by the loss of productive power'.[7] As a result, 'the nation must sacrifice and give up a measure of material property in order to gain culture, skill, and powers of unified production; it must sacrifice some present advantages in order to insure itself future ones' (List 1909: 24). List is aware of the importance of 'broad' NIS when he states that 'we can scarcely conceive of any law or any public legal decision which would not exercise a greater or smaller influence on the increase or decrease of the productive power of the nation' (List 1909: 12).

So, List was the first to develop the idea that the national system of political economy matters for growth. The NIS is a sub-system of the national system of political economy and it embodies the diversity of institutional arrangements, i.e. constraints and opportunities. From the perspective of entrepreneurship we need to explain why NIS represents an institutional opportunity. In order to clarify this point we introduce into the discussion the understanding of institutions in Schmid's (2004) comprehensive work.

Schmid (2004: 1) defines institutions as 'human relationships that structure opportunities via constraints and enablement. A constraint on one person is opportunity for another ... Institutions define the opportunity sets of interdependent transacting parties'.

This understanding of institutions originates from the nature of the technology and products, which generate interdependencies:

> Human interdependence is conditioned by the inherent characteristics of goods (resources). The possibility for one person's actions to affect the welfare of another person is a function of the physics and biology of goods and services. For the purpose of analyzing the impact and change consequences of alternative institutions, these inherent characteristics are given, even though over time, technology may change them.
>
> (Schmid 2004: 90)

From our perspective, it is important to consider Schmid's (2004) point that current interdependencies are determined by technology, but also that technology may change them. While creative destruction does play a role Schmid points out that:

> it is institutions that influence who gets created and who gets benefits and bears the costs. Exclusion costs (appropriability), economies of scale, transaction costs, and uncertainty are among the inherent situation variables relevant to understanding the impact of alternative institutions and both technological and institutional change.[8]
>
> (Schmid 2004: 214)

For Schmid (2004: 199), 'institutions operate on or within other determinants of technical change like relative prices, appropriability, firms and industry specific routines and standard operating procedures, transaction costs, market structure, irreversibility of technological path, and lock-in'. Technological change, Schmid contends, 'is a function of which competing interests can control the process and this is affected by institutions that influence whose preferences count' (Schmid 2004: 207). However, he adds that:

> The point is that this process of creative destruction seldom compensates the would-be losers, and their attempts to avoid losses affect the path of change. Technological change has winners and losers, and institutions affect the outcome of the struggle and the content and timing of that change.
>
> (Schmid 2004: 208)

If we apply this reasoning to our context then institutional opportunities are those that benefit entrepreneurs or innovation constituencies more generally. For example, regulation may hinder the prospects of one entrepreneur, but might open up opportunity for another. Constraints to one's business in terms of protection are opportunities for consumers or workers. We can imagine situations where markets and technological opportunities are visibly present and constraints are primarily due to the absence of institutional opportunities. Also, the structure of market opportunities and entrepreneurial actions is embedded in a web of values, beliefs, norms, traditions, and formal and informal relationships (i.e. in national IS) (Corbetta *et al.* 2004). This points to the importance of common beliefs and mutuality of expectations. For example, Casson (1982: 346) argues that 'equilibrium in the market for entrepreneurs is achieved not by contract but by conjectures. Entrepreneurs must conjecture the behaviour of other entrepreneurs, and only if these conjectures are mutually consistent will equilibrium be achieved'. We would argue that what underpins these conjectures are mutually consistent sets of values, beliefs, norms, traditions and relationships.

As Baumol (1990) points out, the rules of the game influence the payoffs for different kinds of entrepreneurship. Baumol's (1968 and 1990) thesis is that the supply of entrepreneurs in society is constant, but that the societal value of their self-interested ingenuity varies according to the rewards they can expect. This creates a distinction between productive and destructive (unproductive) entrepreneurship.[9]

There is also an argument in the varieties of capitalism literature (bank vs. stock-market based financial systems), that some types of national IS promote better entrepreneurship than others. For a review see Himmelweit *et al.* (2001: Chapter 17).[10] Bowen (2005) explores the institutional factors that influence different types of start-ups (job creating and international).

Metcalfe (2004: 35) succinctly articulates this point:

> an understanding of the entrepreneurial function cannot be separated from the nature of the economic systems in which it is exercised. Its nature and consequences are embedded in the wider system of market and non-market economic institutions. The prevailing features of a market economy produce a particular spectrum of entrepreneurial activities in a different set of institutional arrangements, say of labour managed firms, or of stakeholders capitalism, the entrepreneurial spectrum will take on a different scale because those systems give different meaning and content to entrepreneurial activity and provide different incentive systems from shareholders capitalism.
>
> (Metcalfe 2004: 35)

Shane (2003) elaborates extensively on the effects of the institutional environment on opportunity exploitation. He discusses a variety of variables from the *economic environment* (income, capital gains and property taxes, economic growth and societal wealth, etc.), *political environment* (freedom, rule of law and property rights, and decentralization of power), *socio-cultural environment* (social desirability of entrepreneurship, presence of role models and specific cultural beliefs).

The role of institutional opportunities in the emergence of new industries has been detailed in the sociological and organizational literature. These works argue that new industries based on radical product innovations, require constitutive legitimacy to flourish. Rao (2004) shows that the creation of a new industry is a project in which 'institutional activists' play a central role in securing constitutive legitimacy for the new industry. She points out that the roles of trade associations, professionals and consumer activists in the legitimation process can vary across new industries.

A preponderance of institutional factors hindering and promoting technological opportunities is essential to our understanding of how technologies are introduced and then diffused. Sine and David (2003) use the example of the introduction of energy efficient technologies into the US electric power industry. In this industry the energy crisis forced the abandonment of regulated geographical monopolies and reliance on centralized generators only after a change in 'institutional logic', which created 'a fertile environment for entrepreneurship and ultimately a new set of organizational forms and practices'[11] (Sine and David 2003: 192). Sine and David (2003: 205) also point out that the energy crisis 'did not result in the creation of new technological alternatives so much as the delegitimation of existing institutional logics, and increased awareness of pre-existing technological solutions (e.g. alternative energy and cogeneration)'.

Thus, we can see that institutional opportunities, although not much theorized in relation to entrepreneurship, are essential in matching technological and market opportunities. The variety of the perspectives on institutional opportunities highlighted here suggests that institutions structure interdependencies among agents and thus mediate the coupling between market and technology opportunities. They structure interdependencies among agents.

In the next section, we discuss the linkages between these three dimensions and why their interaction, if favorable, generates entrepreneurship as a systemic phenomenon.

3 What makes entrepreneurship systemic?

Technology is a systemic phenomenon (Fagerberg 2005). Its systemic nature stems from the nature of the interactive innovation model (Kline and Rosenberg 1986), which involves interactions within corporations, but also with knowledge organizations outside the firm (competitors, partners, universities, public R&D institutes, etc.). Linkages and complementarities between different organizations and technologies are key drivers of innovation processes. Institutional set-ups within which these interactions take place are recognized as technological (Carlsson and Stankiewicz 1991; Jacobsson 2005), sectoral (Malerba 2004), regional (Braczyk *et al.* 1998) or national innovation systems (Nelson 1993; Lundvall 1992; Edquist 1997).

While innovation and technology are seen as systemic phenomena, entrepreneurship has traditionally been perceived as primarily an individual act.

Although there is increasing understanding that successful entrepreneurship usually involves more actors than just the entrepreneur, and that its analysis must take account of the behaviour of these other actors (Swedberg 2000: 37), the implications of this fact have not been widely acknowledged by economists and practitioners, despite the fact that historical case studies provide evidence of the relevance of social networks to entrepreneurship (see, for example, Kirby 1993). The network dimension, in the modern meaning of network, has only recently been recognized in analytical terms, primarily through the application of network theory to entrepreneurship (see Burt 1993/2000 and 2005).

To understand the dynamics of entrepreneurship at different levels we need to control for the effects of opportunities. In his analysis of individual entrepreneurship and the opportunities nexus, Shane (2003: 269) points out that:

> if researchers fail to control for the effect of opportunities when measuring the effects of individual differences on the likelihood of opportunity exploitation then the variance attributed to motivation might actually be an artefact of unobserved correlation between the motivation and the expected value of the opportunity.
>
> (Shane 2003: 269)

We would argue that the same logic applies when approaching entrepreneurship from the SI perspective, which is our reason for discussing the role of the three types of opportunities separately. However, unlike Shane (2003) who sees his individual components as being unrelated, we believe that they are closely connected and interact. Entrepreneurial opportunity is a phenomenon that emerges as a result of the interaction between the three types of opportunities, and which develops into a systemic and multi-level event.

Our proposition is that the final outcome, in terms of entrepreneurship, is the result of alignment between these three components. Exploitation of technological opportunities is dependent on market opportunities and on institutional opportunities. On the other hand, if technological opportunities are not present, i.e. the technology has not been commercialized, it may not be possible to exploit market opportunities. Similarly, institutional opportunities may hinder the coupling between technological and market opportunities. Figure 3.1 depicts these relationships.

Metcalfe (2004: 33) articulated this as:

> The dynamics of modern capitalism lies in the combinatorial growth of knowledge [cf. *technological*] and investment [cf. *market*] opportunities combined with the instituted frameworks [cf. *institutional opportunities*] of the market economy that taken together simultaneously stimulate and enable entrepreneurial activity. [My insertions]

In the research on entrepreneurship these individual components have been treated either separately or as substitutes. In our view they are not substitutes,

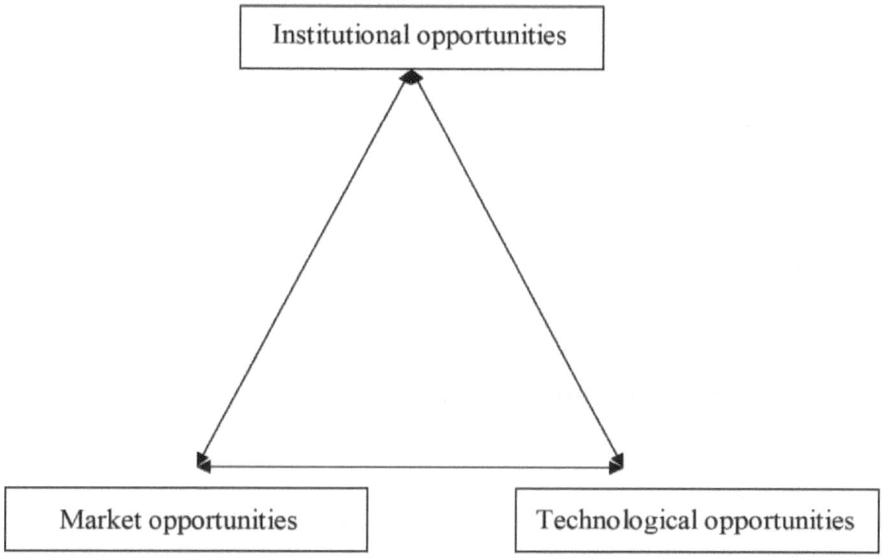

Figure 3.1 Technology-based entrepreneurship as successful matching between three types of opportunities.

but complements. For example, technological opportunities are unlikely to be entrepreneurial unless market opportunities are present, and vice versa. In his interpretation of Schumpeter's theory of motion Oakley (1990) argues that Schumpeter was fully aware of two of these three components: technological opportunity and market (demand).

As already mentioned, Schumpeter underestimated the role of institutional opportunities and the organizational aspects of the innovation process, although he did not ignore them.

Given the linkages between these three forms of opportunity we need to understand what makes them systemic. Their simultaneous presence is enabled by the features of the SI and by pervasive complementarities that operate at the national political economy system. The absence of these complementarities may lead to misalignment or coupling failure, between market, technological and institutional opportunities.

From the SI perspective, the focus is not on the individual components, but on how they relate to each other, i.e. whether they are complementary. Complementarities may be understood as situations where doing (more of) any activity increases the returns to doing (more of) another (Milgrom and Roberts 1995: 181). The essential feature of complementarities is that 'the choice variables tend to move up or down together in a systematic, coherent fashion in response to environmental changes' (Milgrom and Roberts 1995: 185). This phenomenon was first noted by Hirschman (1958), who proposed the theory of unbalanced growth. Hirschman started with the factors and resources that are latent and conditionally available, rather than being completely absent or scarce. This thinking is in line with our view of entrepreneurship as being conditionally available or latent. In other words, entrepreneurship should not be interpreted as the result of scarcity of entrepreneurial talent (Casson 1982) (missing component), but as the result of a (mis)match with other components. The focus is on what Hirschman would call the 'binding agent', or on those elements that can improve the linkages between the three components. Lack of entrepreneurship is the result of mismatches rather than of scarcities. For Hirschman (1958: 6) the solution was to look for '"pressures" and "inducement mechanisms" that will elicit and mobilize the largest possible amounts of these resources'.[12]

This perspective has been rejuvenated through the notion of network alignment (Von Tunzelmann 2004). Network alignment means that 'different elements are pulling in similar directions to one another, even when their purposes differ' (Von Tunzelmann 2004: 25). Network failures arise because:

1 the network relevant to a particular resource flow is missing;
2 the network is present, but anti-developmental; or
3 the networks for different resource flows are mutually inconsistent (Von Tunzelmann 2004).

In our context a mismatch among the three types of opportunities may be due to:

1 one of the components (markets, technology or institutions) being undeveloped;
2 wrong orientation of one of the components (anti-entrepreneurial institutional system; technology lock in; closed markets); or
3 mismatches between technological, market and institutional opportunities, which have developed in isolation and thus are not compatible.

We assume that the rate of mismatches will be smaller in some contexts than in others, in terms of matching of the three types of opportunities. This is a systemic property of SI, which cannot be expressed as a scarce factor, and which is based on compatibilities between different sub-systems of SI and compatibilities between SI and the global economy.

At a more disaggregated level, it is clear that opportunities follow one another rather than being substitutes. For example, East European experience suggests that the price system does not matter so much to the discovery of opportunities, but is important for the exploitation of opportunities (Franicevic 2001). Cheah (1990: 343) argues that Schumpeterian entrepreneurs, who, by innovating and by destroying equilibrium, give rise to uncertainty, are actually widening the scope for Kirznerian entrepreneurs: 'Instead of treating these as contradictory concepts between which we are forced to chose, it would be more fruitful to perceive Schumpeterian and Austrian entrepreneurship ... as opposites and yet complements to each other.' Cheah (1990) thus points to sequential complementarity;

however, the issue is how they can complement one another simultaneously. This highlights the scale of the trade-offs involved among different types of opportunities and the difficulties involved in matching them. This highlights why the emergence of a dynamic NIS is so rare.

In this section, we have outlined the SI approach to entrepreneurship, which is eclectic, integrative and evolutionary. Within our framework entrepreneurship is seen as a function of disequilibria in the market, technological and institutional opportunities. These three types of opportunities are matched and mediated through national systems of political economy or, in modern parlance, through 'broad' SI. This framework is eclectic as it builds on different traditions of research on entrepreneurship in the social sciences. It is integrative because it sees entrepreneurship as a multidimensional phenomenon whose occurrence may be understood through several dimensions (technology, markets and institutions), which should be brought together in empirical research. It is evolutionary in that it interprets entrepreneurship as being a highly context-dependent process, driven by interaction among structural factors (market, technology, institutional opportunities) and how they complement each other (contingent process).

4 Policy implications and policy issues

In this section we discuss the policy implications arising from our analysis. We argue that the implications for policy are highly dependent on the underlying perception of entrepreneurship, which is dependent on the emphasis given to each of opportunities discussed above. The policy options outlined are predominantly focused on one type of opportunity. Policy related to technology opportunities focuses on innovators and on subsidizing R&D to enable entrepreneurs to develop and commercialize new technologies. Market opportunities frameworks are focused on entrepreneurs as arbitrageurs and on the obstacles to market intermediation, entry and exit. The institutional opportunity perspective focuses on the institutional obstacles to entrepreneurship and on institutional support for entrepreneurial activity. We argue that the weaknesses in these approaches lie in their partial treatment of entrepreneurship as being driven by a single factor rather than being determined by their complementing interactions. We argue for a fourth, systemic view of entrepreneurship, which encompasses all three dimensions.

4.1 Market opportunities policies for entrepreneurship

Policy that perceives scarcity of entrepreneurial (market) opportunities as the key issue focuses on deregulation and increased market transparency. In this view, undistorted and transparent markets are seen as essential to the exercise of entrepreneurship. This line of policy thinking is basically grounded in the neoclassical and Austrian economics views. For Hayek (1945: 524), undistorted and transparent markets are essential to enable 'limited individual fields of vision [to] sufficiently overlap so that through many intermediaries the relevant information is communicated to all'.

This market opportunities view of entrepreneurship dominates the reports of Freedom House, GEM, the World Economic Forum (WEF) competitiveness reports and World Bank 'Doing Business', all of which emphasize entry and exit barriers in business, as key impediments to entrepreneurship.[13] For example, the World Bank's 'Doing Business' reports point to high costs, delays and barriers to owning property, starting up businesses, declaring bankruptcy, protecting investors, enforcing contracts and legal rights and in laying off workers as the main barriers to entrepreneurial activity. The 2005 report estimates that improving from the quartile of countries with the worst business regulations to the best would raise annual growth by 2.3 percentage points (Djankov *et al.* 2006; World Bank 2005). This assumes that the market opportunities created by the removal of barriers would automatically generate benefits, irrespective of institutional and technological opportunities. As Hobday and Perini point out, no correlation between over regulation and lack of development is actually established:

> The policy assumption that there is a linear causal relationship between excessive regulation and low rates of development is questionable. It may be that the style, stage and level of development leads to high regulation and low innovative activity, rather than *vice versa.*
>
> (Hobday and Perini 2005: 23)

Exit barriers are somewhat less of a problem, due to the difficulty of monitoring them, and the less well developed understanding of their part in processes of creative destruction. In view of the latest results of business demographics studies, which point to small differences in entry/exit rates across a variety of countries (see Bartelsman and Scarpetta 2004 and Bartelsman *et al.* 2005), an exclusive policy emphasis on this aspect of entrepreneurship seems somewhat misplaced. Also, there is underestimation of sector-specific barriers to entrepreneurship, as opposed to general entry/exit barriers, which seem actually to be a greater hindrance to business. Some of the McKinsey Global Institute reports have highlighted their importance in developed (UK) as well as in less developed economies (India, Russia).[14]

4.2 Institutional support policies for entrepreneurship

This policy perspective assumes that an efficient market economy requires as a minimum an institutional system that effectively enforces property rights and the exchange of property titles. In its minimalist form this view favours strong enforcement of property rights and privatization. In its activist form it favours policies that support entrepreneurship by pointing to market failures in financial and entrepreneurial skills markets. This latter view would justify support to small- and medium-sized enterprises (SMEs) and the venture capital (VC) industry. Such policies are considered to complement market opportunity policies for entrepreneurship.

Hernando de Soto's (2000) ideas on the importance of individual property titles for poor people has been extremely influential in policies promoting individual entrepreneurship, especially in underdeveloped economies. In the absence of secure property rights poor people cannot get ownership over the assets they use, are denied access to finance and cannot legalize their activities. Individual property titles are an essential institutional precondition for entrepreneurship and their effects can be substantial (see de Soto 2000 for evidence). However, there is also evidence that legalization of individual property titles by itself does not solve the problem, and the market value of the newly legalized property generally remains low. In cases where the value of property has dramatically increased it has led either to forced evictions or gradual takeover by richer entrepreneurs.[15] By pushing for one – institutional – solution to the problem of entrepreneurship, the market opportunity aspect of the problem has remained unchanged, which may significantly weaken the effects of this policy.

A possible solution suggested by Geoffrey Payne, a British urban planning consultant, would be to temporarily insulate slums from the commercial land market by granting informal neighbourhood groups land rights for a certain period of time.[16] During this period the value of the property could be increased, and the group land title could be subdivided into individual titles. In terms of our model, the solution involves creating an institutional distortion (group land rights) and then developing market opportunity by investing in improvement to the acquired land assets. However, it is likely that true entrepreneurial opportunity would only arise if the poor people were given decent education and training, providing the necessary third element – technological opportunity – required to grasp the entrepreneurial opportunity. In a nutshell, institutional opportunities (cf. property titles) are not sufficient if market and technological opportunities are not simultaneously present.

In countries where property rights are relatively well enforced, programmes of support for entrepreneurship through support to SMEs are quite widespread. As the OECD (2003a: 9) states, in the OECD group of countries 'the scope, number and growth of entrepreneurship support programs are striking'. In addition, bilateral and multilateral development agencies operating around the globe advocate micro-enterprises as a vehicle for poverty alleviation. Assessment of policies to support entrepreneurship at the local level shows that 'while enterprise creation and development is clearly a critical component of local growth and development it does not constitute a developmental panacea' (OECD 2003a: 13). Problems such as displacement (when competition from new firms causes a loss of output or employment in existing enterprises), small employment effects (at least over the short term), poor quality employment, and deadweight effects (when a programme has effects that would have occurred in its absence) are among those frequently mentioned in evaluations of support programmes.

Aggregate analysis of SMEs as determinants of growth suggest that small- and medium-sized firms do not have a particularly beneficial impact on the incomes of the poor (Beck *et al.* 2003). SMEs are characteristic of fast-growing economies, but they are not determinants of economic growth (Beck *et al.* 2003).

Thus, there are a variety of factors that need to be in place for the successful promotion of entrepreneurship through SMEs. In the absence of market opportunities alongside technological opportunities, which include entrepreneurial capabilities and knowledge, the effects of single-dimensional policies will be limited.

4.3 Supporting technological opportunities for entrepreneurship

Within this policy the lack of technological opportunities is seen as an important barrier to technology based entrepreneurship. It points to a need for funded R&D, and highlights the importance of scarce entrepreneurial talents. This is in line with much EU policy on entrepreneurship, exemplified in various horizontal policies and institutional support for entrepreneurship (S&T parks and innovation centres).

These types of entrepreneurship policies were inspired by the rise of Silicon Valley and the idea that growth of new-technology based firms supported by VC is the model to be aspired to. In countries where new technology sectors have been important this model of support is justified. However, to support technological activities that are behind the frontier, this radical innovation model is not appropriate. In countries where the majority of innovative firms are in non-high-tech industries, this model would skew the policy focus on to a small number of new-technology based firms. A large number of firms involved in small, incremental improvements to existing products and processes based on engineering and technical skills rather than R&D would be overlooked by such policies. Also, a focus on new-technology based firms would exclude large firms and subsidiaries of MNCs, which are more frequent innovators than SMEs.[17]

4.4 Innovation systems based policies towards entrepreneurship

The main problem with the policies outlined above lies in their one-dimensional view of entrepreneurship. By pushing one aspect of entrepreneurship, be it market opportunity, institutional constraints or technological constraints, as the key, they are inevitably constrained by the other two dimensions of entrepreneurship, which are not taken into account. For example, policies to support SMEs that view SMEs in isolation should emphasize the links and synergies between small, medium and large firms. Policies aimed at generating new market opportunities for nascent entrepreneurs through deregulation and liberalization are constrained by insecure property rights (cf. de Soto's point) or by the weak capabilities of SMEs which do not allow them to exploit these newly created opportunities. Policies that are focused on support to new-technology based firms are often constrained by undeveloped markets and regulatory uncertainty. Based on the logic underlying our conceptual model, policies for entrepreneurship should take account of all three dimensions or should try to enhance complementarities between the three types of opportunities – market, institutional and technological.

The fourth policy perspective – the system of innovation perspective – would allow the entrepreneurship problems to be more comprehensively addressed. Within this perspective, the poor connectedness within NIS is often seen as a key policy issue. For an elaboration of this view in the context of the EU see CEC (2004: Chapter 6). Here, the nature of linkages is considered to be dominant over the strength of actors. In a recent critical analysis of EU science, technology and innovation Dosi *et al.* (2005) have rightly pointed to weak European corporations as the major problem, and have criticized the dominance of the so-called 'European paradox', i.e. weak linkages between science and technology, as the problem. In the context of Central and Eastern Europe McGowan *et al.* (2004) pointed to the weakness of domestic firms rather than the poor linkages in the innovation system as the key obstacle to industrial upgrading. However, none of these approaches explicitly looks at the entrepreneurial function as a constitutive element in the analysis of systems of innovation.

The systemic policy view is embodied in the *functional* innovation system view, and is developed the innovation systems research of Bergek *et al.* (2005), Carlsson *et al.* (2004) and Jacobsson (2005). This views leads to systemic policies aimed at remedying poor functionality of innovation systems by strengthening inducement mechanisms and removing blocking mechanisms. From the NIS view, the focus here is on strengthening the entrepreneurial function (entrepreneurial experimentation) and its interaction with other functions in the national or sectoral system of innovation.

From a systems of innovation perspective this policy view is the most appropriate as it departs from the systemic and functional view of entrepreneurship. Other policy perspectives are partial and only tangentially address the key issues of entrepreneurship from a systems-of-innovation perspective. However, the functional approach to NIS–entrepreneurship is also the most demanding of all these approaches in terms of the knowledge required for policy analysis, and the understanding of entrepreneurial functions that is necessary. In this respect, its application is likely to be limited, and confined to sectoral systems of innovation.

So far, we have looked at the policy implications of our analysis through a somewhat technocratic lens by reducing the inherent fuzziness of entrepreneurship to policy implications. However, entrepreneurship emerges as a property of firms and of innovation systems rather than as a result of the policy process. It is the overall social and institutional context, or national system of political economy, of a country that ultimately generates different propensities for entrepreneurship. Direct policies for supporting entrepreneurship will thus inevitably be limited, especially if they contradict the structural features of the system of innovation or if they are incompatible with other functions in the innovation system. Entrepreneurship may be affected much more by indirect policies or other features of the system of innovation. This does not make policies supporting entrepreneurship redundant, but suggests that, almost by definition, entrepreneurship is not an easily directed activity.

Notes

1 An answer to Phan's (2004: 617) question about 'why and when entrepreneurial firms do not emerge even when the condition for the event appear to exist' from Dew *et al.*'s (2004) perspective is that it is entirely due to differences in individuals' perceptions of opportunities.
2 However, inventions or the state of S&T, which is an important basis for innovation, are exogenous to Schumpeter's model.
3 For example, see North's (1993: 2) definition of institutions set out in his Nobel Prize lecture:

> Institutions are the humanly devised constraints that structure human interaction. They are made up of formal constraints (e.g. rules, laws, and constitutions), informal constraints (e.g. norms of behaviour, conventions, and self-imposed codes of conduct) and their enforcement characteristics. Together they define the incentive structure of societies and specifically economies.
>
> (North 1993: 2)

4 Oakley (1990: 117) argues that Schumpeter saw potential entrepreneurial agents as facing a more or less given and unlimited 'pool' of inventions and relevant knowledge:

> They [possibilities] are always present, abundantly accumulated by all sorts of people. Often they are also generally known and even discussed by scientific or literary writers. In other cases, there is nothing to discover about them, because they are quite obvious.
>
> (Oakley 1990: 117)

5 For a state-of-the-art review see Fagerberg (2005). For the theory of localized technical change see Antonelli (1995).
6 Institutional opportunities relate to 'broad' NIS, i.e. they consists of institutions, norms and rules (formal and informal) that affect (directly or indirectly) the innovation process.
7 As quoted in an introductory essay by J. Shield Nicholson.
8 Discussions on the income distribution effects of IT and whether global income inequalities are trade or technology driven can be better interpreted within this framework.
9 Aidis (2005) discusses the relevance of Baumol's thinking in the context of transition economies.
10 For a rejection of the relevance of this dichotomy see Thorsten and Levine (2002).
11 Sine and David (2003: 188) define institutional logics as 'socially-constructed assumptions, values, and beliefs that define formal and informal rules of behaviour and guide interpretation about why certain structures and practices exist'.
12 This view is in line with the functional perspective on systems of innovation as developed in Bergek *et al.*'s (2005) methodology.
13 A completely different policy stance, but based on similar concerns, would be to create market opportunities through unequal access to the market. Hikino and Amsden (2004) argue that this pattern of initial asymmetries, which were then corrected, was essential for East Asian entrepreneurship.
14 See www.mckinsey.com/knowledge/mgi/reports.
15 For an interesting debate on the effects of de Soto's inspired policies see John Gravois 'The De Soto Delusion', updated 29 January 2005 at: www.slate.com/id/2112792 as well as replies from his supporters at www.techcentralstation.com/020305D.html. See also Hobday and Perini (2005).
16 See www.slate.com/id/2112792.
17 See also Hobday and Perini (2005) on this point.

References

Bartelsman, E.J.H. and Scarpetta, S. (2004) 'Microeconomic evidence of creative destruction in industrial and developing countries', Tinbergen Institute Discussion Paper no. TI 2004–114/3.

Bartelsman, E.J.H., Scarpetta, S. and Schivardi, F. (2005) 'Comparative analysis of firm demographics and survival evidence from micro-level sources in OECD countries', *Industrial and Corporate Change*, vol. 14 (3): 365–91.

Barth, F. (1978/2000) 'Economic spheres in Darfur', in Firth, R. (ed.) *Themes in Economic Anthropology*, London: Tavistock Publications, reprinted in Swedberg, R. (ed.) *Entrepreneurship. The Social Science View*, Oxford: Oxford University Press, pp. 139–60.

Baumol, J.W. (1968) 'Entrepreneurship in economic theory', *The American Economic Review*, vol. 58 (2): 64–71.

Baumol, J.W. (1990) 'Entrepreneurship: productive and destructive', *Journal of Political Economy*, vol. 98 (5): 893–921.

Baumol, J.W. (2002) *The Free-market Innovation Machine: Analyzing the Growth Miracle of Capitalism*, Princeton: Princeton University Press.

Beck, T., Demirguc-Kunt, A. and Levine, R. (2003) 'Small and medium enterprises, growth and poverty: cross country evidence', World Bank Policy Research Working Paper no. 3178, December, Washington, DC.

Bergek, A., Jacobssoni, S., Carlsson, B., Lindmarki, S. and Rickne, A. (2005) 'Analyzing the dynamics and functionality of sectoral innovation systems – a manual', paper presented at the DRUID Tenth Anniversary Summer Conference 2005 on Dynamics of Industry and Innovation: Organizations, Networks And Systems, Copenhagen, Denmark, 27–29 June.

Bowen, P.H. (2005) 'National institutions and the allocation of entrepreneurial effort', Vlerick Leueven Gent working paper series no. 2005/15.

Braczyk, H.J., Cooke, P. and Heidenreich, M. (eds) (1998) *Regional Innovation Systems*, London and Bristol, PA: University College London Press.

Breschi, S., Malerba, F. and Orsenigo, L. (2000) 'Technological regimes and sectoral patterns of innovation', *Economic Journal*, vol. 110 (463): 388–410.

Burt, S.R. (1993/2000) 'The network entrepreneur', in Swedberg, R. (ed.) *Explorations in Economic Sociology*, Russell Sage Foundation, reprinted in Swedberg, R. (ed.) (2000) *Entrepreneurship. The Social Science View*, Oxford: Oxford University Press, pp. 281–307.

Burt, S.R. (2005) *Brokerage and Closure: an Introduction to Social Capital*, Oxford: Oxford University Press.

Carlsson, B. and Stankiewicz, R. (1991) 'On the nature, function, and composition of technological systems', *Journal of Evolutionary Economics*, vol. 1 (2): 93–118.

Carlsson, B., Jacobsson, S. and Bergek, A. (2004) 'Dynamics of innovation systems: policy-making in a complex and non-deterministic world', mimeo.

Casson, M. (1982) *The Entrepreneur. An Economic Theory*, Oxford: Martin Robertson.

CEC (2004) European Economy, no. 6, Office for Official Publications of the EC, Luxembourg, KC–AR–03–006–EN–C.

Cheah, H.B. (1990) 'Schumpeterian and Austrian entrepreneurship: unity within duality', *Journal of Business Venturing*, vol. 5 (6): 341–7.

Corbetta, G., Huse, M. and Ravasi, D. (2004) 'Crossroads of entrepreneurship research: an introduction', in Corbetta, G., Huse, M. and Ravasi, D. (eds) *Crossroads of Entrepreneurship Research*, Dordrecht: Kluwer Academic Publishers.

de Soto, H.D. (2000) *The Mystery of Capital: why Capitalism Triumphs in the West and Fails Everywhere Else*, New York: Basic Books.

Dew, N., Velamuri, S.R. and Venkataraman, S. (2004) 'Dispersed knowledge and an entrepreneurial theory of the firm', *Journal of Business Venturing*, vol. 19 (5): 659–79.

Djankov, S., McLiesh, C. and Ramalho, R. (2006) 'Regulation and growth', *Economic Letters*, vol. 92 (3): 395–401.

Dosi, G., Llerena, P. and Sylos Labini, M. (2005) 'Evaluating and comparing the innovation performance of the United States and the European Union', expert report prepared for the TrendChart Policy Workshop 2005, 29 June.

Ebner, A. (2000) 'Systems of innovation between globalisation and transformation: policy implications for the support of Schumpeterian entrepreneurship', in Kuklinski, A. (ed.) *The Knowledge-Based Economy – The Challenge of the 21st Century*, Warsaw: State Committee for Scientific Research of the Republic of Poland, pp. 80–99.

Edquist, C. (ed.) (1997) *Systems of Innovation: Technologies, Institutions, and Organizations*, London: Pinter.

Fagerberg, J. (2003) 'Schumpeter and the revival of evolutionary economics: an appraisal of the literature', *Journal of Evolutionary Economics*, vol. 13 (2): 125–59.

Fagerberg, J. (2005) 'Innovation: a guide to the literature', in Fagerberg, J., Mowery, D.C. and Nelson, R.R. (eds) *The Oxford Handbook of Innovation*, Oxford: Oxford University Press.

Franičević, V. (2001) 'Some considerations on post-socialist entrepreneurship', paper presented at the international conference 'Post-Communist Transition then Years later. Challenges and outcomes for Europe and the role of CEI', Forle, 2–3 February.

Freeman, C. and Perez, C. (1988) 'Structural crises of adjustment: business cycles and investment behavior', in Dosi, G., Freeman, C., Nelson, R., Silverberg, G. and Soete, L. (eds) *Technical Change and Economic Theory*, London: Pinter.

Gartner, W.B. (1988) '"Who is an entrepreneur?" is the wrong question', *American Journal of Small Business*, vol. 12 (4): 11–32.

GEM (2004) 'Global Enterpreneurship Monitor', Report, Babson College and London Business School, authors: Acs, Z.J., Arenius, P., Hay, M. and Minniti, M.

Gravois, J. (2005) 'The de Soto delusion', *Slate.com*, 29 January, online, available at: www.slate.com/id/2112792.

Harvey, M. and Metcalfe, S. (2005) 'The ordering of change: Polany, Schumpeter and the nature of the market mechanism', CRIC University of Manchester Discussion Paper no. 70, March.

Hayek, F.A. (1945) 'The use of knowledge in society', *American Economic Review*, vol. 35 (4): 519–30.

Hikino, R.T. and Amsden A.H. (2004) 'Entrepreneurship in late developing countries: what is unique about East Asia?', in Corbetta, G., Huse, M. and Ravasi, D. (eds) *Crossroads of Entrepreneurship Research*, Dordrecht: Kluwer Academic Publishers, pp. 209–23.

Himmelweit, S., Simmonetiu, R. and Trigg, A. (2001) *Microeconomics: Neoclassical and Institutionalist Perspective on Economic Behaviour*, London: The Open University, Thomson.

Hirschman, O.A. (1958) *The Strategy of Economic Development*, New Haven: Yale University Press.

Hoang, H. and Antoncic, B. (2003) 'Network based research in entrepreneurship. A critical review', *Journal of Business Venturing*, vol. 18 (2): 165–87.

Hobday, M. and Perini, F.A.d.B. (2005) 'Entrepreneurship and bureaucracy in developing countries: a policy perspective', rough draft, SPRU, Freeman Centre, University of Sussex, March.

Holcombe, R.G. (1999) 'Entrepreneurship and economic growth: reply', *The Quarterly Journal of Austrian Economics*, vol. 2 (2): 73–8.

Jacobsson, S. (2005) '"Functional analysis" as a tool for policy makers in identifying policy issues in an emerging sectoral innovation systems', Chalmers University of Technology, mimeo.

Kirby, M. (1993) 'Quakerism, entrepreneurship and the family firms in north-east England, 1870–1860', in Brown, J. and Rose, M.B. (eds) *Entrepreneurship, Networks and Modern Business*, Manchester and New York: Manchester University Press, pp. 105–26.

Kirzner, I.M. (1973) *Competition and Entrepreneurship*, Chicago and London: University of Chicago Press.

Kirzner, I.M. (1980) 'The primacy of entrepreneurial discovery', in IEA (The Institute of Economic Affairs), *The Prime Mover of Progress. The Entrepreneur in Capitalism and Socialism. Papers on the Role of the Entrepreneur*, Lancing, Sussex: Goron_Pro-Print Co. Ltd, pp. 5–28.

Kline, S. and Rosenberg, N. (1986) 'An overview of innovation', in Landau, R. and Rosenberg, N. (eds) *The Positive Sum Strategy: Harnessing Technology for Economic Growth*, Washington, DC: National Academy Press, pp. 275–305.

Langlois, R.N. (2002) 'Schumpeter and the obsolescence of the entrepreneur', University of Connecticut Working Paper no. 2002–19, August.

List, F. (1909) *The National System of Political Economy*, London: Longmans Green and Co. Printed from www.econolib.org.

Lundvall, B.A. (ed.) (1992) *National Systems of Innovation: Towards a Theory of Innovation and Interactive Learning*, London: Pinter.

McGowan, F., Radosevic, S. and von Tunzelmann, N. (2004) *The Emerging Industrial Structure of the Wider Europe*, London: Routledge.

Malerba, F. (2004) *Sectoral Systems of Innovation: Concepts, Issues and Analysis of Six Major Sectors in Europe*, Cambridge: Cambridge University Press.

Metcalfe, S. (2004) 'The entrepreneur and the style of modern economics', in Corbetta, G., Huse, M. and Ravasi, D. (eds) *Crossroads of Entrepreneurship Research*, Dordrecht: Kluwer Academic Publishers, pp. 33–52.

Milgrom, P. and Roberts, J. (1995) 'Complementarities and fit strategy, structure, and organizational change in manufacturing', *Journal of Accounting and Economics*, vol. 19 (2): 179–208.

National Commission on Entrepreneurship (NCOE) (2001) *Five Myths about Entrepreneurs*, Washington, DC: NCOE.

National Commission on Entrepreneurship (NCOE) (2002) *Entrepreneurship: a Candidate's Guide*, Washington, DC: NCOE.

Nelson, R.R. (ed.) (1993) *National Innovation Systems: a Comparative Analysis*, Oxford: Oxford University Press.

North, D.C. (1993) 'Economic performance through time', lecture to the memory of Alfred Nobel, 9 December. Online, available at: http://nobelprize.org/nobel_prizes/economics/laureates/1993/north-lecture.html.

Oakley, A. (1990) *Schumpeter's Theory of Capitalist Motion. A Critical Exposition and Reassessment*, Aldershot: Edward Elgar.

OECD (2003a) *Entrepreneurship and Local Economic Development: Programme and Policy Recommendations*, Paris: OECD.

OECD (2003b) *The Sources of Economic Growth in OECD Countries*, Paris: OECD.

Perez, C. (1983) 'Structural change and the assimilation of new technologies in the economic and social systems', *Futures*, vol. 15 (5): 357–75.

Perez, C. (2003) *Financial Capital and Technological Revolutions*, Cheltenham: Edward Elgar.

Phan, P.H. (2004) 'Entrepreneurship theory: possibilities and future directions', *Journal of Business Venturing*, vol. 19 (5): 617–20.

Rao, H. (2004) 'Institutional activism in the early American automobile industry', *Journal of Business Venturing*, vol. 19 (3): 359–84.

Schmid, A. (2004) *Conflict and Cooperation. Institutional and Behavioral Economics*, Malden: Blackwell.

Schumpeter, J.A. (1934) *The Theory of Economic Development*, London: Oxford University Press.

Schumpeter, J.A. (1942) *Capitalism, Socialism and Democracy*, New York: George Allen and Unwin.

Schumpeter, J.A. (1949) 'Economic theory and entrepreneurial history. Change and the entrepreneur', reprinted in Schumpeter, J.A. (1989) *Essays on Entrepreneurs, Innovations, Business Cycles and Capitalism*, edited by Clemence, R.V., New Brunswick: Transaction, pp. 253–71.

Shane, S. (2003) *A General Theory of Entrepreneurship. The Individual – Opportunity Nexus*, Cheltenham: Edward Elgar.

Shane, S. and Venkataraman, S. (2000) 'The promise of entrepreneurship as a field of research', *Academy of Management Review*, vol. 25 (1): 217–26.

Shane, S. and Venkataraman, S. (2003) 'Guest editors' introduction to special issue on technology entrepreneurship', *Research Policy*, vol. 32 (2): 181–4.

Sine, W.D. and David, D.J. (2003) 'Environmental jolts, institutional change, and the creation of entrepreneurial opportunity in the US electric power industry', *Research Policy*, vol. 32 (2): 185–207.

Spinosa, C., Flores, F. and Dreyfus, H.L. (1997) *Disclosing New Worlds, Entrepreneurship, Democratic Action and the Cultivation of Solidarity*, Cambridge, MA: MIT Press.

Swedberg, R. (2000) 'Introduction', in Swedberg, R. (ed.) *Entrepreneurship. The Social Science View*, Oxford: Oxford University Press.

Tsoukas, H. (1996) 'The firm as distributed knowledge system: a constructionist approach', *Strategic Management Journal*, vol. 17: 11–25.

von Tunzelmann, N. (2004) 'Network alignment in the catching-up economies of Europe', in McGowan, F., Radosevic, S. and von Tunzelmann, N. (eds) *The Emerging Industrial Structure of Wider Europe*, London: Routledge, pp. 23–37.

Wennekers, S. and Thurik, A.R. (1999) 'Linking entrepreneurship and economic growth', *Small Business Economics*, vol. 13 (1): 27–55.

World Bank (2005) *Doing Business*, Washington, DC: World Bank.

4 Academic patents, spin-offs and beyond

The many faces of scientific entrepreneurship

Chiara Franzoni[1] *and Francesco Lissoni*[2]

1 Introduction

The present chapter discusses the notion of 'academic entrepreneurship', as it emerges from a number of recent theoretical and applied contributions to the economics of innovation and science and technology studies.

In Section 2 we put forward the proposition that contemporary science is the result of an 'entrepreneurial' effort, undertaken both by individual scientists and by the academic institutions that host them. The intensity and specific features of the entrepreneurial effort depends very much on the institutional characteristics of national academic systems, which we outline by looking briefly at the history of the US and French systems, the last one taken as the extreme example of the European case (Section 3).

In Sections 3 and 4 we examine the recent literature on university patenting and spin-off firm creation, and suggest that both commercial activities result from the broad entrepreneurial agendas of individual scientists and universities, and not merely from a few individual scientists' profit-seeking attitudes.

In Section 5 we propose a few conclusions and outline a few directions for future research.

2 'Entrepreneurship' in science

'Entrepreneurship' is quite a popular word in a number of studies dealing with the philosophy, sociology, economics and history of science. A common thread can be traced across the different authors and approaches that have used it.

2.1 Entrepreneurship as an individual features: the sociology of scientist–entrepreneurs

The contemporary sociology and economics of science describe the organization of scientific research, especially in experimental sciences, as necessarily 'entrepreneurial'. Scientists at the head of large laboratories perform a number of activities which are typical of the modern entrepreneur, such as setting up and managing increasingly complex organizations, and providing them with adequate funding and human capital. More generally, scientists with innovative research agendas have to relate frequently with agents outside the universities (especially policy-makers and industrialists), in order to find political and material support for that agenda.

The starting point of many recent essays is Robert Merton's portrait of academic scientists as individuals engaged in careers based upon peers' recognition of their contributions to the advancement of scientific knowledge (Merton 1973). In Merton's view, such recognition takes primarily the form of 'scientific credit', which is given to any scientist whose 'priority claim' over a discovery is acknowledged by the scientific community at large. Philosophers and economists of science have gone a long way in exploring how the quest for scientific credit may shape social relationships in science, and have reinforced the notion that being credited with one or more 'discoveries' (through the mechanism of bibliographic citations and, possibly, eponymy) is essential to a scientist's career (Kitcher 1993; Dasgupta and David 1994).

An even more complex view of how scientists manage their careers according to entrepreneurial criteria comes from the sociological tradition of 'science studies' (Callon 2002), and a number of related contributions to the history of science and technology (Latour 1988; Lenoir 1997).

This literature explores in greater depth the relational aspect of the scientific enterprise. Scientific facts are not merely 'discovered' by the first scientist who solves a theoretical puzzle or creates an innovative experimental routine (and thus wins the priority race). Rather, they are established by obtaining social consensus on the relevance of the topic, on the legitimacy of the theoretical assumptions, and on the solidity of experimental routines. Such consensus has to be gained both from fellow scientists (especially within one's own disciplinary field) and from other relevant actors, such as businessmen and policy-makers.

Fellow scientists can validate the contents of a scientific paper or programme by citing it as a legitimate source of information, or neglect it as irrelevant or poorly conceived. Their consensus has to be elicited either by indirect means (e.g. by choosing the best publication outlet or through a perfunctory use of paper citations) or by more direct ones, by establishing social ties through research cooperation, conference invitations and joint lobbying for economic resources from state and industry.

In this respect, businessmen and policy-makers can be instrumental in providing funds, data, scientific materials and instruments, as well as ethical validation. Participation to science policy forums, policy and ethical committees, and scientific boards of large companies are all necessary activities for senior scientists to support the activity of their laboratories.

If seen within this context, IPR management, consultancy and equity participation to spin-off companies are not simply market and market-like alternative activities, but indeed necessary steps, conditional to the scientific entrepreneur's chief goal of setting up or expanding her lab, and promoting her academic career (OECD 1999: 37).

This perspective suggests that technology transfer activities such as patenting or joining a start-up company may have deeper motivation than the immediate quest for profits. Ease of patenting and access to resources for setting up a company are welcome insofar they are instrumental in widening and thickening the scientist's network. They will be disregarded if they do not fit in the research agenda. At the same time, more research funds, or more career opportunities, although totally unrelated to any technology transfer target or firm creation objective, may naturally push more scientists to pursue the ambition of setting up their own laboratory; and it may well be that, by doing so, those scientists will reach out of the academic walls anyway.

Lenoir's (1997) portrait of German physiologists, physicists and chemical scientists in the nineteenth century confirms this view. In particular, Lenoir contrasts scientists engaged in academic careers within the boundaries of established disciplines with those whose research agendas foresee the birth of a new discipline, or require disciplinary boundaries to be re-drawn, either to allow for interdisciplinary work or to establish new hierarchies between disciplines.

Lenoir's scientist–entrepreneur first aims at acquiring superior skills and technical expertise in handling complex experimental procedures and equipment, so that other scientists will find it hard to disprove his experimental results, and will require his approval or help to validate their own findings. Then he will promote a wider agenda, which aims at proving the social benefits bestowed by the new disciplinary programme for society. Practical applications of the new scientific discipline (were it nineteenth-century organic chemistry or twentieth-century nuclear physics) are proved through patenting, licensing, consulting and the encouragement of start-ups by young colleagues and students.[3]

Thus, academic entrepreneurship, as part of the more general phenomenon of scientific entrepreneurship, proves to result from a complex bundle of strategies and incentives and career planning.

2.2 *Academic entrepreneurship as an institutional feature: US vs. European universities*

One of the best-known papers on entrepreneurship in academy is Henry Etzkowitz's (1983) essay on 'Entrepreneurial scientists and entrepreneurial universities in American academic science'. The 'entrepreneurial university' is there portrayed as the outcome of a revolutionary process started in the United States with the Big Science programmes launched in the aftermath of the Second World War. In a later paper, Etzkowitz (2003) suggests that, slowly but inevitably, European research-oriented universities will leave room to profit motives for their research, and turn into entrepreneurial ones such as their overseas counterparts.

On closer inspection, however, the American model of the 'entrepreneurial university' appears to be rooted much more deeply in the gradual evolution of US universities from teaching colleges of divinity and liberal arts to modern research institutions. By contrast, many contemporary efforts to promote entrepreneurial attitudes in universities outside the United States are at odds with an institutional history of central planning and control.

The US university system has been, since its early boom in the first half of the nineteenth century, a heterogeneous collection of a large number of autonomous institutions cherished by their local communities or by religious groups and individual philanthropists (Rudolph 1990). Their faculty members were neither subject to their students' control (as in Italian medieval universities) nor ever served as civil servants paid by the state, as in most European countries. Since their inception, the president and board of trustees of US colleges exercised a degree of local control which federal and state governments never managed to overcome. Attempts to centralize the university system have always been overthrown (both by the oldest private colleges and the more recent state universities), even at times when financial distress could have advised otherwise (Trow 2003). Nowadays, autonomy is one of the greatest strength of the US universities, and this is also the main background reason for their transformation into 'entrepreneurial' organizations.

Since right after Second World War, and well through the 1960s, the extraordinary success achieved by basic science applications to military technology legitimized the well-known exponential increase of federal funding of academic research. However, Vannevar Bush's famous report 'Science, the endless frontier', while presenting the case for large public funding of research universities convincingly, failed to persuade US lawmakers of the need to set up a centralized body for the administration of all funds (Graham and Diamond 1997).

Nowadays, a number of concurrent institutions still provide research grants in various fields. The National Institute of Health, the National Science Foundation, the Ministry of Defence and other governmental bodies all run projects both in separate scientific fields and in a few overlapping ones. The possibility to be financed, in certain fields, by different agencies has helped to keep alive a healthy heterogeneity of research targets (also within the same scientific field) and administration models.

All of these programmes rely on the so-called Principal Investigator (PI) principle, by which individual scientists (neither their departments nor their institutions) are made entirely responsible for a project. The PI's research record is instrumental to win the grant in order to set up or expand her own laboratory or research group. Individual scholarship, and not any political objective of equal distribution of resources, becomes the key allocation criteria for research funding. As a consequence, universities have been engaging in a race to recruit the most talented scientists, whose contribution is decisive to get public funds.

At the same time, individual scientists engage in self-promotion activities in order to win and then manage the grants. They devote their efforts not only to publishing, teaching and attending scientific meetings, but also to establishing relationships with one or more funding agencies, aiming at influencing the choice of the research topics to be funded, as well as networking at the academic level for recruiting brilliant young scientists to the ever-increasing needs of their laboratories. As employees of their universities, and not of the federal government nor of the individual states, the academic scientists are let free and possibly encouraged to engage in these typical entrepreneurial activities, as long as this enhances their university's reputation and financial health. As a result, the US academic system has witnessed in recent years an overall tendency of research teams to increase in size and complexity in all scientific fields, albeit at different rates (Adams *et al.* 2005).

Etzkowitz (1983) describes this pattern as one of diffusion of 'quasi-firms' (laboratories and research groups), whose survival and expansion depends upon chasing funds, recruiting skilled employees, managing funds, delivering results and moving up to higher-level funding agencies. PIs provide the necessary entrepreneurial effort and skills to do the job, in exchange for a large bite of the credit for the success of the scientific enterprise. Stephan and Levin (2002) offer a similar view. One cannot fail to see here a strong parallel with the redefinition of academic entrepreneurship we have proposed in Section 2.1.

The history and organization of the US system differs in many ways from the European one. These are well exemplified by French universities, whose history is one of abrupt termination and slow recreation under tight centralized control.

French medieval universities were abolished by the Revolution, as part of the wider effort to reduce the influence of the Catholic Church in education. The task of educating the technical and administrative elites was then assigned to the so-called Grandes Ecoles, modelled after the Ecole des Ponts Chaussées, founded in 1775.

After that, in 1806–8, a brand new institution was set up in order to educate a new generation of teachers, lawyers, medical doctors and the rank-and-file of public administration needed for the renovated state: the Imperial University. Its lecturers were asked to act as civil servants, organized along rigid disciplinary lines and within regional faculties under the state's control. It was not until 1896 that the regional faculties were transformed into local universities, and not until the 1970s that they gained a substantial degree of organizational (but not yet financial) autonomy (Neave 1993).

Research tasks were increasingly assigned to specialized institutes, often founded around a new discipline, such as the Institute Pasteur (1887), or under the direct control of a ministry, as in the case of various agricultural agencies. In 1939 the National Centre for Research (CNRS) was founded. Still well into the 1990s, CNRS employed over 14,000 full-time researchers, and even more were employed by the other public research organizations (PROs), as opposed to 45,000 university professors, whose research engagement was, at best, on a part-time basis.

Although since the late 1980s more and more CNRS labs have been moved within academic walls, a vertical hierarchy of labs exist, starting from those staffed exclusively by CNRS personnel (funded directly by CNRS and the Ministry of Education), followed by those staffed both by CNRS and university personnel, and down to those staffed entirely by university faculty, with no access to CNRS funds (Neave 1993; Laredo and Mustar 2001).

In recent times, French policy-makers have actively encouraged academic staff to engage in patenting and even in entrepreneurial ventures, but have also been unwilling to grant more autonomy either to the universities (which still cannot manage freely their personnel, real estate and finance) or to their researchers (whose contacts with industry are regulated in great detail). While the supposed lessons of the Bayh–Dole Act (see below) are clearly upheld, very little is retained of the lessons derived from the long US history of generous support to fundamental research, faculty mobility and university autonomy, nor from the role these features play both in promoting technology transfer and in shaping scientific entrepreneurship.

Rigidities such as those described for France are common throughout continental Europe, starting with large countries such as Germany and Italy (Clark 1993). And the cross-country similarities extend to the policy-makers' wish to imitate the US university patenting boom, and their reluctance to go beyond university IPR reform.

3 Academic entrepreneurship and university patents

Over the past 20 years, university patents have come to the attention of the academic community, of policy-makers, and more recently of the general public, as a consequence of the impressive growth registered in the number of patent applications by US universities after the introduction of the Bayh–Dole Act in 1980.

In particular, USPTO patent applications by universities have been growing at a much faster rate than those by business companies and individuals. Above all, there has been an impressive growth in the number of academic institutions entering for the first time in the patent system. Henderson *et al.* (1998) calculate that from 1965 to 1991 the number of universities patenting in the United States increased from 30 to 150. Most patents, however, remained concentrated in the hands of the major research universities: in 1991, the top 20 universities held 70 per cent of patents. Biotechnology and, later, software have been the fields where university patenting has thrived most.

A close inspection of the recent literature, however, reveals once again that this particular form of academic entrepreneurship (AE) has deep roots in the history and institutional profile of the US academic system; and that such roots are as important as the patent-based incentive scheme created by the Bayh–Dole Act, if not more.

The lack of comprehensive data on Europe makes any comparison effort extremely arduous, and the need to collect data with imaginative criteria the most urgent. The few available studies, however, tend to confirm the importance of historical conditions and institutional frameworks.

3.1 University patents before the Bayh–Dole Act

University patenting was common in US academia well before the introduction of the Bayh–Dole Act, but for a long time it was hardly associated with a profit motive, at least on the part of the university. The earliest university patents, and the first organizations dedicated to manage them, were due to academic inventors who wished to make their inventions accessible to the public at large, and return any commercial proceedings to science in the form of research grants.

Mowery and Sampat (2001) remind us of the importance, in the history of academic patenting, of Frederick Cottrell, professor at UC Berkeley, who in 1912 founded the Research Corporation, a non-profit company he endowed with his own patents, with the specific purpose of preventing the rampant monopolistic firms of the time from gaining control of the technological field of his interest, and used any proceeds to finance further academic research in a variety of fields. Apple (1989) offers a similar story for Wisconsin's Professor Harry Steenbock, who in 1925 founded the Wisconsin Alumni Research Foundation (WARF), and trusted it with his own patents in the food-processing field.

The Research Corporation, after receiving other patents by Cottrell's colleagues in various universities, went on to become the most important broker of academic patents. WARF similarly allowed the University of Wisconsin to be, for a long time, the top patenting university in the United States.

The Research Corporation and WARF have been instrumental in diffusing IPR management expertise in the US academic system. In the 1970s the Research Corporation, facing increasing costs for the management of a scarcely profitable patent portfolio, started decentralizing a number of tasks to the client universities, to which it offered training courses in IPR management and consultancy services. When, in the 1980s, the Research Corporation sold its university business to a for-profit spin-off, most universities had learnt enough to start their own technology transfer offices and exploit effectively the new opportunities opened up for them by the Bayh–Dole Act.

In Europe, only Britain had a similar experience with the British Technology Group (BTG), which was founded in 1948 (originally with the name of the National Research Development Corporation) with the specific aim of commercializing the results of British public research and re-investing the proceedings in the university system (Clarke 1985; Gee 1991).

3.2 Academic patents as a by-product of entrepreneurial science

Assessing the impact of the Bayh–Dole Act has been a major line of research in the United States. Among the many questions investigated, two are of particular interest here: did the Act really increase the number of university patents, or is it that progress in biotechnology and software (and the concurrent strengthening of IPR laws) would have led to the observed growth anyway? Did the economic incentives attached to patenting change the research pattern of universities, either by increasing the overall research effort, or by addressing it to more applied fields?

Research addressed to these questions has first investigated the kind of inventions patented by universities. The Act aimed at creating a marketplace for proofs of concepts and prototypes, to be acquired, developed and finally placed on the market. Granting IPRs to universities was necessary to overcome any potential market failure.

As argued by many studies we review below, clearly stated IPRs, along with proper contractual arrangement, can help overcoming agency problems.

Case studies by Zucker *et al.* (1998a) suggest that prominent US biotech scientists whose patents were licensed either to established or new companies, play a prominent role after licensing for the precise reason that their expertise and skills are needed to further develop the inventions. Jensen and Thursby (2001) argue that this is also the case in other fields. On a similar vein, Thursby and Thursby (2002) suggest, on the basis of survey data, that growth in university patenting and licensing may be explained by 'universities becoming more entrepreneurial' at all levels: scientists are more willing to disclose their inventions, while the university administrations have increased the rate at which disclosed inventions are patented. Academic research has not shifted from basic to applied, but commercialization efforts have become so aggressive that also inventions of minor importance have been patented. This may also explain early results by Henderson *et al.* (1998), who observe a decline in the importance of university patents (as measured by citations received) after the Bayh–Dole Act.

However, there is no complete agreement on the importance of agency problems, and the related IPR solutions, in the history of US technology transfer.

Studying the cases of Stanford, University of California, and Columbia University, Mowery *et al.* (2001) reach the conclusion that one can hardly attribute the recent historical developments to the Bayh–Dole Act. A greater influence has been possibly played by legislative changes allowing the general strengthening and broadening of IPRs, and in particular the increasing freedom to patent the results of biomedical research. Mowery *et al.* (2001) also find that academic research has not been diverted from basic targets, and criticize the methodology behind the above-mentioned papers whose findings go in the opposite directions.

Colyvas *et al.* (2002) study in depth the history of 11 blockbuster patents from Columbia and Stanford, and confirm that they do not originate from applied research, but rather from basic research aimed at the solution of practical problems. In contrast to Thursby and Thursby (2002), Colyvas *et al.* (2002) suggest that most inventions, while in need of further development for a number of applications, were of immediate use in industry for other purposes; and that the academic research leading to these inventions was, from its very beginning, already sponsored by industry or closely monitored (and influenced) by 'strategically placed people in industry'. According to this view, the Bayh–Dole Act not only did not divert academic scientists from their research interests, but also did little to promote additional technology transfer, which was already in place *before* any patent was taken.

Even the work by Zucker and colleagues, at a closer look, suggests that the latter might have been the case. The academic scientists who have exploited more effectively the opportunities offered by the new patenting regime are the so-called 'superstars', that is scientists with an outstanding publication record and an influential position in the local academic and business communities, as well as a long record in managing grants and personnel for their labs.

More recently, a number of empirical contributions have tested the hypothesis of a trade-off between commitment to scientific (basic) research and patenting by academic scientists, more often than not coming to negative conclusions.

In a cross-sectional perspective, scientists with a superior patenting performance also exhibit superior publication scores, while longitudinal analysis has revealed that inventorship of a patent has the tendency to be associated with a temporary increase in the number of publications, and not to a decrease in their quality. The most productive scientists are also those more likely to become academic inventors, as soon as they tap into a promising set of scientific opportunities.

In addition, institutional factors seem to matter a lot: other things equal, scientists whose colleagues are also academic inventors, or whose universities have a longer tradition of IPR management, are more likely to match a good publication record with some patenting activity (Azoulay *et al.* 2005).

Once again it is confirmed that, as far as individual scientists are concerned, patenting adds a further dimension to traditional forms of scientific entrepreneurship, but it does not change more traditional motivations and behaviours. Rather than pursuing patenting only for personal enrichment, scientists may disclose their research results because they hope to access further funds for their ongoing research, and support their careers within the academy. University administrators, on the contrary, are more prone to exploitative behaviour, as they chase royalties to strengthen the budgets of their institutions. Technology transfer offices may have often the additional problem to legitimize themselves by proving to be useful, as when their performance is evaluated on the basis of the number of disclosures, patents and licences they obtain.

This complexity is confirmed by studies on licensing schemes. As we noted above, licensee companies often require the commitment of scientists to develop further their inventions, so that the contract provides for output-based payments such as royalties. Thursby *et al.* (2005) explore licensing contracts more in depth, and find that they often involve explicit consulting duties for the academic scientists. The higher the importance of tacit knowledge and learning-by-doing in the application of an invention, the higher the chance that commercialization will require a direct involvement of the inventor (Lowe 2002). However, in a survey, Jensen and Thursby (2001) find also that technology transfer offices regard themselves as mediating between the requests of the university administration, which point at maximizing returns for the institutions via fees and royalties and therefore push for binding licensing-cum-consulting contracts, and those of the inventors, whose natural preferences go towards sponsored research.

3.3 Academic patenting in Europe

Research on academic patenting in Europe has been much less substantial than in the United States. The most part of it has dealt with the institutional differences between the European and the US academic systems. Discussion of these differences has served two different purposes:

1 As a possible explanation of quantitative differences in the patenting patterns by universities in the two systems.
2 As a justification for adopting different measurement criteria of academic patenting activity in the two systems.

The three most important institutional differences between Europe and the United States relate, or related until recently, to:

1 The legal ownership of IPRs over academic research, epitomized by the so-called 'professor's privilege', i.e. the exemption of academic personnel from attributing the rights over their inventions to their employers.
2 The balance between research conducted by universities and research conducted by other PROs in countries such as Germany and France.
3 The comparatively little autonomy and competencies of the European university administrations in matters of IPRs.

Until 2002, the academic (or professor's) privilege was a typical institution of the German patent law, which reflected the power achieved by academic scientists in the late 1800s. Over the twentieth century, it was also adopted by many of the countries which imitated the German academic system and science policies. In the UK, until recently, it was also a peculiar arrangement of Cambridge University.

According to a simplistic application of microeconomics, this arrangement should be expected to provide scientists with the highest possible incentives to exploit their research results via patenting. In practice, it has always achieved little, at least compared to the US tradition, as it was the result of an academic system that granted little managerial power to universities and empowered greatly individual professors as opposed to departments.

With no or little administrative support by the university, the costs and complexities of the patent system have traditionally discouraged academic scientists to actively manage their own IPRs. Lacking financial autonomy, most European universities saw little reason to obviate to this situation by arranging a transfer of the professors' IPRs to them, in exchange of royalties. Being forbidden or heavily constrained to conduct commercial activities by governmental regulations, European universities could not manage effectively a patent portfolio. And, with the exception of the BTG in Britain, no European country has ever hosted an organization such as the Research Corporation or WARF.

The academic privilege has been recently abolished both by Germany and Denmark, and Sweden is also considering abandoning it (PVA–MV 2003). Currently, IPRs from publicly funded research belong to universities in Austria, Belgium, France, Germany, the Netherlands, Norway, Portugal and Spain (OECD 2003).

However, no matter whether the national legislation imposed the academic privilege, most European universities have for long lacked the autonomy and administrative skills required in order to take advantage of their professors' patenting activities. They traditionally resisted being involved in such activities, and took the shortcut of allowing scientists engaged in cooperative or contract research with various business companies or PROs to sign blanket agreements that left all IPRs in their partners' hands.

This suggests that a large part of academic patents in Europe may simply escape the most commonly available statistics, which classify the origin of the patent according to the identity of the grantees or applicants, rather than of the inventors. If this methodological remark were true, traditional comparisons with the United States may be proved to be misleading; insofar as they exaggerate the scarcity of academic patents in Europe.

Following this clue, Meyer (2003) for Finland, Iversen *et al.* (2007) for Norway and Lissoni *et al.* (2008) for Italy, France and Sweden have re-classified patents by inventor, and matched the inventors' names with available datasets on university faculties. They found out that in both countries a significant percentage of the business companies' patents originate from academic inventors (from 3 per cent of EPO patents in Italy to 8 per cent in Finland). CNR, CNRS and VTT (the most prominent PROs of France, Italy and Finland, respectively) also hold many patents signed by academic inventors.

Attempts to measure the number of academic patents in Germany have relied on a thinner tactic, namely that of looking for the academic title 'Professor' in the inventor's field of patent applications, given that the title, in Germany, is awarded only to academics with tenured positions. Schmiemann and Durvy (2003) suggest that, according to this kind of calculation, 5 per cent of German patents at EPO can be attributed to universities. Gering's and Schmoch's (2003) calculations suggest that academic inventors' patents at the German patent office have grown from about 200 to almost 1,800 between 1970 and 2000.

Data of this kind have also been collected for individual institutions in Spain, Belgium, Denmark and, on a sample basis, for the Netherlands, the UK and a score of other European countries.

Many European countries' public research systems, however, rely extensively not just on universities, but also on a large web of public laboratories, which have often enjoyed more financial autonomy than universities and whose patenting record, as a consequence, is more open to scrutiny. Germany is a well-documented case.

Gering and Schmoch (2003) estimate that the number of German institutes' patents compare favourably to that of academic professors', especially if one takes into account their smaller budget as opposed to the overall R&D budget of universities.

A cursory look at patent statistics also confirms the importance of French PROs' patents, especially the CNRS (which also retains the IPRs over the research results achieved by teams working within universities, even when they include university staff).

All together, these early measurement attempts suggest that academic patenting in Europe is far from negligible. However, it seems due either to individual scientists' efforts to link up with business companies or to default arrangements between scientists and funding agencies or PROs. University administrations seem to play a lighter role, in line with their lack of autonomy and entrepreneurial chances and attitudes.

4 Academic entrepreneurship and firm creation

As university technology managers became more experienced in technology commercialization, it soon became clear that in many cases licensing of potentially valuable patents was not easily achievable for a number of reasons (Thursby *et al.* 2001; Jensen *et al.* 2003). First, because in many cases academic inventions were disclosed at a proof-of-concept stage, it was hard to convince a firm to take on the long and risky development work needed to bring a final product to the market. Second, in many cases, this work could not be effectively done by an external firm alone, because the tacit and know-how dimension of the knowledge involved was too high. Third, many of the most promising cutting-edge and disrupting technologies are of no interest for large incumbents and would make a good investment only for venture capitals and high-risk equity market.

At the same time, with the development of biotech companies in the United States, several successful examples of superstar scientists that had raised huge amount of capitals in the market by selling the equity of their start-ups were impressing the public opinion, and seem to suggest that academia and industry could join their effort to leverage a new generation of high-tech companies, characterized by a strong research focus.

Business angels and venture capitals were starting to knock on the universities' doors, in search not only of promising business ideas, but also of qualified consulting and peered opinions to evaluate and manage the strategic choices of their biotech portfolios.

These new opportunities were taken in good favour by US and European university administrators, who soon adapted their regulations to allow giving equity capital and branding to start-ups and to ensure job security and institutionalized temporary leave to professors on 'entrepreneurial duties'. Many technology managers saw academic spin-offs as a sort of advanced solution to technology transfer that would help finding viable commercialization strategies to growing patent portfolios (Franklin *et al.* 2001).

4.1 Early studies: academic knowledge as a non-tradable asset

Since the beginning of 1990s, many scholars have investigated the individual motivations and the rationale behind the claim for a proactive role of university-based scientists in the generation of new high-technology applications for nascent industries.

Early contributions to academic spin-off company creation tended to stress that university-based scientists own a specific set of knowledge and information, enabling them to spot valuable opportunities of investment, which would remain hidden to other people. Hence a scientist or a novel PhD student may have a comparative advantage vis-à-vis other potential entrepreneurs in the recognition of promising businesses, thanks to the idiosyncratic knowledge gained while working at a scientific discovery.

This view was supported by several pieces of empirical evidence, especially with regard to emerging high-tech industries. For instance, Zucker and Darby (1996) suggested that the most successful biotech companies were co-publishing with university professors and showed that their commercial success, in terms of the number of products developed and commercialized, was positively associated to the scientific eminence of researchers participating in the scientific board and holding equity stakes. In a later study, co-publications were also shown to explain a firm's patent citations rate, suggesting the idea that a stronger technological base would produce higher quality patent applications in fields characterized by a high strategic value of IPR assets (Zucker *et al.* 1998b). Shane and Stuart (2002) studied the probability of success of 134 new ventures exploiting MIT inventions and found that both the academic rank of the inventor and the number of MIT patents in the company portfolio were likely to increase the probability of an IPO and decrease the failure rate.

The idea underlying those early studies is deeply rooted in a linear model of science–technology relationships: in which cutting-edge science is produced by the universities and highly attached to individual researchers, hence, professors and personnel trained in the academic environment have easier and quicker foresight of technological solutions and a preferred access to business.

Besides, the attention of early studies was especially focused upon the growing US biotechnology industry and on its innovative potential, as compared to more traditional drug industry and market incumbents. Certainly the idiosyncratic features of that industry makes any generalization hard and does not justify some excess of emphasis placed on early result.

With regard to the choices on the structure of ownership, those contributions stressed that cutting-edge science is naturally attached to individuals and, because of the poor absorptive capacity of the environment, transfer could not occur through simple licensing and requires aligning professors' remuneration to the success of the venture (Audretsch 1995; Audretsch and Stephan 1999). This seemed particularly the case of newly created firms, which can be shaped on the emerging scientific culture and may be better suited to the exploitation of new and radical technologies (Henderson 1993).

In such a context, because the intellectual capital was seen as the true key asset, the founding of a firm looked like a unique means for the scientist to extract private gains from her idiosyncratic knowledge. Additionally, because the diffusion of this knowledge is naturally bounded by face-to-face interactions, many authors foresaw a lower need of engaging in enforcement and protection of IPRs (Audretsch 1995). Hence, the mantra goes that best scientists enjoy both a superior access to high-value knowledge and a stronger natural excludability leading to higher-value entrepreneurial opportunities in the selection phase and sustainable competitive advantages later on (Zucker *et al.* 1998b).

Besides, in highly incomplete informational contexts, the scientific reputation of the academic entrepreneur, or the rank of the related institution could have been used by the stakeholders as an indirect signal of the high perspective value of the venture (Shane and Khurana 2003; Stuart and Ding 2004). In the absence of more accurate information, a researcher's eminence could serve to proxy the strength of a company's technological base whereas the star scientist's research specialties would signal the future technology strategies that the company would have been undertaken (Audretsch and Stephan 1996). In a study of biotechnology IPOs, Stephan and Everhart (1998) found that the amount of funds raised and the initial stock evaluation of firms were positively linked with the reputation of the university-based scientist associated with the firm. *Ceteris paribus*, Di Gregorio and Shane (2003) found that spin-off companies from peer-universities were more likely to attract venture capitals than less prestigious ones, whereas Franklin *et al.* (2001), in a survey of key competitive factors conducted among UK technology managers, reported that the researcher's reputation was ranked immediately after their scientific preparation and that this was especially true for higher performing and more experienced universities.

4.2 Incentive problems rediscovered

Following this line of thought, at the beginning of 1990s, most academic administrations, technology managers and venture capitalists were especially stressing the technical content of university applications, which they expected to be more radical and broader in scope than innovations with purely industrial backgrounds. Nevertheless, this emphasis on the knowledge capital and on the alleged superior technological endowments eventually fades at the end of 1990s, when broader studies reported mixed evidence. For instance, Nerkar and Shane (2003) found that the top technological level of MIT start-ups reduced failure rates only in low-concentration industries. The same study also re-established the importance of industry differences in terms of patent effectiveness and appropriability regimes in explaining venture success.

In the meantime, with the help of policy-makers, an increasing number of universities had invested in (often unprofitable) technology transfer activities (Thursby and Thursby 2002), and it started to be questioned if the importance of firm creation from academia had been over-sold, possibly beyond any true economic advisability, both in terms of economic gains and of professors' intentions, which brought into play an entirely new set of problems.

As soon as profit started to become a concern of universities at the institutional level, technology managers discovered that a good technological endowment or the expectations of business profits in many cases were not enough to justify or to convince a scientist to take part in a venture, as ultimately entrepreneurship also meant risk-taking, a strategic vision and possibly a life change. Indeed 'entrepreneurial type' scientists were hard to find, a considerable mismatch of objectives between faculties, technology managers and investors was affecting transactions and several aspects of the academic culture after all seemed to be at odds with the rules of market competition (Siegel *et al.* 2003).

Despite their technological strengths, newborn firms were frequently reported to be unsuccessful because of a failure in complying with the market needs. Field studies and extensive interviews with technology managers portray scientists as individuals with a good taste for science, but with relatively naive ideas about the pursuit of market goals (Thursby and Thursby 2003c).

The knowledge-endowment argument and its related theory of entrepreneurship hence lost much of their appeal, as a stronger trade-off between scientific and market concerns was brought back to the forefront of analysis.

What falls over in the notion of AE applied to spin-off policies and strategies is not the capacity of scientists to offer a valuable pool of technological opportunities to market investors, nor really the role of the 'knowledge entrepreneur' in chasing market opportunities. Rather, the focus is shifted towards the alignment of a scientist's objectives to the goals of a nascent firm, where the expected gains of a scientific entrepreneur are not seen only as those of profit in case of firm success, but also come in the form of increased availability of funds for complementary research.

To scientists concerned with their academic careers, research funds made available through the firm's R&D activities may be particularly appealing insofar as they may serve to buy instruments and data, hire additional personnel, pay for travel to conferences, and generally enlarge the professor's budget for research. Hence, the decision to start up a company would depend for a good part on the researcher's expectations of engaging in stimulating, fruitful and possibly generously funded development activities, which comes along with the creation of a new venture, rather than on expectations of profit and growth, especially when she is not required to put in a big share of the equity upfront.

In addition, because the gains to earn from big research budgets vary with the different stages of a career, the propensity of faculty members to engage in interchanges with industry was seen to be also depending on lifecycle effects and on the choices of investigative pathways (Thursby and Thursby 2003a). Whenever the contiguity of scientific and industrial effort fade, monetary incentives should be raised to compensate for the time taken by purely commercial activities with unclear effects on the academic career (Thursby and Thursby 2003b).

The idea that, in many cases, market goals as such simply fail to produce a set of incentives compatible with the day-to-day life of the entrepreneurial scientists has been a finding of many surveys. Jensen *et al.* (2003) report that scientists may voluntary retain disclosures of potentially marketable technologies and suggest that the opportunity cost of development activities was stronger for higher quality scientists, whose inventions arise typically at a very embryonic stage. Franklin *et al.* (2001) report that technology managers indeed regard the academic founders of their spin-off companies as entrepreneurial individuals with good commitment on the research projects, but they signal a stronger mismatch of perceived goals as the most common cause of venture failure.

The researcher's attitude towards pure scientific investigation, the privilege of having her own lab and the ability to enlarge her group of graduate students frequently clashes with reward schemes based upon commercialization. Not surprisingly, many scholars report that the problem arises most often when the development stage was nearly completed and the firm had to promote a general shift of goals towards the industrialization of the product and to cope with marketing and financial pressures (Shane 2004; Vohora *et al.* 2004). It is at that stage that financial constraints challenge the availability of funds for further development and laboratory work and the appeal of having sponsored additional research fades.

In the follow-up of a survey conducted on 62 US universities in the 1990s, Jensen *et al.* (2003) describe the relationship linking university administration, technology managers and individual scientists as an agent–principal game theoretic model. Scientists are seen as positively reacting to both monetary incentives, and to the share of sponsored research they may obtain for their labs, but, because high-quality faculties would disclose inventions at a more embryonic stage, willingness to disclose would depend more substantially on the latter than on the former.

Besides, the opportunity costs faced by scientists would not just depend on exogenous preferences and personal interests, but also on the availability of other funds, on other appointments and on purely lifecycle effects. In this respect, older scientists may be more willing to cash-in the market gains of their knowledge assets than their younger colleagues because they have already achieved the highest academic ranks (Audretsch and Stephan 1996). This could also be the case for professors of continental European countries, where the academic environment is characterized by lower competition and by job security. For instance, Audretsch (2000) found that the probability for an individual scientist to create a private venture is higher for older professors, suggesting the idea that academic entrepreneurship becomes a more viable option when career pressures have cooled down and the scientist has coped with the concern of establishing her scientific position in academia. This can be especially true within the contexts in which social rules discourage for-profit activities, in which case, only older and highly reputed scientists may dare to undergo non-traditional academic pathways (Stuart and Ding 2004). For younger scientists, such as novel PhD students and research assistants, the founding of a venture may rather become appealing as a viable strategy to exit academia (Roberts 1991; Franklin *et al.* 2001).

4.3 Business creation vs. patent licensing. Do we really need academic spin-offs?

Although university patents, spin-off company creation, consulting and joint research agreements are often addressed as separate, alternative transfer mechanisms, in practice, commercializing a piece of university research may require a variable mix of all those instruments. For instance, in a recent survey on commercialization of US academic research, it emerged that licensing contracts made by technology transfer offices in the majority of cases involve royalties, annual fees, equity, milestones and consulting agreements (Thursby *et al.* 2005). The question of what instrument is best suited to transfer different pieces of knowledge has been the focus of many recent contributions, whose central argument goes that market inefficiencies in the transfer of knowledge claim for an ownership structure with some risk-taking positions from the party that possesses the most idiosyncratic assets.

Because scientists' knowledge is characterized by natural excludability, it resists codification in a fully transmittable form and tends to stick to individuals, even after a patent has been filed or an article published. At the same time, we have already stressed that many academic inventions are no more than a proof of concept at the frontier of knowledge. It follows that, in order to take up the nutshell technology and undertake the final development stage on their own, firms need to recruit the scientist as a partner or stakeholder: in the absence of her personal involvement, they would not be able to profit from the innovation (Jensen and Thursby 2001). Therefore, it may be that academic scientists face a stronger need to become entrepreneurs, the higher the degree of sophistication of their technology compared to that of the outside business world (Shane 2004).

Besides, the decision of whether or not the exploitation of a technology is best achieved by patent licensing or by a start-up depends on the technological regime and on the appropriability of the innovation. In low-appropriability patent regimes, licensing may be hard and innovations may not be commercialized because of a lack of incentives, but if the knowledge is also characterized by natural excludability, the creation of a company exploiting a scientist's idiosyncratic knowledge may become the only viable transfer option (Shane 2004).

Some empirical evidence in support of this thesis has been provided both in case studies and empirical analyses. Shane (2001b and 2002) found that the probability of an MIT invention to result in of a patent application was higher in strong appropriability regimes. In a related study he also found that the spin-off rate increased with the novelty and importance of the technology behind it (Shane 2001a). In a study of the technology transfer activities at University of California, Lowe (2002) found that patents characterized by a stronger scientific base and a higher degree of tacitness were significantly more likely to be licensed to their original inventors, thus supporting the idea that spin-off creation is necessary when the scientist's knowledge is highly uncodified and idiosyncratic.

Finally, Feldman *et al.* (2002) report that the willingness of US universities to take equity in a new venture was generally higher among longer-experienced technology offices, which suggests that equity positions of university-administrations may offer a second-best solution to the problem of achieving higher transfer of knowledge to the market, one that perhaps involves a lower risk to divert good scientists from their original tasks.

4.4 Consequences (intended and unintended) of academic entrepreneurship

The argument that an academic entrepreneurship may do a non-replaceable job in fostering the emergence of new generations of high-tech firms and the renovation of local economic systems, in recent decades, has been widely popularized by policy-makers in many European countries. In addition, common arguments in favour of academic spin-off creation normally emphasized that the core attitude of spin-off companies for experimentation would resist the start-up phase and result in a superior propensity of the firm to deliver continuous innovation later on.

However, if one considers the widespread consensus of those claims from scholars and policy-makers alike, it comes as a surprise that very few pieces of evidence have been made available so far that assess the actual performances and the contributions of academic venturing to technological change and local development. As we look at the empirical evidence, notwithstanding the problem of the reliability of field analyses in the absence of a clear-cut definition of academic spin-off (Pirnay *et al.* 2002), we have little more than anecdotes on success stories of university-based inventions that were incorporated into a firm, developed a successful application, grew big and eventually clustered with other firms (see Roberts 1991). Research on the biotechnology sector suggests that the presence of a scientist has a positive effect over start-up success. Nevertheless, those results have hardly been extended to different industries (Nerkar and Shane 2003) and to institutional frameworks other than in the United States.

When it comes to appreciating the actual contribution of academic ventures, only some very preliminary evidence is available that proves the supposed higher performances of spin-off companies either in terms of innovativeness, or in terms of employment created and new products developed and sold. Mustar (1997) reports that R&D intensity of French academic spin-offs was higher than that of other new-technology based start-ups. Similar results were found for samples of UK firms.

Perhaps some stronger, though highly industry-specific, evidence has been provided in support of the claim that companies founded by academic personnel were likely to locate around universities (Audretsch and Stephan 1996; Zucker *et al.* 1998b). This can be a somehow desirable feature from the point of view of policy-makers, concerned with fostering economic development locally, and for university administrators alike, to the extent that spin-off companies may serve as good partners for joint research and technology licensing later on. The clustering choices observed in many research spin-offs may reflect the initial need for part-time scientists to locate closer to their academic jobs and to a hard-science environment, in order to comply with their multi-task careers (Audretsch and Stephan 1996). However, it is dangerous to push this observation further and take it as a confirmation of a higher-than-average focus on innovation and high-technology content of spin-off firms (see Shane 2004), as ultimately location entails a strong path dependent component.

Overall, as we look at successful case studies, it is worth asking what was the fundamental contribution offered to firms by the origin in the academic environment: if it was the unique knowledge offered by universities, the support given in terms of credibility and networking in a context of jeopardized information or, finally, if it was simply good training and qualified scientific consulting that proved to be critical.

This is a question worth asking because spin-off activities also bring several downsides and costs even beyond the general costs and risk of the investments.

Major opportunity costs faced by university administrators, irrespective of their civil service mission, are of at least two kinds. First, universities may lose good scientists or may simply divert them from high-quality publications and teaching. Second, at a more fundamental level, they should be afraid of losing their long-lasting reputation of reliable and non-opportunistic agents, which is functional to their capability to act as a broker for the market of technology, as well as for their more traditional goals. This concern seems to have been understated more in the literature than in practice. For instance, some institutions, such as Cambridge University in the UK, although proactive in business creation, refuse to commit their commercialization activities to a purely profit-oriented mission and describe their role as one of facilitators in the diffusion of knowledge for the benefit of society. In practice, concerns have been expressed that professors may use students as low-paid employees and indiscriminately re-sell the effort of collective commitments. Shane (2004) reported that, in order to cope with the problem of moral hazard, many US faculties have also introduced a general prohibition for a scientist to work at the same research project both in their internal unit and in their external private ventures, after a person died at University of Pennsylvania Medical School during the test of a therapy developed by an academic spin-off.

5 Conclusions

Economic incentives moulding academic entrepreneurship are much more affected by the long-standing institutional features of national university systems, than by any ad hoc legislation affecting IPRs in university. These institutional features do not simply influence the intensity of patenting and firm-creation activities. More generally, they explain to what extent commercial activities may or may not help scientific entrepreneurs to progress in their careers. Among those institutional features, university autonomy, personnel mobility, and the principal investigator principle stand out as the most prominent. Patent-based and spin-off based technology transfer is, by and large, the product of a specific institutional history, that of the US research universities, where these features have been prominent. Every introduction of those issues within the various European university systems should require first and foremost strong reflections and adjustments that take into account institutional, organizational and environmental characteristics of academic research at the national level.

Academic entrepreneurs who are active in patenting, firm-founding (and more generally in technology transfer) come disproportionately from the ranks of scientific entrepreneurs with a brilliant scientific record, possibly oriented to fundamental research. These scientists' economic agenda is centred upon entrepreneurial efforts within the university, aimed at gaining reputation through discipline building, creation and management of laboratories and research teams, and an appetite for the economic resources necessary to pursue those goals.

To those scientists, patent licensing and spin-off creation are appealing not just because of the expectation of profits, but also because they offer valuable opportunities to enlarge their sphere of influence, to empower their internal and external consensus, and inflate the budgets available for their research. Hence, any wise policy of technology transfer in academia should move from a broad consideration of the overall personal incentives faced by scientists and framed within the context of academic careers.

The complexity of academic scientists' incentives to commercialize their discoveries suggests an immediate policy conclusion, albeit a speculative one (at this stage of research): the two objectives of promoting academic entrepreneurship and restraining public expenditures for academic science (which are often found to go hand-in-hand in Europe) are largely incompatible. Starving academic science does not push 'unruly' scientists to apply more thoroughly their knowledge to technologically relevant issues; it merely stifles the entrepreneurial spirits of the younger and more dedicated researchers, from whose ranks we expect to emerge one day the most active producers of patents, companies and any other form of technology transfer effort. Additionally, when the goals of science and market diverge, the cost of convincing good scientists to take part in commercial activities grows, and technology managers may end up to with only untalented scientists.

The biggest absence for the literature we reviewed in this chapter is the demand side of the market for academic inventions. Future theoretical efforts to properly conceptualize AE will have to take it into account.

As for empirical research, this will have to be directed towards a better measurement of entrepreneurial activities taking place in universities, without drawing any preconceived distinction between the industrial exploitation of research results, and more traditional efforts to build academic careers within the university via breakthroughs into new research fields and the creation of new research groups, labs and departments.

Acknowledgements

Comments from Nicoletta Corrocher, Nicola Lacetera, Fabio Montobbio and Grid Thoma are gratefully acknowledged, as well as comments received by participants in: the KEINS project; ESSID, the European Summer School of Industrial Dynamics held in Cargese (Corsica), September 2005; the CEMI-EPFL retreat in Saillon (CH), November 2005; and seminars at Bocconi and Brescia universities. Suggestions for readings that proved enlightening came from Paul David, Christian Zellner and Mario Biagioli. Francesco Lissoni accomplished much of his contribution to the chapter while visiting the Sloan School of Management, MIT, thanks to Fullbright grant. Usual disclaimers apply.

Notes

1 DISPEA–Politecnico di Torino.
2 Università di Brescia and KITeS–Cespri–Università Bocconi.
3 We find other striking accounts of this breed of scientists in Latour's (1988) portrait of Louis Pasteur, and in Mowery and Sampat's (2001) and Apple's (1989) biographical notes on Frederick Cottrell and Harry Steenbock.

References

Adams, J.D., Black, G.C., Clemmons, J.R. and Stephan, P.S. (2005) 'Scientific teams and institutional collaborations: evidence from U.S. universities, 1981–1999', *Research Policy*, vol. 34 (3): 259–85.

Agrawal, A. and Henderson, R. (2002) 'Putting patents in context: exploring knowledge transfer from MIT', *Management Science*, vol. 48 (1): 44–60.

Apple, R.D. (1989) 'Patenting university research: Harry Steenbock and the Wisconsin Alumni Research Foundation', *Isis*, vol. 80 (3): 374–94.

Audretsch, D.B. (1995) *Innovation and Industry Evolution*, Cambridge: MIT Press.

Audretsch, D.B. (2000) 'Is university entrepreneurship different?', mimeo.

Audretsch, D.B. and Stephan, P.E. (1996) 'Company–scientists locational links: the case of biotechnology', *American Economic Review*, vol. 86 (3): 641–52.

Audretsch, D.B. and Stephan, P.E. (1999) 'Knowledge spillovers in biotechnology: sources and incentives', *Journal of Evolutionary Economics*, vol. 9 (1): 97–107.

Azoulay, P., Ding, W. and Stuart, T. (2005) 'The determinants of faculty patenting behavior: demographics or opportunities?', NBER Working Paper no. 11348.

Callon, M. (2002) 'From science as an economic activity to socioeconomics of scientific research', in Mirowski, P. and Sent, E.M. (eds) *Science Bought and Sold. Essays in the Economics of Science*, Chicago: University of Chicago Press.

Clark, B.R. (ed.) (1993) *The Research Foundations of Graduate Education: Germany, Britain, France, United States, Japan*, Berkeley: University of California Press.

Clarke, M. (1985) 'British Technology Group – UK technology transfer grows', *Nature*, vol. 316: 385.

Colyvas, J., Crow, M., Gelijns, A., Mazzoleni, R., Nelson, R.R., Rosenberg, N. and Sampat, B.N. (2002) 'How do university inventions get into practice?', *Management Science*, vol. 48 (1): 61–72.

Dasgupta, P. and David, P.A. (1994) 'Toward a new economics of science', *Research Policy*, vol. 23 (5): 487–521.

Di Gregorio, D. and Shane, S. (2003) 'Why do some universities generate more start-ups than others?', *Research Policy*, vol. 32 (2): 209–27.

Etzkowitz, H. (1983) 'Entrepreneurial scientists and entrepreneurial universities in American academic science', *Minerva*, vol. 21 (2): 198–233.

Etzkowitz, H. (2003) 'Research groups as "quasi-firms": the invention of the entrepreneurial university', *Research Policy*, vol. 32 (1): 109–21.

Feldman, M., Feller, I., Bercovitz, J. and Burton, R. (2002) 'Equity and the technology transfer strategies of American research universities', *Management Science*, vol. 48 (1): 105–21.

Franklin, S.J., Wright, M. and Lockett, A. (2001) 'Academic and surrogate entrepreneurs in university spin-out companies', *Journal of Technology Transfer*, vol. 26 (1): 127–41.

Gee, H. (1991) 'Privatization ahead for BTG', *Nature*, vol. 349.

Gering, T. and Schmoch, U. (2003) 'Management of intellectual assets by German public research organisations', in OECD, *Turning Science into Business. Patenting and Licensing at Public Research Organizations*, Paris: Organisation for Economic Co-operation and Development.

Graham, H.G. and Diamond, N. (1997) *The Rise of American Research Universities. Elites and Challengers in the Postwar Era*, Baltimore: Johns Hopkins University Press.

Henderson, R. (1993) 'Underinvestment and incompetence as a responses to radical innovation: evidence from the photolithographic alignment equipment industry', *Rand Journal of Economics*, vol. 24 (2): 248–70.

Henderson, R., Jaffe, A. and Trajtenberg, M. (1998) 'Universities as a source of commercial technology: a detailed analysis of university patenting 1965–1988', *Rev. of Economics and Statistics*, vol. 80 (1): 119–27.

Iversen, E.J., Gulbrandsen, M. and Klitkou, A. (2007) 'A baseline for the impact of academic patenting legislation in Norway', *Scientometrics*, vol. 70 (2): 393–414.

Jensen, R. and Thursby, M.C. (2001) 'Proofs and prototypes for sale: the licensing of university inventions', *American Economic Review*, vol. 91 (1): 240–59.

Jensen, R., Thursby, J.G. and Thursby, M.C. (2003) 'Disclosure and licensing of university inventions: "the best we can do with the s**t we get to work with" ', *International Journal of Industrial Organization*, vol. 21 (9): 1271–300.

Kitcher, P. (1993) *The Advancement of Science*, Oxford: Oxford University Press.

Larédo, P. and Mustar, P. (eds) (2001) *Research and Innovation Policies in the New Global Economy: an International Comparative Analysis*, Cheltenham: Edward Elgar.

Latour, B. (1988) *The Pasteurization of France*, Cambridge, MA: Harvard University Press.

Lenoir, T. (1997) *Instituting Science. The Cultural Productivity of Scientific Disciplines*, Stanford: Stanford University Press.

Lissoni, F., Llerena, P., McKelvey, M. and Sanditov, B. (2008) 'Academic patenting in Europe: new evidence from the KEINS database', *Research Evaluation*, vol. 16: 87–102.

Lowe, R.A. (2002) 'Entrepreneurship and information asymmetry: theory and evidence from the University of California', mimeo.

Merton, R.K. (1973) *The Sociology of Science: Theoretical and Empirical Investigations*, edited by N.W. Storer, Chicago: University of Chicago Press.

Meyer, M. (2003) 'Academic patents as an indicator of useful research? A new approach to measure academic inventiveness', *Research Evaluation*, vol. 12 (1): 17–27.

Mowery, D.C. and Sampat, B.N. (2001) 'Patenting and licensing university inventions: lessons from the history of the research corporations', *Industrial and Corporate Change*, vol. 10 (2): 317–55.

Mowery, D.C., Nelson, R.R., Sampat, B.N. and Ziedonis, A.A. (2001) 'The growth of patenting and licensing by US universities: an assessment of the effects of the Bayh–Dole act of 1980', *Research Policy*, vol. 30 (1): 99–119.

Mustar, P. (1997) 'Spin-off enterprises. How French academics creates high tech companies: conditions for success or failure', *Science and Public Policy*, vol. 24 (1): 37–43.

Neave, G. (1993) 'Séparation de corps. The training of advanced students and the organization of research in France', in Clark, B.R. (ed.) *The Research Foundations of Graduate Education: Germany, Britain, France, United States, Japan*, Berkeley: University of California Press.

Nerkar, A. and Shane, S. (2003) 'When do start-ups that exploit patented academic knowledge survive?', *International Journal of Industrial Organization*, vol. 21 (9): 1391–410.

OECD (1999) *University Research in Transition*, Paris: Organisation for Economic Co-operation and Development.

OECD (2003) *Turning Science into Business. Patenting and Licensing at Public Research Organizations*, Paris: Organisation for Economic Co-operation and Development.

Pirnay, F., Surlemont, B. and Nlemvo, F. (2002) 'Toward a typology of university spin-offs', *Small Business Economics*, vol. 21 (4): 355–69.

PVA-MV (2003) *Report on the Abolition of the German Professors' Privilege: Overview of Changes and Challenges*, Stockholm: Vinnova – Swedish Agency for Innovation Systems.

Roberts, E.B. (1991) *Entrepreneurs in High Technology*, Oxford: Oxford University Press.

Rudolph, F. (1990) *The American College and University: a History*, Athens, GA: University of Georgia Press, 1st edition in 1962.

Schmiemann, M. and Durvy, J.N. (2003) 'New approaches to technology transfer from publicly funded research', *Journal of Technology Transfer*, vol. 28 (1): 9–15.

Shane, S. (2001a) 'Technology opportunities and new firm formation', *Management Science*, vol. 47 (2): 205–20.

Shane, S. (2001b) 'Technology regimes and new firm formation', *Management Science*, vol. 47 (2): 1173–90.

Shane, S. (2002) 'Selling university technology: patterns from MIT', *Management Science*, vol. 48 (1): 122–37.

Shane, S. and Khurana, R. (2003) 'Bringing individuals back in: the effect of career experience on new firm founding', *Industrial and Corporate Change*, vol. 12 (3): 519–43.

Shane, S. and Stuart, T (2002) 'Organizational endowments and the performance of university start-ups', *Management Science*, vol. 48 (1): 154–70.

Siegel, D., Westhead P. and Wright, M. (2003) 'Assessing the impact of university science parks on research productivity: explanatory form-level evidence from the United Kingdom', *International Journal of Industrial Organization*, vol. 21 (9): 1357–69.

Stephan, P.E. and Everhart, S.S. (1998) 'The changing rewards to science: the case of biotechnology', *Small Business Economics*, vol. 10 (2): 141–51.

Stephan, P.E. and Levin, S.G. (2002) 'The importance of implicit contracts in collaborative scientific research', in Mirowski, P. and Sent, E.M. (eds) *Science Bought and Sold. Essays in the Economics of Science*, Chicago: University of Chicago Press.

Stuart, T.E. and Ding, W.W. (2004) 'When do scientists become entrepreneurs? The social structural antecedents of commercial activity in the academic life sciences', mimeo.

Thursby, J.G. and Thursby, M.C. (2002) 'Who is selling the ivory tower? Sources of growth in university licensing', *Management Science*, vol. 48 (1): 90–104.

Thursby, J.G. and Thursby, M.C. (2003a) 'Are faculty critical? Their role in university–industry licensing', NBER Working Paper no. 9991, National Bureau of Economic Research.

Thursby, J.G. and Thursby, M.C. (2003b) 'Pattern of research and licensing activity of science and engineering faculty', mimeo.

Thursby, J.G. and Thursby, M.C. (2003c) 'Industry/university licensing: characteristics, concerns and issues from the perspective of the buyer', *Journal of Technology Transfer*, vol. 28 (3): 207–13.

Thursby, J.G., Jensen, R. and Thursby, M.C. (2001) 'Objectives, characteristics and outcomes of university licensing: a survey of major U.S. universities', *Journal of Technology Transfer*, vol. 26 (1): 59–72.

Thursby, M.C., Thursby, J.G. and Dechenaux, E. (2005) 'Shirking, shelving, and sharing risks: the role of university license contracts', mimeo.

Trow, M. (2003) 'In praise of weakness: chartering, the University of the United States, and Dartmouth College', Center for Studies in Higher Education Research Occasional Paper CSHE.2.03, University of California, Berkeley, CA.

Vohora, A., Wright, M. and Lockett, A. (2004) 'Critical junctures in the development of university high-tech spin-out companies', *Research Policy*, vol. 33 (1): 147–75.

Zucker, L.G. and Darby, M.R. (1996) 'Star scientists and institutional transformation: patterns of invention and innovation in the formation of biotechnology industry', Proceedings of the National Academy of Sciences of the United States of America, Colloquium Paper no. 93: 12709–16.

Zucker, L.G., Darby, M.R. and Armstrong, J. (1998a) 'Geographically localized knowledge: spill–overs or markets?', *Economic Inquiry*, vol. 36 (1): 65–86.

Zucker, L.G., Darby, M.R. and Brewer, M.B. (1998b) 'Intellectual human capital and the birth of US biotechnology enterprises', *American Economic Review*, vol. 88 (1): 290–306.

Part III

Knowledge-intensive entrepreneurship and innovation systems

The European case

5 Patterns of technological entry in different fields

An analysis of patent data

Roberto Camerani and Franco Malerba

1 Introduction

The aim of this chapter is to study the patterns of technological entry of organizations into different industries through the analysis of patent data. The chapter examines the emergence of new innovators in different technologies. The study is based on patents applications at the European Patent Office. The basic starting point is that very often organizations that patent for the first time are also new organizations.

The chapter intends to answer questions such as: which are the most common patterns of technological entry? Are there significant differences among countries (France, Germany, Italy, Sweden, the United Kingdom and the United States) and technological fields? Do technological entrants differ among countries and sectors? Which are the characteristics of those organizations that managed to persist in patenting activity over time?

The chapter is divided in three parts. First, we provide a summary of the main themes about entry and innovation, as well as a description of the KITeS–Cespri patent database and the methodology used (Section 2). Then we use patent data to examine technological entry in different technological fields. We focus on organizations from six countries, that patented in 13 technological fields during the period 1990–2003. In particular, we describe the patterns of technological entry and the differences among sectors and countries (Section 3), and then we concentrate on the post-entry behaviour, analysing the survival of entrants, in terms of persistence in patenting activity and patents citations (Section 4).

2 Overview and methodology

2.1 Entry and the innovation: a review of main themes

Traditional theories about entry of new organizations into sectors claim that the entry process is driven by the existence of supernormal profits in the markets. Entry of new organizations will eventually drive the sector to a steady state equilibrium, by eroding market profits and making less efficient organizations exit from the market. Therefore, entry represents a threat for those organizations already operating in the markets. Since Bain (1956), the first empirical studies about entry concentrated on the existence and relevance of the above-mentioned barriers impeding the entry of new companies into the market, especially focusing on the relationship between barriers and the presence of supernormal profits in the sector. On the contrary, more recent studies challenged the traditional view of entry, and concentrated on effective rates of entry in different industries and on the relationship between entry and exit of organizations. The results of these studies did not confirm the main assumptions of the traditional model of entry. A primary and remarkable result was that all industries are characterized by a non-negligible rate of entry, even if there are high barriers to entry protecting them. Furthermore, market profitability does not seem to be the primary source of attraction of new entrants (Marsili 2001: 18). In addition, entrants' ability to erode the market shares of incumbents has not been demonstrated; even in sectors characterized by high entry rates, entry is not followed by a decrease in average profits (Geroski and Jacquemin 1988; Geroski and Schwalbach 1991). Some of the most important and widespread empirical results about entry have been summarized by Geroski (1995). First of all, entry is quite common. However, entry rates are also associated with low levels of market penetration and low survival rates. Entry is not particularly related to profitability and the entry barriers and does not have particularly strong effects on average industry price or cost margins. Finally, one of the most interesting results is that entry is often associated with high rates of innovation and increases in efficiency. This is confirmed by other studies (such as Acs and Audretsch 1990; Geroski 1989a and 1989b) that have reported the existence of a positive correlation between entry and innovation. To these empirical results one may add the more general considerations by Nelson and Winter (1982) who describe the entry of new organizations into markets as a process triggered by the perception of technological opportunities to innovate or imitate.

Despite these results, for many years the relationship between entry and innovation – and particularly the role of new innovators – has been neglected. More recently, studies on industrial dynamics gave a renewed interest to the role of new innovators (Malerba and Orsenigo 1999), especially of new entrants in different stages of the lifecycle of an industry. In particular, new entrants have a major role in the early stages of the product lifecycle, when a dominant design has not yet emerged: new firms enter into the market bringing new ideas and product innovations (Agarwal and Gort 1996; Gort and Klepper 1982; Klepper 1996; Utterback 1994).

Despite the extensive empirical literature on the entry of new organizations into markets, no comprehensive cross-country and cross-sector comparisons about new innovators have been provided. Some efforts in this direction have been attempted by using patent statistics in order to study technological entry. A broad picture of the phenomenon has been provided by Malerba and Orsenigo (1999), who studied technological entry and exit of organizations in 49 technological fields from 1978 to 1990. Other studies considered technological entry as a measure for the turbulence in technological fields, in order to test the presence of technological regimes and Schumpeterian patterns of innovation in different fields (Breschi *et al.* 2000; Malerba and Orsenigo 1995 and 1996). Finally some other studies analysed post-entry behaviour, by focusing on the persistence in the patenting activity (Cefis 2003; Cefis and Orsenigo 2001).

The present chapter aims to analyse the technological entry of organizations into sectors, using and consolidating the methodology used by Malerba and Orsenigo (1999). This chapter considers as technological entrants those organizations that patent for the first time in a specific technological field. More specifically, this chapter analyses patterns of technological entry and the characteristics of those organizations which patent for the first time in different sectors and in different countries. The main questions that the chapter intends to answer are the following. Are the patterns of technological entry different across countries and sectors? Do technological entrants manage to be persistent innovators and therefore patent again?

2.2 Description of the patent dataset

The data used in this work is based on patents and patent citations dataset. The dataset includes all patent applications to the European Patent Office (EPO), from 1978 to 2005. The dataset includes more than 1.7 million patents, applications to firms, research organizations, universities and individuals from 174 different countries.

Furthermore, for each patent application, the dataset includes the following set of bibliographic variables:

- Priority, application and publication number.
- Priority, application and grant dates.
- Title and abstract.
- Designated states for protection.
- Status of application.
- Main and secondary International Patent Classification (IPC) codes.
- Applicant's name and address.
- Inventors' names and addresses.

In addition, the dataset also includes all the citations made and received by each EPO patent. The data set contains 834,310 citing patents and 642,217 cited patents, corresponding to a total of 1,621,329 citations.

In particular, bibliographic data on patent applications are periodically obtained by the Espace Bulletin CD–R provided by the EPO, while data about patent citations are retrieved from the REFI tape also supplied by the EPO. The information provided by EPO goes through a process of names cleaning and standardization conducted by KITeS–Cespri (concerning misspelled names, applicants who patent with different names, or legal status of organizations changed over time). This process has led to the identification of 169,557 unique organizations, starting from 215,236 different original applicants' names.

Table 5.1 Classification of patents into different technological classes

Code	*Technological classes*	*Macro classes*
1	Electrical engineering	Electrical engineering; electronics
2	Audiovisual technology	
3	Telecommunications	
4	Information technology	
5	Semiconductors	
6	Optics	Instruments
7	Technologies for control/measures/analysis	
8	Medical engineering	
9	Nuclear technology	
10	Organic chemistry	Chemicals; materials
11	Macromolecular chemistry	
12	Basic chemistry	
13	Surface technology	
14	Materials; metallurgy	
15	Biotechnologies	Pharmaceuticals; biotechnology
16	Pharmaceuticals; cosmetics	
17	Agricultural and food products	
18	Mechanical engineering (excluding transport)	Industrial processes
19	Handling; printing	
20	Agricultural and food apparatuses	
21	Materials processing	
22	Environmental technologies	
23	Machine tools	Mechanical engineering; machines; transport
24	Engines; pumps; turbines	
25	Thermal processes	
26	Mechanical elements	
27	Transport technology	
28	Space technology; weapons	
29	Consumer goods	Consumer goods; civil engineering
30	Civil engineering	

For the purposes of this work, we have adopted the INPI–OST–FhG–ISI classification, developed by Institut National de la Propriété Industrielle (INPI, Paris), Observatoire des Sciences and des Techniques (OST, Paris) and Fraunhofer Gesellschaft–ISI (Karlsruhe). This classification, reported in Table 5.1, is technology-oriented and groups patents in 30 technological classes and seven macro-classes, based on their IPC[1] codes. Figure 5.1 shows the total number of patents assigned by EPO in each year. Except for the beginning of the 1990s and the year 2001, the number of patents grows considerably from 1978 to 2003. European and US applicants, including both organizations and individual inventors, hold a large majority of EPO patents and, as reported in Figure 5.1, European patents always outnumber the US ones.

The number of applicants follows approximately the same pattern of the number of patents (see Figure 5.2).

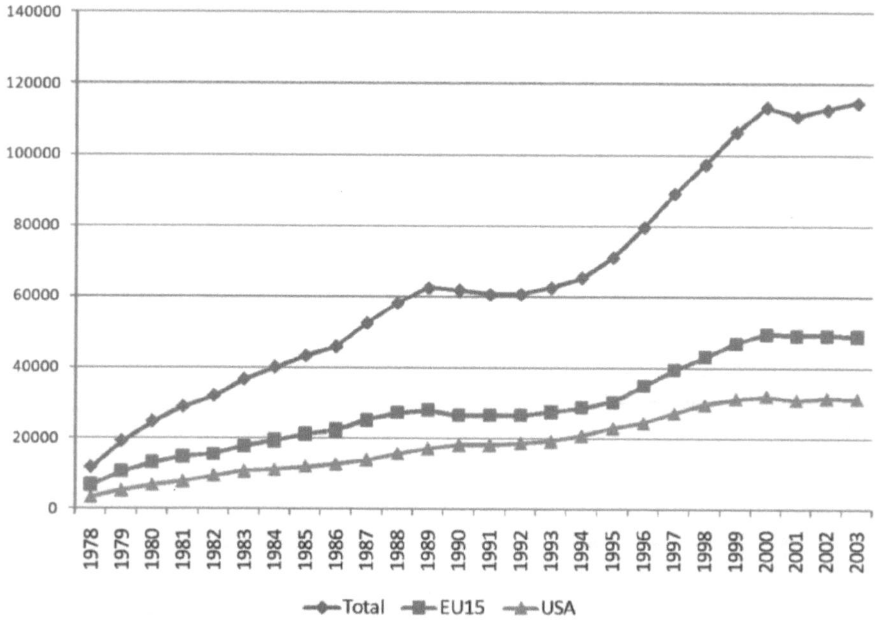

Figure 5.1 Patents assigned by EPO from 1978 to 2003, broken down by year and geographic area.

This number increased during the whole period, except for the first half of the 1990s and the year 2001. As it happened for the patents, the number of applicants from the European Union is systematically higher than those from the United States. Figure 5.3 reports the average number of patents per applicant in every year. On average, it is about four at the beginning of the time series, it reaches a peak during the 1990s, and it stabilizes at the end of 1990s, reaching around 11 patents per applicant in 2003. While US applicants always patent less than European ones, they have a higher average number of patents.

2.3 Methodology

This chapter aims to analyse patterns of technological entry and the characteristics of the technological entrants across different countries and technological fields. Technological entrants are defined as those organizations and other institutions that patent for the first time in a certain technological field. Although individual inventors provide a significant contribution to the patenting activity, those patents will not be taken into account. In this chapter we will use the term 'organization' for both firms and other organizations.

The definition of economic entry classifies as an entrant a firm that offers a product within a market for the first time. This definition includes *de novo*

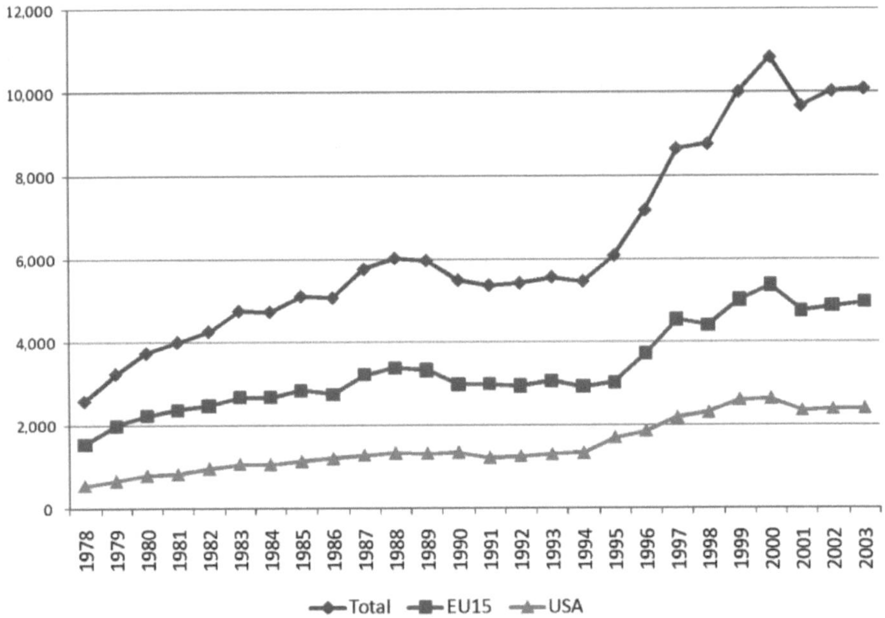

Figure 5.2 Number of applicants from 1978 to 2003, broken down by year and geographic area.

Note
The number of applicants includes organizations, universities, research centres, etc., as well as individuals.

entrants, such as new established organizations, or entrants by diversification, such as incumbents that establish new plants in the same sector, or that start to sell new products in different sectors. In a similar way, a technological entrant – i.e. a company or another organization patenting for the first time in a certain sector – could be a *de novo* entrant or a technological diversifier. It is important to remark again that technological entry does not necessarily imply economic entry or the establishment of a new company. Although case studies and analyses in progress confirm that the date of the first patent of a firm is highly correlated with its founding date, in general a firm can patent shortly after its foundation or after many years.

Another consideration regards the use of patents as a measure of innovation. Critiques about patent analyses point out that not all innovations are patented or that different organizations show different propensities to patent. Moreover, a patent is not necessarily an indicator of innovation, and many patents will never have a commercial value. However, patents represent a uniform and unique measure of inventions across countries and industries (Griliches 1990). Furthermore, a patent has a definite technological connotation, and it represents a firm's technological effort to produce an invention that it believes to be relevant,

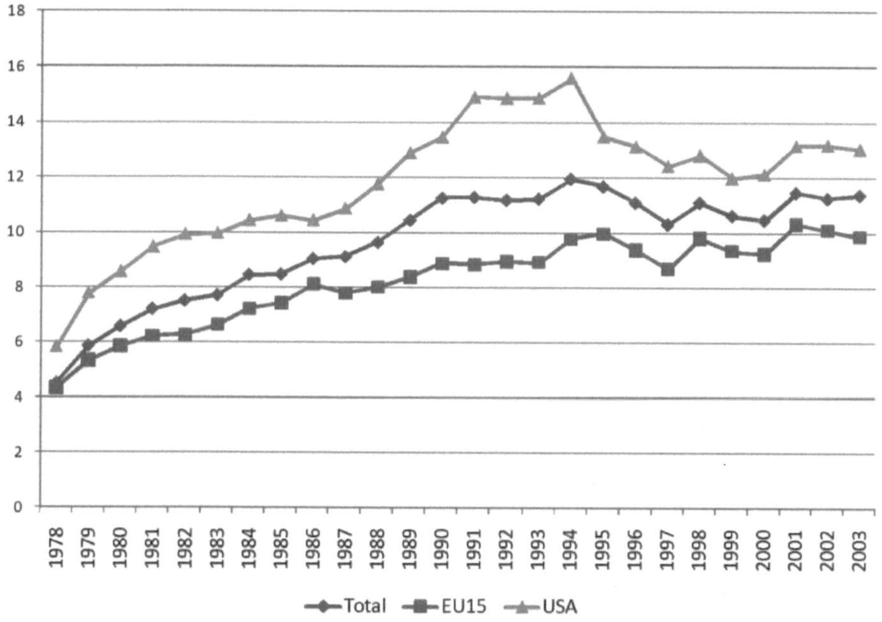

Figure 5.3 Average number of patents per applicant, broken down by year and geographic area.

Note
The number of applicants includes organizations, universities, research centres, etc., as well as individuals.

original and at the technological frontier. For this reason, we prefer to define the organizations that patent for the first time as technological entrants, rather than innovative entrants.

Technological entrants will be divided in *de novo* entrants or entrants *by diversification*. More specifically, the latter are organizations that patent for the first time in a technological field after having already patented at least in another field, while *de novo* entrants are those organizations that patent for the very first time in a certain technological field without having patented before.

The analysis is limited to six countries (France, Italy, Germany, Sweden, the United Kingdom and the United States) for two main reasons. First, because the data about the organizations from those countries have been thoroughly cleaned by KITeS–Cespri so that possible mistakes (such as misspelled names or multiple names for unique organizations) could be avoided. The second reason is that those six countries represent a very high percentage of the overall number of patents applications at EPO. In particular, during the period 1990–2003, the organizations from those six countries hold 65 per cent of the total number of patents.

The sectors taken into account are reported in Table 5.2. The 13 sectors selected have been chosen from the 30 technological fields classified by the

Table 5.2 Sectors taken into account in the analysis

Class	*Description*	*Type*
2	Audiovisual technology	Electrical engineering; electronics
3	Telecommunications	
4	Information technology	
5	Semiconductors	
6	Optics	Instruments
7	Technologies for control/measures/analysis	
8	Medical engineering	
11	Macromolecular chemistry	Chemicals; materials
14	Materials; metallurgy	
15	Biotechnologies	Pharmaceuticals; biotechnology
16	Pharmaceuticals; cosmetics	
23	Machine tools	Mechanical engineering; machines; transport
28	Space technology; weapons	

INPI–OST–FhG–ISI nomenclature. These selection criteria have been twofold. On one side, the selection reflects the purpose of providing a broad picture of the technological entry phenomenon. On the other side, high-tech sectors, which have experienced an important growth in their innovative activity during the last decade, have been given preference over others. It is important to point out that the classification used is technology-based and not market-based. As a result, sectors' names represent specific technologies in which the patent has been classified by the Patent Office, and not the sector in which the firm operates. Some organizations may patent in some technologies and then use those patents in complete different markets.

Furthermore, another methodological issue is related to the period of time of analysis. In fact, the patent date represents a crucial issue for our analysis, since our purpose is to determine when organizations patent for the first time in their history. In our analysis we will consider the *priority date* of the patent, since that date is assumed to be the formal date of invention for patent law purposes. The period of analysis is 1990–2003. We chose to start from 1990 for two reasons. First, because our main interest is in recent innovative activities. The second reason is related to the methodology used. Since organizations will be classified as technological entrants when they patent for the first time, a conspicuous number of patents and patenting organizations before the starting date of the analysis is required. Since the patent dataset begins with the year 1978, the 13-year period before the first date of analysis of this chapter appears enough to ensure reliable information. In addition, although more recent data could be available, we decided to limit the analysis to 2003. The reason is the presence of a time lag between the *application date* and the *publication date* (date in which patent data become public and available for researches and analysis). In fact, the EPO publishes all the data about patent applications only 18 months after their application date. In addition, if the application is submitted first to a particular national patent office a further delay occurs, since the national office can take up to another 12 months to extend the patent protection to the European office. Furthermore, those organizations that apply for an International Patent Application through the WIPO (World Intellectual Property Organization) from other patent offices, such as the USPTO (United States Patent and Trademark Office) will extend these delays to another 18 months. In brief, the time lag between the publishing date by EPO and the application date can be up to 48 months long. Because of these time lags, too recent data are not considered reliable for this study. Taking into account too recent patents will therefore bias the analysis against those European organizations that applied first to their national patent office, and even more against US organizations. For these reasons, the year 2003 has been considered as the most recent one to which the analysis could be extended, in order to base the study on reliable data.

Besides the simple count of the number of total technological entrants, *de novo* and diversifier, other indicators will be proposed. The first is the rate of technological entry, calculated in terms of organizations or in terms of patents. The *entry rate* in terms of organizations is the ratio between the number of entrants in a certain sector divided by the total number of organizations patenting in the same sector. The same ratio can be expressed in terms of patents. The *entrants' share* is the ratio between the patents held by new entrants and those held by all the organizations patenting in the same sector. This measure can be used to test the traditional assumption that entrants are in general smaller than incumbents. Other measures of size are the average number of patents held by entrants and the entrants' relative size (calculated as the entrants' average size in terms of patents divided by the average size of all the organizations patenting in the same sector and period). Finally, other measures used in the chapter take into account the performance of organizations after their entry. In particular, the post-entry behaviour will be analysed in term of persistence (if organizations patent occasionally or persistently) and differentiation in the patenting fields (if organizations keep patenting in the same field after entry or differentiate their activity). Ultimately, also patent citations will be taken into consideration as a measure of the 'success' of the organizations (see, for example, Jaffe and Trajtenberg 2002). Specifically, the average number of citations will be provided in order to analyse how successful the entrants were compared with the incumbents.

3 The technological entry of organizations

3.1 Patterns of technological entry

Technological entry is a very common process. Table 5.3 reports some general figures about technological entry in all the sectors and all the countries included in the analysis. The first figure is the technological entry rate, calculated as the ratio between the number of technological entrants in the period and the total number of firms patenting in the same period. The share is calculated both for the total number of entrants (*de novo* and diversifiers) and for the *de novo* entrants only.

Table 5.3 Total technological entry

	1990–1	*1992–3*	*1994–5*	*1996–7*	*1998–9*	*2000–1*	*2002–3*
Technological entry in terms of number of organizations							
Total tech. entry rate	0.42	0.41	0.41	0.45	0.47	0.47	0.43
De novo technological entry rate	0.25	0.24	0.25	0.28	0.29	0.28	0.25
Technological entry in terms of number of patents							
Total tech. entrants' share	0.16	0.16	0.16	0.17	0.18	0.19	0.17
De novo technological entrants' share	0.09	0.09	0.09	0.10	0.11	0.11	0.10
Average size of technological entrants in terms of patents relative to the average size							
Tech. entrants' relative size	0.38	0.38	0.38	0.39	0.39	0.40	0.40
De novo entrants' relative size	0.38	0.38	0.38	0.38	0.39	0.40	0.40

The total entry rate ranges from 41 to 47 per cent, meaning that around 45 per cent of the patenting organizations in each period have not patented before in the same technological field. Excluding entrants by diversification, and considering only those organizations that in every period patent for the very first time (*de novo* entrants), the ratio drops to 24–9 per cent, meaning that *de novo* entry represents the most important form of technological entry of organizations into sectors.

This simple figure provides a quite remarkable result: more than one-quarter of the organizations that patent in every period have not patented in the past. Furthermore, technological entry is quite stable during the period. In fact, by looking at Figure 5.4 it is possible to see that the total number of organizations, the total and *de novo* entrants follow a very similar trend during the period 1990–2003.

If entry in terms of organizations is a very relevant phenomenon, however, entrants' size is much lower. Table 5.3 reports statistics about this issue. The first is the technological entry in terms of patents, calculated as the number of patents held by all the entrants (or *de novo* entrants) divided by the total patents in the period. It is quite clear that while total entrants and *de novo* entrants are numerous in terms of number, they represent a much smaller percentage of patents. In fact, entrants hold around 16–19 per cent of the patents and *de novo* entrants only 9–11 per cent. This indicates that the average size of entrants is much smaller than the average incumbent. This result is confirmed by another measure: the average size of entrants in terms of patents divided by the average size of organizations in the same period. In this case, entrants and *de novo* entrants appear very similar and much smaller than the average. In fact, their size (in terms of patents) is only around 40 per cent of the average size of patenting organizations. It is interesting to point out again that all those measures appear quite stable during the whole period. This is because the growth in the number of patenting organizations and entrants coincided with a similar growth of patents (see Figure 5.5).

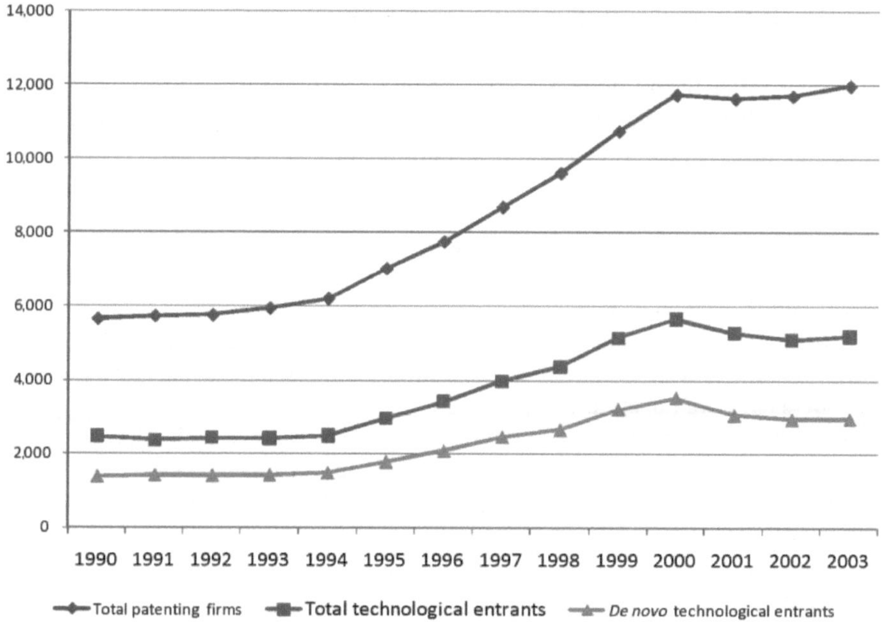

Figure 5.4 Total number of patenting organizations, total and *de novo* technological entrants.

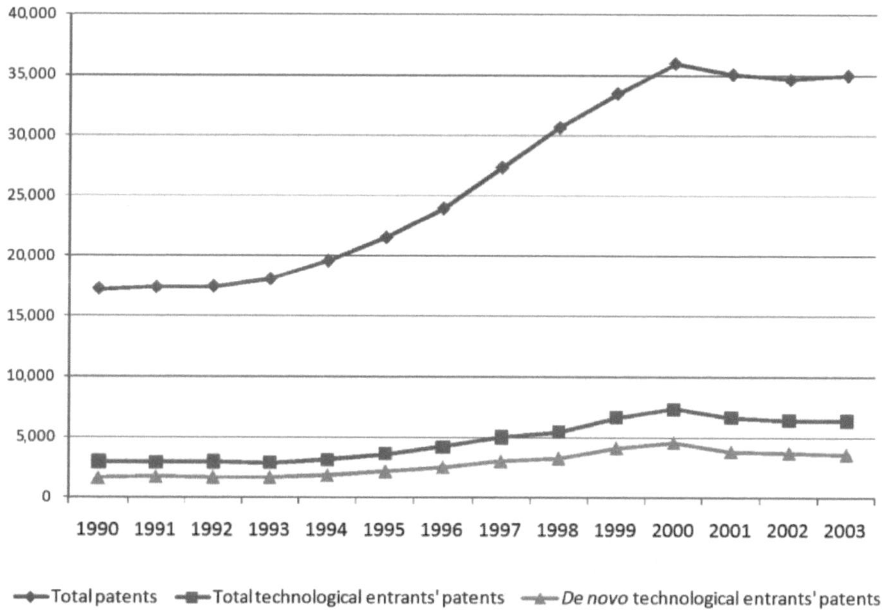

Figure 5.5 Total number of patents, entrants' and *de novo* entrants' patents.

In summary, entrants appear to be a quite homogeneous group and stable over time. If another dimension (country) is taken into account, the picture does not change significantly.

In general, as from Figure 5.6, the United States holds a much higher amount of patents than any other European country considered singularly, and also presents a remarkable increasing trend during time. The European country with the highest amount of patents is Germany, which is also the only country that shows a consistent increase in the number of patents.

Table 5.4 shows the average number of patents held by each firm by country. On average, French, German, Swedish and US organizations patent more than Italian and British ones. However, these geographic differences in terms of total patents and average number of patents do not affect the average size of entrants and *de novo* entrants, since their size is almost the same in every country (around 1.2 patents per organization).

The first considerations that it is possible to draw are that technological entry is a relevant phenomenon, very stable over time. However entrants are much smaller than incumbents in terms of patents. Furthermore, entrants appear to

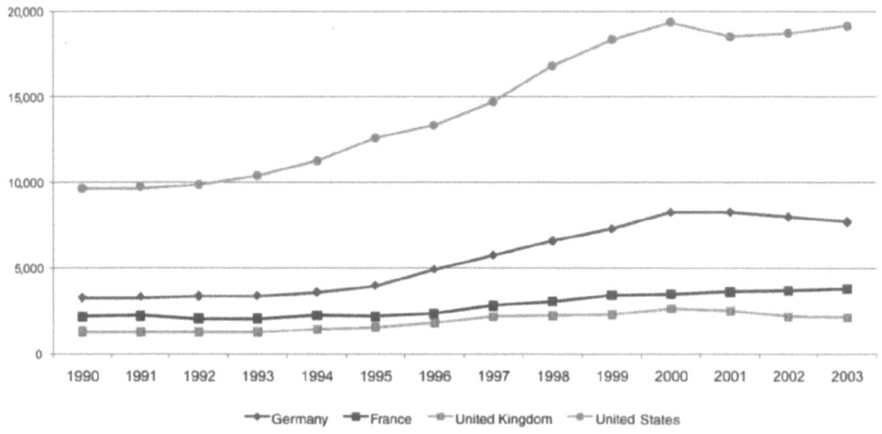

Figure 5.6 Total number of patents per country and period.

Table 5.4 Average number of patents held by each organization per country

Area	*All organizations*	*Total technological entrants*	De novo *technological entrants*
France	3.3	1.2	1.2
Germany	3.2	1.2	1.2
Italy	1.6	1.1	1.1
Sweden	3.4	1.2	1.2
United Kingdom	2.2	1.2	1.2
United States	3.2	1.3	1.3

have the same characteristics regardless of the period of time and the geographic area in which they are established.

3.2 Technological entry in different fields

Sectors differ significantly regarding the number of patenting organizations, patents and technological entrants. Figure 5.7 shows the number of patenting organizations in five sectors from 1990 to 2003. The sector that has the largest number of patenting organizations every year is technologies for measure/control/analysis, followed by medical engineering. Information technology is the sector that shows the largest increase in the number of organizations during the last part of the 1990s, but then they decline after the year 2000. Also in pharmaceuticals, telecommunications and biotechnology the number of patenting organizations has increased (especially from the end of the 1990s).

Also in terms of patents, the differences between technological fields persist. Figure 5.8 shows, however, that to a large number of patenting organizations do not necessarily correspond a large number of patents. The biggest sector in terms of patents is telecommunications, followed by information technology and technologies for measure/control/analysis. Other sectors that show an increase in patenting activity are medical engineering, pharmaceuticals and audiovisual technology. The other industries are quite stable during the whole period. The trend of patents in biotechnologies is quite puzzling, as it constantly grew until the year 2000 and sharply declined after that date. This trend is not confirmed by the data in Figure 5.7 about the number of patenting organizations, which does not decline after 2000.

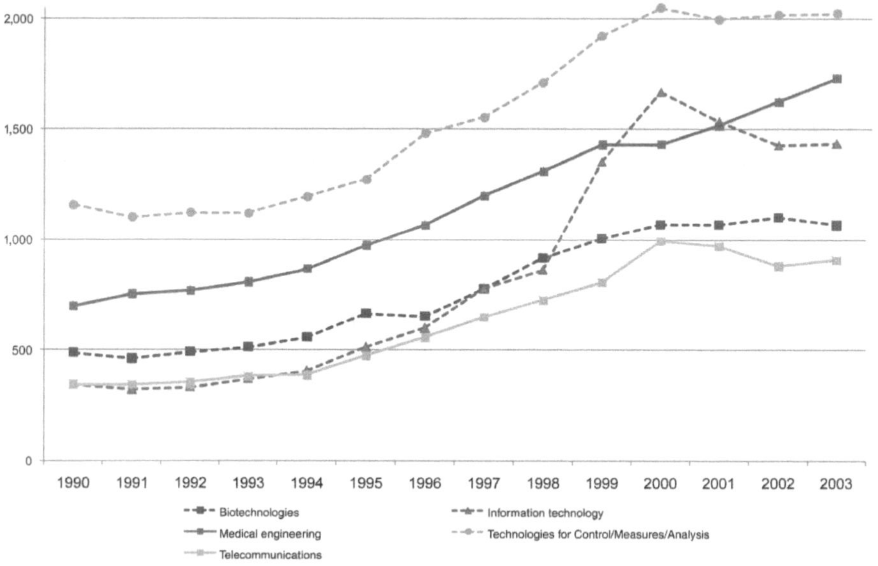

Figure 5.7 Total number of patenting organizations per year and sector (1990–2003).

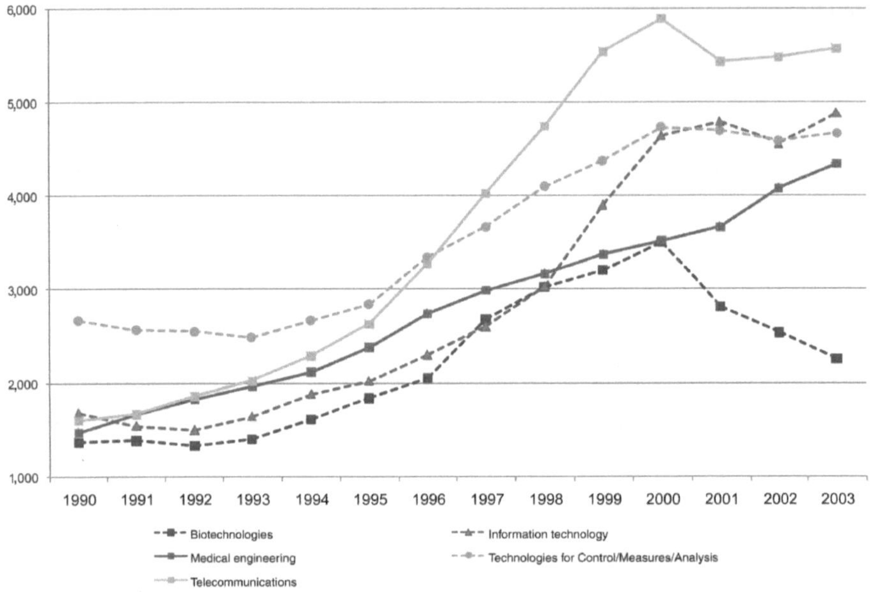

Figure 5.8 Total number of patents per year and sector (1990–2003).

In summary, the sustained increase of patent applications reported in Figure 5.1 becomes much more differentiated by breaking down the data by technological fields, since different sectors show different patterns in patenting activity. In principle, one would expect that such trends are also reflected in the technological entry by sector, for instance indicating a correspondence between high growth of patents and high technological entrance.

However, the sectoral differences in the patenting activity only partially reflect technological entry in those fields. In fact, regardless of the differences among sectors in the growth of patents, entry seems to be very frequent in every field. Table 5.5 reports the entry rate (in terms of number of organizations) for each period and sector.

Some interesting differences among sectors emerge. In general total entry rates are always between 32 and 64 per cent, while the *de novo* entry rates are between 11 and 45 per cent. The sector with the highest entry rate is information technology, in which the total and *de novo* entry rates reach 64 per cent and 45 per cent, respectively, in the period 2000–1. Other sectors with high entry rates are medical engineering, telecommunications and audiovisual technology. On the contrary, macromolecular chemistry, materials/metallurgy, optics and semiconductors have lower (but still remarkable) entry rates.

Table 5.6 reports the entry rates in terms of patents per period and sector. Some sectoral differences are noticeable. The sectors with highest shares of patents held by entrants or *de novo* entrants are machine tools, medical engineering,

Table 5.5 Technological entry rates per sector and period (number of organizations)

Sector	*Entry rate*	*1990–1*	*1992–3*	*1994–5*	*1996–7*	*1998–9*	*2000–1*	*2002–3*
Audiovisual technology	Total entrants	0.47	0.48	0.46	0.53	0.53	0.53	0.48
	De novo	0.28	0.31	0.25	0.33	0.31	0.28	0.27
Biotechnologies	Total entrants	0.33	0.33	0.33	0.36	0.39	0.38	0.33
	De novo	0.18	0.19	0.20	0.23	0.24	0.23	0.19
Information technology	Total entrants	0.50	0.51	0.54	0.58	0.62	0.64	0.55
	De novo	0.29	0.29	0.34	0.41	0.44	0.45	0.36
Machine tools	Total entrants	0.49	0.47	0.45	0.48	0.47	0.45	0.45
	De novo	0.28	0.28	0.27	0.27	0.27	0.24	0.23
Macromolecular chemistry	Total entrants	0.36	0.38	0.34	0.40	0.39	0.40	0.38
	De novo	0.18	0.16	0.14	0.17	0.18	0.15	0.15
Materials; metallurgy	Total entrants	0.38	0.38	0.36	0.41	0.42	0.41	0.41
	De novo	0.20	0.20	0.16	0.20	0.20	0.20	0.19
Medical engineering	Total entrants	0.46	0.43	0.42	0.44	0.43	0.42	0.43
	De novo	0.35	0.33	0.32	0.33	0.32	0.31	0.29
Optics	Total entrants	0.42	0.36	0.39	0.42	0.44	0.43	0.41
	De novo	0.22	0.19	0.19	0.21	0.23	0.21	0.20
Pharmaceuticals; cosmetics	Total entrants	0.35	0.36	0.35	0.38	0.40	0.39	0.37
	De novo	0.18	0.19	0.21	0.22	0.23	0.23	0.21
Semiconductors	Total entrants	0.32	0.35	0.36	0.42	0.43	0.42	0.40
	De novo	0.15	0.11	0.14	0.14	0.18	0.16	0.14
Space technology; weapons	Total entrants	0.34	0.34	0.35	0.40	0.42	0.40	0.39
	De novo	0.21	0.19	0.23	0.27	0.26	0.19	0.25
Tech. for control/measures/analysis	Total entrants	0.45	0.44	0.45	0.46	0.49	0.47	0.46
	De novo	0.25	0.25	0.26	0.27	0.29	0.27	0.25
Telecommunications	Total entrants	0.42	0.43	0.44	0.51	0.52	0.53	0.46
	De novo	0.23	0.24	0.31	0.33	0.35	0.35	0.27

Table 5.6 Technological entrants share per sector and period (patents)

Sector	*Entrants' share*	*1990–1*	*1992–3*	*1994–5*	*1996–7*	*1998–9*	*2000–1*	*2002–3*
Audiovisual technology	Total entrants	0.16	0.18	0.16	0.23	0.19	0.21	0.17
	De novo	0.09	0.12	0.09	0.14	0.11	0.11	0.10
Biotechnologies	Total entrants	0.14	0.15	0.15	0.13	0.16	0.16	0.19
	De novo	0.08	0.09	0.09	0.09	0.10	0.10	0.10
Information technology	Total entrants	0.13	0.14	0.17	0.21	0.26	0.28	0.21
	De novo	0.07	0.08	0.11	0.14	0.18	0.19	0.14
Machine tools	Total entrants	0.35	0.31	0.30	0.32	0.31	0.30	0.27
	De novo	0.20	0.18	0.18	0.17	0.18	0.16	0.14
Macromolecular chemistry	Total entrants	0.07	0.09	0.08	0.10	0.10	0.11	0.13
	De novo	0.04	0.04	0.03	0.04	0.05	0.04	0.05
Materials; metallurgy	Total entrants	0.21	0.22	0.21	0.25	0.23	0.24	0.24
	De novo	0.11	0.11	0.09	0.12	0.11	0.11	0.11
Medical engineering	Total entrants	0.27	0.22	0.21	0.22	0.23	0.22	0.20
	De novo	0.20	0.17	0.16	0.17	0.17	0.16	0.14
Optics	Total entrants	0.10	0.09	0.11	0.14	0.16	0.15	0.18
	De novo	0.05	0.05	0.05	0.07	0.08	0.08	0.09
Pharmaceuticals; cosmetics	Total entrants	0.15	0.15	0.14	0.14	0.16	0.17	0.17
	De novo	0.07	0.08	0.08	0.08	0.09	0.10	0.09
Semiconductors	Total entrants	0.11	0.10	0.10	0.12	0.13	0.15	0.15
	De novo	0.05	0.03	0.04	0.04	0.05	0.06	0.05
Space technology; weapons	Total entrants	0.17	0.18	0.18	0.22	0.25	0.23	0.22
	De novo	0.10	0.11	0.11	0.15	0.14	0.10	0.14
Tech. for control/measures/analysis	Total entrants	0.23	0.22	0.23	0.24	0.25	0.24	0.24
	De novo	0.12	0.12	0.13	0.14	0.15	0.13	0.13
Telecommunications	Total entrants	0.11	0.10	0.11	0.11	0.10	0.13	0.10
	De novo	0.06	0.05	0.07	0.07	0.07	0.09	0.06

space technology, technologies for control/measure/analysis and information technology (at least in the last periods). On the contrary, audiovisual technology, macromolecular chemistry, optics, semiconductors and telecommunications have lower rates. The general decrease both in the number of entrants and patents during the final period (2002–3) is partially due to the above-mentioned lag between the application date and publication date of the patents by the European Patent Office.

The previous analysis leads to the individuation of some interesting patterns. Sectors can be classified according to two measures: patents technological growth (high growth of patents) and technological turbulence (high *de novo* entry rates) leading to the classification reported in Table 5.7. In most of the cases the sectors with the highest growth of patents experienced a higher entry of new patenting firms (e.g. telecommunications, information technology, medical engineering and technologies for control/measures/analysis). On the contrary, other sectors are more stable (i.e. they have lower growth and entry): optics, materials, space technology, macromolecular chemistry. Some sectors follow a different pattern: for example biotechnology, pharmaceutical and semiconductors grow at a high pace, but at the same time do not show remarkable entry rates. Finally, other sectors characterized by a high degree of turbulence experienced a relative low growth in terms of patents (machine tools, audiovisual technology).

The classification of sectors in terms of technological turbulence and growth seems to be consistent with recent history and the evolution of each sector. The only puzzling result regards biotechnology. In fact, this sector is classified as low turbulent (which means a lower *de novo* entrance than the average), while in recent years this sector has been considered one of the most dynamic, attracting new entrants and generating intense innovations. One of the possible reasons for this discrepancy could be related to the fact that biotechnology and pharmaceuticals overlap in our data. This means that the organizations that patent in one sector are probably also patenting in the other. The two sectors (at least regarding the patenting activity) appear to be very similar and dominated by large

Table 5.7 Classification of technological fields by levels of growth and turbulence

	Technological turbulence	
	High	*Low*
Technological growth		
High	Telecommunications Information technology Medical engineering Tech. for contr./meas./analysis	Biotechnology Pharmaceuticals Semiconductors
Low	Machine tools Audiovisual technology	Optics Materials, metallurgy Space technology, weapons Macromolecular chemistry

corporations in which patenting for the first time is relatively more difficult than in other technological sectors.

However, as it happened with the aggregate data, entrants and *de novo* entrants appear to be very similar across sectors in terms of average patents held by *de novo* entrants and total entrants. Table 5.8 shows the average number of patents per organizations in every sector. In some sectors, such as telecommunications and chemistry, organizations have a higher number of patents, while in others, like machine tools, materials and space technology, the number is smaller. Despite these differences, in every sector both entrants and *de novo* entrants have always the same average number of patents: between 1.2 and 1.4. This remark is still valid when differences among countries are considered. Table 5.9 reports the number of patents held by each firm by sector and country. Despite many other differences among countries and sectors, entrants and *de novo* entrants are always very homogeneous.

In summary, when differences among sectors and countries are considered, the analysis of technological entry becomes more varied; however, some common patterns are identifiable. First of all, some homogeneity in the characteristics of entrants regardless of sector, country and period becomes evident. Furthermore, some of the empirical results that traditional studies reported about entry seem to be confirmed by the analysis of technological entry: technological entry is a very common phenomenon, persistent in time, and entrants are generally smaller than incumbents. The next step is analysing the survival of entrants, hence their post-entry behaviour, in order to understand if there are difficulties in being persistent in the patenting activity.

Table 5.8 Average number of patents held by each firm, total entrants and *de novo* entrants

Sector	*All organizations*	*Total technological entrants*	De novo *technological entrants*
Audiovisual technology	3.4	1.3	1.3
Biotechnologies	2.9	1.3	1.3
Information technology	3.4	1.3	1.3
Machine tools	1.8	1.2	1.2
Macromolecular chemistry	4.7	1.2	1.2
Materials; metallurgy	2.1	1.2	1.2
Medical engineering	2.4	1.2	1.3
Optics	3.9	1.3	1.3
Pharmaceuticals; cosmetics	3.0	1.3	1.2
Semiconductors	3.9	1.3	1.3
Space technology; weapons	2.2	1.2	1.2
Technologies for control/ measures/analysis	2.3	1.2	1.2
Telecommunications	5.9	1.4	1.4

4 Post-entry behaviour

4.1 Persistence and specialization

The post-entry behaviour of technological entrants is analysed in two ways: persistence and specialization. The first measure indicates the persistency in the patenting activity of technological entrants. This statistic takes into consideration only *de novo* entrants, since entrants by diversification can be considered by definition in some way persistent, as they already patented before entering into the sector. *De novo* entrants are divided in three groups: *occasional*, if they have only one patent; *sporadic*, if they have more than one patent but less than the average; *persistent*, if they have more than one patent and they patent more than the average. For instance, an occasional technological entrant in the period 1992–3 is a firm that patented for the first time in that period and then did not patent anymore. On the contrary, a sporadic or persistent entrant in the period 1992–3 is a technological entrant patenting for the first time in that period and that managed to patent at least another time in the following period. Table 5.10 resumes persistence in patenting activity of *de novo* entrants by period and geographic area.

It is important to point out that, as technological entry is not related to the date of foundation of the firm, in a similar way the fact that a firm does not keep patenting does not mean that it has exited the sector. A firm could not patent again because of a series of reasons, including: lack of innovativeness, exit from the market, acquisition by another firm, or the decision to protect its inventions in other ways. With the available patent data it is impossible to determine why a firm does not patent anymore. However, the analysis of persistence gives very useful insights that complement those coming from the analysis of technological entry.

From Table 5.10 one can notice that a large amount of organizations do not patent again after their entry. In particular more than 50 per cent of the organizations have only one patent (i.e. they are occasional entrants). This means that if technological entry is very frequent, survival is more difficult. Then, among the non-occasional organizations, only 13–15 per cent are persistent (i.e. they patent more than the average), while 30–7 per cent are sporadic. This seems to indicate that even those organizations that managed to survive find it very hard to continue to be highly innovative. The last period shows in general a lower persistency. This is due to the right truncation of the data, which gives less time to the organizations entered in the period 2002–3 to keep patenting and hence being classified as persistent.

These results are generally confirmed in every country. However, some evident differences emerge. For instance, Italian, Swedish and British organizations are in general more persistent than the others. The differences in persistence are more evident by looking at Table 5.11, that provides the measures of persistence by sector. On average, the sectors in which persistence is most difficult are information technology and technology for control/measure/analysis, while the

Table 5.9 Average number of patents held by each firm, total entrants and *de novo* entrants by sector and country

Sector		*France*	*Germany*	*Italy*	*Sweden*	*United Kingdom*	*United States*
Audiovisual technology	All organizations	6.5	2.8	1.4	2.8	1.8	3.5
	Total tech. entrants	1.3	1.2	1.1	1.1	1.3	1.3
	De novo tech. entrants	1.2	1.2	1.1	1.1	1.3	1.3
Biotechnologies	All organizations	3.1	3.0	1.3	1.5	2.7	3.0
	Total tech. entrants	1.2	1.3	1.1	1.2	1.2	1.3
	De novo tech. entrants	1.2	1.3	1.1	1.2	1.2	1.3
Information technology	All organizations	3.1	4.3	1.6	3.4	1.8	3.6
	Total tech. entrants	1.3	1.2	1.1	1.3	1.3	1.3
	De novo tech. entrants	1.2	1.2	1.1	1.2	1.3	1.3
Machine tools	All organizations	1.4	1.8	1.4	2.2	1.6	2.0
	Total tech. entrants	1.1	1.2	1.1	1.2	1.1	1.2
	De novo tech. entrants	1.1	1.2	1.1	1.3	1.1	1.1
Macromolecular chemistry	All organizations	4.2	6.2	2.7	1.2	2.8	4.8
	Total tech. entrants	1.2	1.3	1.2	1.1	1.2	1.2
	De novo tech. entrants	1.1	1.2	1.2	1.0	1.3	1.2
Materials; metallurgy	All organizations	2.2	2.1	1.5	2.1	1.6	2.1
	Total tech. entrants	1.2	1.2	1.1	1.2	1.2	1.2
	De novo tech. entrants	1.2	1.1	1.1	1.2	1.2	1.1
Medical engineering	All organizations	1.5	2.2	1.5	2.7	1.7	2.8
	Total tech. entrants	1.2	1.2	1.1	1.2	1.2	1.3
	De novo tech. entrants	1.2	1.2	1.1	1.2	1.2	1.3

Optics	All organizations	3.9	2.9	1.9	2.3	2.0	4.9
	Total tech. entrants	1.2	1.2	1.2	1.1	1.3	1.3
	De novo tech. entrants	1.2	1.2	1.2	1.1	1.3	1.3
Pharmaceuticals; cosmetics	All organizations	4.6	3.9	1.7	1.4	3.8	2.7
	Total tech. entrants	1.3	1.3	1.2	1.2	1.2	1.3
	De novo tech. entrants	1.2	1.4	1.1	1.2	1.2	1.2
Semiconductors	All organizations	2.9	5.4	2.1	2.8	1.4	4.0
	Total tech. entrants	1.2	1.3	1.2	1.1	1.2	1.3
	De novo tech. entrants	1.2	1.3	1.1	1.1	1.3	1.3
Space technology; weapons	All organizations	2.6	2.6	1.3	2.1	1.4	2.1
	Total tech. entrants	1.3	1.3	1.1	1.3	1.1	1.2
	De novo tech. entrants	1.2	1.3	1.1	1.1	1.1	1.2
Technologies for control/ measures/analysis	All organizations	2.2	2.8	1.3	1.6	1.7	2.4
	Total tech. entrants	1.2	1.2	1.1	1.2	1.2	1.2
	De novo tech. entrants	1.1	1.2	1.1	1.2	1.1	1.2
Telecommunications	All organizations	9.1	7.8	2.0	16.5	3.0	4.9
	Total tech. entrants	1.4	1.2	1.2	1.3	1.3	1.4
	De novo tech. entrants	1.3	1.3	1.2	1.4	1.3	1.4

Table 5.10 Persistence in patent activity after *de novo* entry by period and country (%)

Area		*1990–1*	*1992–3*	*1994–5*	*1996–7*	*1998–9*	*2000–1*	*2002–3*
France	Occasional	56.6	58.1	61.0	59.7	56.4	60.0	73.3
	Sporadic	28.6	25.3	23.5	28.4	27.4	27.8	17.0
	Persistent	14.9	16.6	15.5	11.9	16.2	12.2	9.7
Germany	Occasional	45.1	49.2	49.9	45.1	46.7	55.1	65.2
	Sporadic	38.9	35.8	37.8	41.3	37.5	33.4	24.9
	Persistent	16.0	15.0	12.3	13.7	15.8	11.5	9.9
Italy	Occasional	50.9	50.4	57.9	55.7	55.9	64.0	73.4
	Sporadic	31.6	28.9	23.7	27.3	27.5	20.6	15.9
	Persistent	17.5	20.6	18.4	17.0	16.6	15.4	10.6
Sweden	Occasional	49.0	59.4	50.5	51.8	52.0	66.7	75.0
	Sporadic	24.5	21.7	29.7	29.0	30.0	19.0	10.8
	Persistent	26.5	18.9	19.8	19.2	18.0	14.3	14.2
United Kingdom	Occasional	55.4	49.8	54.2	51.2	55.1	62.4	79.1
	Sporadic	28.0	32.4	31.4	33.7	30.2	24.9	13.6
	Persistent	16.6	17.8	14.4	15.0	14.7	12.7	7.3
United States	Occasional	44.7	45.2	44.9	44.5	52.0	52.7	71.3
	Sporadic	42.4	42.5	43.0	40.0	35.8	34.2	20.5
	Persistent	13.0	12.3	12.1	15.5	12.2	13.2	8.2
Average	Occasional	48.2	48.9	49.3	47.8	51.9	56.3	71.5
	Sporadic	36.7	36.3	37.2	37.2	34.2	30.8	19.5
	Persistent	15.1	14.8	13.5	15.0	14.0	12.9	9.0

Note
Only *de novo* entrants are considered.

Table 5.11 Persistence in patent activity after *de novo* entry by sector (%)

Area		*1990–1*	*1992–3*	*1994–5*	*1996–7*	*1998–9*	*2000–1*	*2002–3*
Audiovisual technology	Occasional	59.5	53.8	50.6	52.3	53.0	52.4	69.2
	Persistent	16.2	13.8	13.5	11.7	15.3	12.2	11.2
Biotechnologies	Occasional	33.5	35.1	36.1	33.5	35.7	46.7	69.1
	Persistent	21.4	16.8	16.2	18.0	18.9	18.2	11.7
Information technology	Occasional	56.5	58.4	57.6	57.8	61.0	64.9	74.4
	Persistent	8.8	12.9	10.9	10.9	9.7	9.2	6.4
Machine tools	Occasional	51.9	56.1	53.6	49.0	56.2	59.2	77.6
	Persistent	16.0	14.5	12.6	12.1	14.4	11.1	7.7
Macromolecular chemistry	Occasional	44.8	39.1	50.5	58.8	40.2	55.4	64.9
	Persistent	20.0	18.3	16.2	14.2	19.5	16.2	11.9
Materials; metallurgy	Occasional	47.3	54.9	59.2	46.6	57.7	66.2	67.3
	Persistent	15.9	14.3	12.2	15.0	15.9	12.0	13.3
Medical engineering	Occasional	47.0	44.4	49.4	41.9	48.5	53.1	73.4
	Persistent	11.7	14.0	13.8	19.8	14.9	13.3	8.1
Optics	Occasional	41.0	49.5	48.1	46.5	43.7	51.4	66.3
	Persistent	22.1	11.7	16.7	17.4	18.8	15.8	9.6
Pharmaceuticals; cosmetics	Occasional	34.3	45.2	39.9	41.9	46.8	50.5	71.2
	Persistent	19.3	20.2	18.5	18.8	17.7	17.0	7.9
Semiconductors	Occasional	25.0	25.7	44.2	44.4	37.0	45.5	59.0
	Persistent	31.3	34.3	23.1	28.6	22.0	22.3	16.2
Space technology; weapons	Occasional	61.4	45.9	45.8	59.7	50.0	53.3	67.2
	Persistent	22.7	24.3	22.9	19.4	25.8	22.2	23.9
Tech. for control/measures/analysis	Occasional	52.8	52.4	51.7	48.8	55.3	59.9	72.9
	Persistent	10.3	12.4	9.2	12.4	10.2	11.1	8.5
Telecommunications	Occasional	49.4	46.6	45.5	51.2	55.2	48.7	67.3
	Persistent	19.0	14.8	14.8	12.9	12.2	13.6	8.8

Note
Only *de novo* entrants are considered.

most persistent are semiconductors, space technology, pharmaceuticals and biotechnology. It is interesting to point out that there is a quite clear relationship between turbulence (hence high rates of entry, see Table 5.7) and persistency. In fact, all the sectors that present a high level of technological turbulence (telecommunications, information technology, medical engineering, technologies for control/measure/analysis, machine tools and audiovisual technology) also show a low degree of persistency. On the contrary (with the only exception of materials/metallurgy), those sectors that are more stable in terms of technological entry are also characterized by a higher level of persistency (biotechnology, pharmaceuticals, semiconductors, optics, space technology, macromolecular chemistry). In particular, those sectors that attract more technological entrants are those in which the entrants are less successful in terms of survival and persistency.

The second measure of post-entry behaviour regards the specialization of organizations. The technological entry of a firm in a certain sector does not guarantee that the firm will keep patenting in the same sector. This could be because the firm may diversify its inventive efforts, or may be simply related to the fact that some sectors are technologically more related than others and hence the same company can patent in both of them. The measures of specialization will be based only on *de novo* entrants. This is because an entrant by diversification is by definition not specialized in a single sector. Furthermore, the organizations with only one patent (occasional) will be excluded, because in this case the entry sector is obviously also the most important. The remaining cases are divided into *deepening* organizations, if their technological field of entry continues to be the most important (in term of patents), or *widening* organizations, if their most important technological field is not the one of entry. Table 5.12 reports the differences between deepening and widening organizations by period and geographic area. The most important result is that in general *de novo* entrants keep patenting in the same sector of entrance (around 80 per cent). Only a relative small percentage of organizations are not specialized. Furthermore, this percentage drops from 16.9 per cent at the beginning of the 1990s to 12.3 per cent in 2000–1. The last period suffers from the above-mentioned right-truncation issue. European and US organizations are quite similar regarding their degree of specialization. However, German organizations seem to have in general a higher level of widening organizations than the average.

Table 5.13 provides a breakdown of the specialization measure by sector and period. The sectors with more specialized organizations are medical engineering, machine tools and information technology. On the contrary, the sectors where organizations follow a widening strategy are semiconductors, macromolecular chemistry and materials. It is interesting to point out that all the above-mentioned sectors with a high percentage of widening organizations are also characterized by low turbulence. On the contrary, the more specialized sectors are also those with high turbulence. It seems that in some cases, operating in a less turbulent technological field enhances the possibility to survive and also to follow a widening strategy.

Table 5.12 Deepening and widening organizations by period and country (%)

Area		*1990–1*	*1992–3*	*1994–5*	*1996–7*	*1998–9*	*2000–1*	*2002–3*
France	Deepening	83.6	84.3	83.5	85.0	84.4	84.4	89.4
	Widening	16.4	15.7	16.5	15.0	15.6	15.6	10.6
Germany	Deepening	80.4	78.3	79.9	79.7	84.0	86.9	90.9
	Widening	19.6	21.7	20.1	20.3	16.0	13.1	9.1
Italy	Deepening	82.6	81.4	85.0	83.6	87.2	92.4	87.3
	Widening	17.4	18.6	15.0	16.4	12.8	7.6	12.7
Sweden	Deepening	84.6	83.7	85.6	84.9	85.8	90.1	88.3
	Widening	15.4	16.3	14.4	15.1	14.2	9.9	11.7
United Kingdom	Deepening	84.3	86.5	84.2	84.4	85.2	89.4	91.9
	Widening	15.7	13.5	15.8	15.6	14.8	10.6	8.1
United States	Deepening	83.7	82.1	83.1	85.5	88.2	87.5	90.3
	Widening	16.3	17.9	16.9	14.5	11.8	12.5	9.7
Average	Deepening	83.1	82.0	82.9	84.1	86.6	87.7	90.2
	Widening	16.9	18.0	17.1	15.9	13.4	12.3	9.8

Note
It considers only *de novo* entrants and it does not consider those organizations that have only one patent.

Table 5.13 Deepening and widening organizations by period and sector (%)

Sector		*1990–1*	*1992–3*	*1994–5*	*1996–7*	*1998–9*	*2000–1*	*2002–3*
Audiovisual technology	Deepening	84.3	78.9	85.7	81.0	82.3	82.9	86.8
	Widening	15.7	21.1	14.3	19.0	17.7	17.1	13.2
Biotechnologies	Deepening	81.7	71.0	76.0	82.4	83.3	86.9	84.7
	Widening	18.3	29.0	24.0	17.6	16.7	13.1	15.3
Information technology	Deepening	81.0	82.1	82.6	89.0	91.6	89.7	90.2
	Widening	19.0	17.9	17.4	11.0	8.4	10.3	9.8
Machine tools	Deepening	85.9	85.0	87.1	85.4	88.0	88.0	94.9
	Widening	14.1	15.0	12.9	14.6	12.0	12.0	5.1
Macromolecular chemistry	Deepening	68.1	70.0	67.3	72.1	82.7	87.9	92.5
	Widening	31.9	30.0	32.7	27.9	17.3	12.1	7.5
Materials; metallurgy	Deepening	85.4	78.5	85.0	78.2	78.5	75.3	89.9
	Widening	14.6	21.5	15.0	21.8	21.5	24.7	10.1
Medical engineering	Deepening	91.1	94.2	90.3	92.7	93.8	93.5	96.5
	Widening	8.9	5.8	9.7	7.3	6.2	6.5	3.5
Optics	Deepening	84.7	86.5	75.0	78.3	90.0	88.9	87.1
	Widening	15.3	13.5	25.0	21.7	10.0	11.1	12.9
Pharmaceuticals; cosmetics	Deepening	83.5	80.7	81.6	85.1	89.3	87.5	95.2
	Widening	16.5	19.3	18.4	14.9	10.7	12.5	4.8
Semiconductors	Deepening	69.4	65.4	65.5	74.3	79.4	77.3	81.4
	Widening	30.6	34.6	34.5	25.7	20.6	22.7	18.6
Space technology; weapons	Deepening	76.5	70.0	88.5	88.9	87.9	90.5	90.9
	Widening	23.5	30.0	11.5	11.1	12.1	9.5	9.1
Tech. for control/measures/analysis	Deepening	78.6	79.6	79.6	77.5	83.5	86.5	86.6
	Widening	21.4	20.4	20.4	22.5	16.5	13.5	13.4
Telecommunications	Deepening	83.8	83.0	88.2	85.7	79.0	85.6	88.1
	Widening	16.3	17.0	11.8	14.3	21.0	14.4	11.9

Note
It considers only *de novo* entrants and it does not consider those organizations that have only one patent.

In conclusion, the analysis of post-entry behaviour provides some interesting insights. First of all, while technological entry is very common, innovative persistence is always difficult. A very high percentage of the new patenting organizations do not patent anymore after their first patent, and even those that manage to keep patenting face significant difficulties. Second, the classification in turbulent and stable sectors matches with the classification in persistent and occasional/sporadic sectors. As a result, technological entry and exit from the pool of innovators seem to be correlated. Finally, the less turbulent sectors are also those in which the organizations are significantly diversifying their patenting activity, while the more turbulent ones are also those in which organizations are more specialized.

4.2 Patent citations made and received by technological entrants

The final measures considered in this chapter regard the citations made and received by organizations. The number of citations received by a patent represents a measure of its success, since it measures how many later patents were at least partially based on it. However, the number of citations received by a patent does not represent necessarily a proxy for its economic value, but only for the technological relevance of the invention that it represents. Figures 5.9 and 5.10

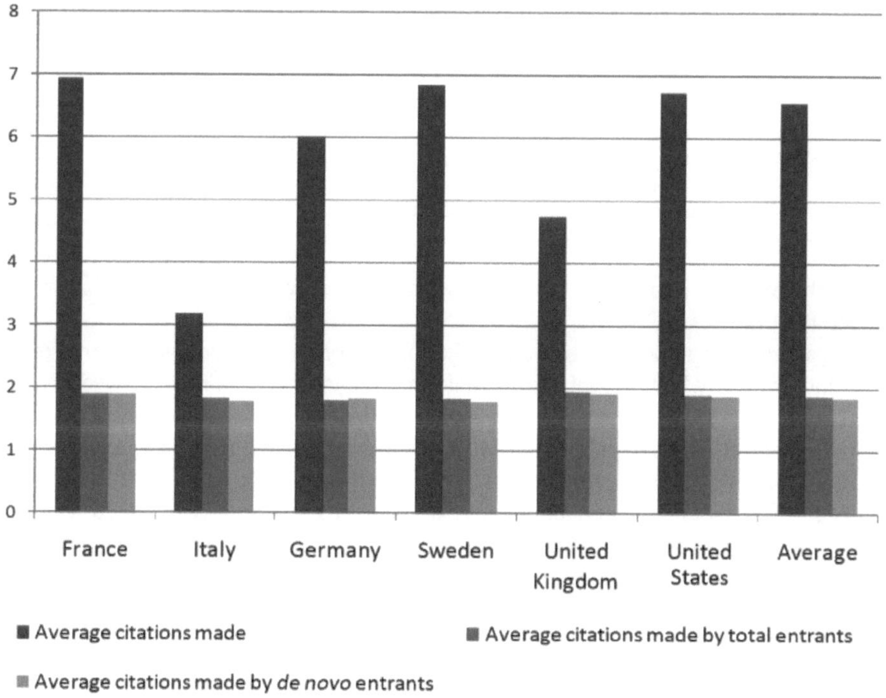

Figure 5.9 Average citations made by country (1990–2003).

Note
Average number of citations made by each citing firm.

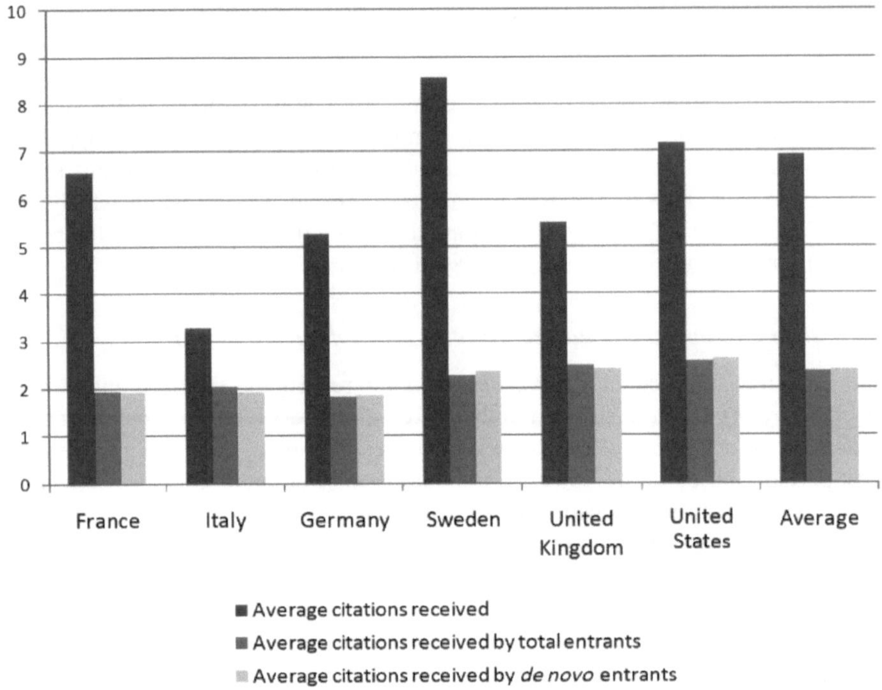

Figure 5.10 Average citations received by country (1990–2003).

Note
Average number of citations received by each cited firm.

show the average number of citations made and received by organizations from different countries.

On average organizations make 6.5 citations and receive almost seven citations. The countries in which organizations make more citations are France, Sweden and the United States. The ones in which organizations receive less citations are Italy, Germany and the UK. Total entrants and *de novo* entrants make and receive fewer citations than the average (with no differences among those two groups).

The differences between countries for citations made and received vanish when we analyse technological entrants and *de novo* entrants. Total technological entrants and *de novo* technological entrants appear much more homogeneous in terms of citations made and received. In particular, every new entrant receives on average 2–2.5 citations. The figures on citations received by entrants are necessarily lower than the average, because most of the technological entrants have only one patent. For this reason, it would be interesting to check whether the average number of citations received is higher for those organizations that managed to survive after entry. Figure 5.11 shows the citation received by *de novo* entrants from different geographic areas depending on their level of

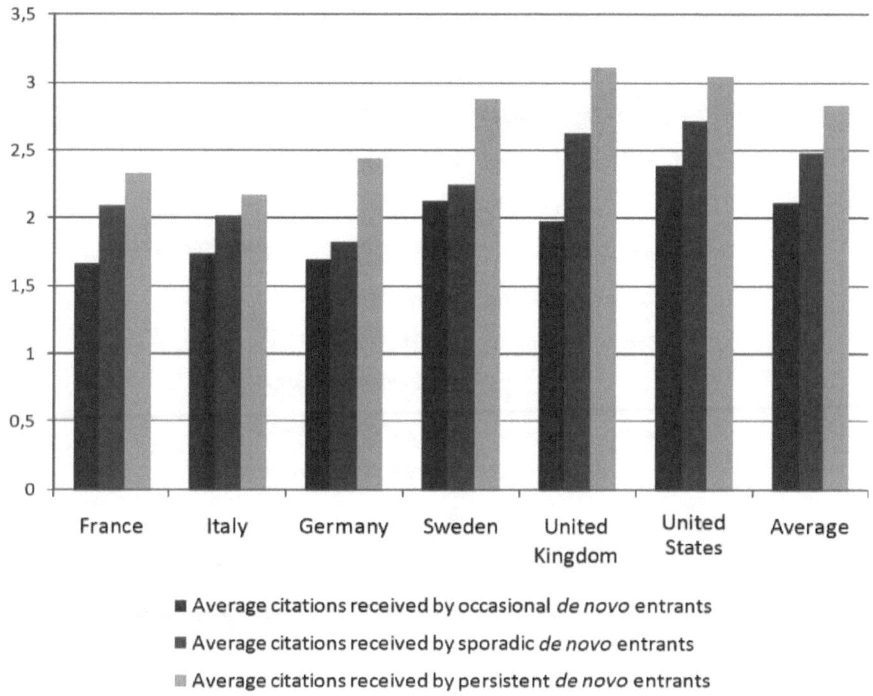

Figure 5.11 Average citations received by *de novo* entrants' level of persistency (1990–2003).

Note
Average number of citations received by each cited firm.

persistency. Persistent organizations are always more cited than occasional and sporadic ones. In other words, persistency and citations received are positively linked; those two measures can be considered as measures of the success of *de novo* entrants. In particular, persistent entrants are not only successful in terms of number of patents (as they patent more than the average), but also in qualitative terms, since their patents are more cited than the other type of entrants. However, even if organizations are persistent, they receive much less citations than the average. In fact, while the average number of citations received by all organizations is seven (Figure 5.10), persistent entrants receive only 2.8 citations. The data confirm that entrants are in general smaller (in terms of patents and citations) than incumbents. The breakdown by level of persistency shows some interesting differences by country. In fact, while the UK receives on average fewer citations than the other countries, its sporadic and persistent entrants are much more cited than the other countries. Similarly, the *de novo* entrants from Sweden and the United States are much more cited.

Figure 5.12 reports the citations made and received by all the organizations patenting in different sectors. Sectors are very different in terms of number of

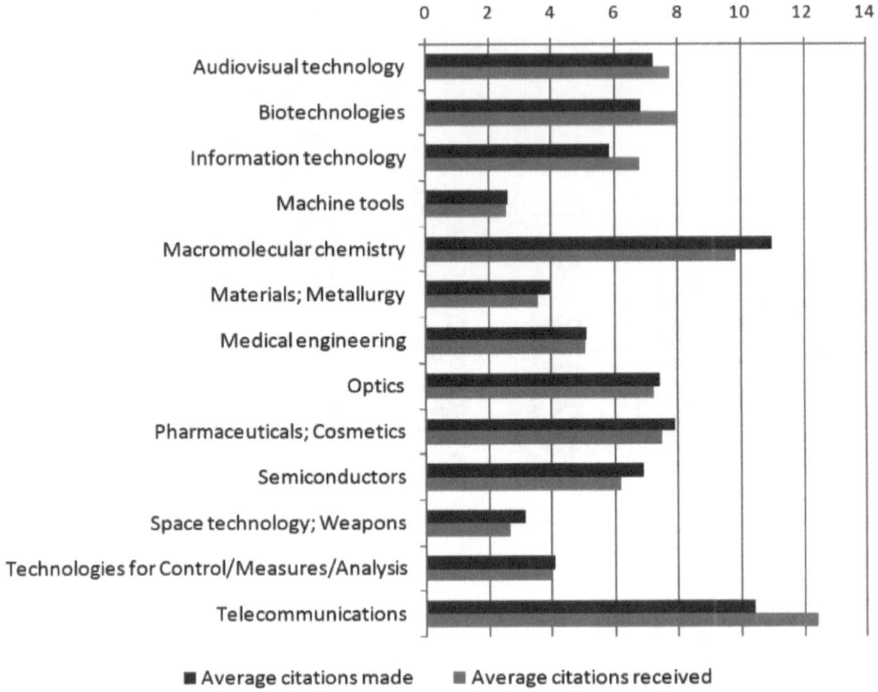

Figure 5.12 Average citations made and received by sector (1990–2003).

citations made and received. In general the sectors that make a lot of citations also receive a lot of them. The sectors with higher levels of citations made and received are telecommunications, macromolecular chemistry, pharmaceuticals, biotechnology and audiovisual technology. It is interesting to point out that these levels are not generally correlated with the growth in the number of patents.

If we take into consideration only entrants and *de novo* entrants, the differences between sectors become smaller, but are still considerable. Figure 5.13 reports the citations received by all entrants and *de novo* technological entrants in each sector. In some sectors, such as machine tools and space technology, entrants receive on average fewer citations (around 1.5). In others, such as telecommunications and biotechnologies, they receive almost the double amount of citations (around three).

In sum, if persistency could be considered as a measure of success for a technological entrant in terms of *quantity* of patents, the number of citations received can be assumed to be a measure of success in terms of *quality*. These measures of success seem to be positively related. However, even if persistent entrants are more cited, incumbents still receive always more citations than entrants. However, even if at an aggregate level entrance and *de novo* entrance appear very similar, important differences among sectors persist.

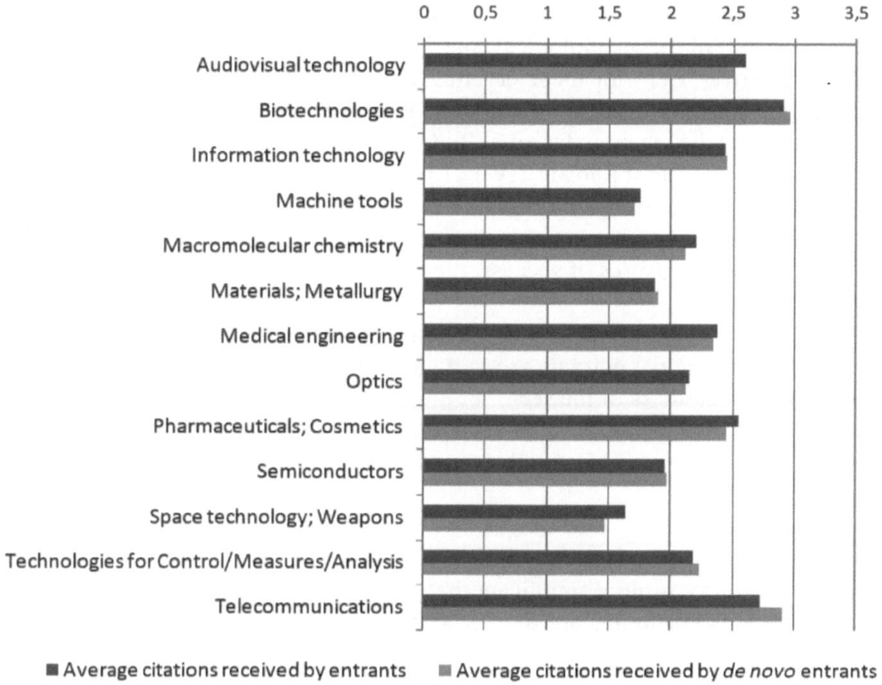

Figure 5.13 Average citations received by entrants and *de novo* entrants in every sector (1990–2003).

5 Conclusions

This chapter provides a first broad picture of the phenomenon of technological entry which is closely related to knowledge-based entrepreneurship. The analysis is descriptive and does not attempt to reach any causal explanation. In this respect much further detailed and finer grained work is needed in terms of case studies, sectoral analyses and links to the literature on learning (technological) regimes (Breschi *et al.* 2000) that affect innovation in sectors and on national innovation systems that differentiate technological entry across countries.

The analysis has been carried out through the analysis of patent data in 13 technological fields by organizations from six countries over the period 1990–2003. From the analysis of technological entry across countries and sectors and the post-entry behaviour, four conclusions emerge.

First of all, technological entry is a very frequent and common phenomenon over all the period examined. Around 25 per cent of the patenting organizations in each period are organizations that have not patented before. However, important differences emerge among countries, and even more among sectors. In particular, sectors can be classified in terms of growth of patents and of technological turbulence.

Second, the post-entry behaviour analysis indicates that if technological entry is a very common process, innovative survival is much more different. In fact, around one half of the technological entrants do not patent anymore after the first patent. Moreover, even for those organizations that managed to survive, growth in patenting appears even more difficult. However the classification based on persistency matches the one based on turbulence. Those sectors with higher turbulence have in general a lower level of persistency. The citations received can be used as a measure of the success of a patent: also in this case entrants receive on average much fewer citations than incumbents. Persistence and citations appear to be two measures of success, the first in terms of quantity, the second in terms of quality.

Third, while several differences among countries and sectors have been reported, technological entrants are very similar in terms of share of patents or average size, regardless of the sector or the country.

In sum, the patent analysis of technological entry confirms several results coming from studies of industrial dynamics concerning entry of organizations into sectors:

1 Technological entry is common.
2 *De novo* entry is more significant than entry by diversification.
3 Technological entrants are smaller than incumbents.
4 Persistence in patenting by technological entrants is difficult.
5 High rates of technological entry (technological turbulence) are related with low levels of persistence in patenting and vice versa.

Note

1 The IPC (International Patent Classification) is an international patent classification system developed and administered by the World Intellectual Property Organization (WIPO). The current classification (8th edition) entered into force on 1 January 2006 (WIPO 2005).

References

Acs, Z. and Audretsch, D. (1990) *Innovation and Small Organizations*, Boston: MIT Press.

Agarwal, R. and Gort, M. (1996) 'The evolution of markets and entry, exit and survival of organizations', *The Review of Economics and Statistics*, vol. 78 (3): 489–98.

Bain, J.S. (1956) *Barriers to New Competition*, Cambridge, MA: Harvard University Press.

Breschi, S., Malerba, F. and Orsenigo, L. (2000) 'Technological regimes and sectoral patterns of innovation', *Economic Journal*, vol. 110 (463): 388–410.

Cefis, E. (2003) 'Is there persistence in innovative activities?', *International Journal of Industrial Organization*, vol. 21 (4): 489–515.

Cefis, E. and Orsenigo, L. (2001) 'The persistence of innovative activities. A cross-countries and cross-sectors comparative analysis', *Research Policy*, vol. 30 (7): 1139–58.

Geroski, P.A. (1989a) 'Entry and the rate of innovation', *Economics of Innovation and New Technology*, vol. 1 (3): 203–14.

Geroski, P.A. (1989b) 'Entry, innovation and productivity growth', *Review of Economics and Statistics*, vol. 71 (4): 572–8.

Geroski, P.A. (1995) 'What do we know about entry?', *International Journal of Industrial Organization*, vol. 13 (4): 421–40.

Geroski, P.A. and Jacquemin, A. (1988) 'The persistence of profits: a European comparison', *Economic Journal*, vol. 98 (391): 375–89.

Geroski, P.A. and Schwalbach, J. (eds) (1991) *Entry and Market Contestability: an International Comparison*, Oxford: Basil Blackwell.

Gort, M. and Klepper, S. (1982) 'Time paths in the diffusion of product innovations', *Economic Journal*, vol. 92 (367): 630–53.

Griliches, Z. (1990) 'Patent statistics as economic indicators: a survey', *Journal of Economic Literature*, vol. 28 (4): 1661–707.

Jaffe, A. and Trajtenberg, M. (2002) *Patents, Citations, and Innovations. A Window on the Knowledge Economy*, Cambridge, MA and London: MIT Press.

Klepper, S. (1996) 'Entry, exit and growth, and innovation over the product life cycle', *American Economic Review*, vol. 86 (3): 562–83.

Malerba, F. and Orsenigo, L. (1995) 'Schumpeterian patterns of innovation', *Cambridge Journal of Economics*, vol. 19 (1): 47–65.

Malerba, F. and Orsenigo, L. (1996) 'Schumpeterian patterns of innovation are technology-specific', *Research Policy*, vol. 25: 451–78.

Malerba, F. and Orsenigo, L. (1999) 'Technological entry, exit and survival: an empirical analysis if patent data', *Research Policy*, vol. 28 (6): 643–60.

Marsili, O. (2001) *The Anatomy and Evolution of Industries: Technological Change and Industrial Dynamics*, Cheltenham: Edward Elgar.

Nelson, R.R. and Winter, S.G. (1982) *An Evolutionary Theory of Economic Change*, Cambridge, MA: Harvard University Press.

Utterback, J. (1994) *Mastering the Dynamics of Innovation*, Boston: Harvard University Press.

WIPO (2005) *International Patent Classification, 8th edition (2006)*, core level, volume 5 Guide.

6 Survival of innovative entrants in knowledge-based sectors

Stefano Breschi, Franco Malerba and Maria Luisa Mancusi

1 Introduction

The aim of this chapter is to explore how the survival likelihood of *innovative entrants* in highly dynamic and knowledge-intensive industries is affected by the characteristics of such firms at the time of their founding. By innovative entrants here we mean companies that are new to the industry (i.e. they are entrants) *and* that enter into the market by introducing an innovation. Our methodology to identify such companies has been based on information contained in patent data for two technological fields, i.e. lasers and data-switching networks. In particular, for each technological field examined, we have identified all those firms whose first patented invention occured within a relatively short time after their entry into the industry. For this sample of innovative start-ups, we have collected further information regarding the date of exit (if any), the background of founders, the background of the inventing team and the characteristics of the patented invention. Our ultimate aim is to examine the impact of both founders' and inventing team background on start-ups' performance as captured by their survival.

It is worth remarking that our methodology differs from the prevailing literature in this field in two main respects. First, while most recent studies investigating the performance of new entrants are based on samples identified using either *product* definitions or *industry* classifications, the sample of firms examined here has been extracted starting from a *technological* classification. Second, our sample includes only innovative start-ups that enter by patenting. Therefore, it excludes both innovative start-ups that enter by innovating, but that adopt different appropriability strategies, and truly non-innovative start-ups, namely firms that are new to the industry, but whose products and technologies do not contain novel elements.

The main result of our analysis is that the importance of the firm's innovation at entry, as measured by the number of citations it receives within three years of the priority date, positively affects the firm's survival. We claim that this variable could proxy for the founder's previous market-related experience and hence the firm's ability to exploit knowledge about market opportunities and customer needs. Therefore, our interpretation of these results is that the firms in our sample, being all innovative firms, are all well endowed with technological knowledge and that only those better positioned to learn about market opportunities enjoy a survival advantage.

The chapter is organized as follows. Section 2 provides a short of review of the background literature. Section 3 describes the basic methodology used in the chapter, while Section 4 illustrates the main empirical findings from a model estimating the likelihood of survival of innovative start-ups. Section 5 concludes.

2 Background literature

The focus of this chapter is on the relationship between the characteristics of a start-up at the time of its founding and its subsequent performance, in particular its survival. This is an issue that has recently attracted the attention of many scholars. Given the importance of entrepreneurial activity in fostering the exploration of emerging technologies and in tapping into non-commercialized technologies and unexploited market opportunities, start-up companies are increasingly seen as a key means to sustain economic growth and employment creation. However, the empirical observation that a high proportion of new business organizations fail within a short time after establishment and that the failure rate of new ventures has remained substantial over time suggests the need for a better understanding of the motives driving new firm founding and of the determinants of new firms' survival. In this respect, recent works suggest that entrants are quite heterogeneous at the time of birth and that such differences may explain a good deal of their survival chances. The basic argument that may be found in several studies is that start-up firms enter the industry with differential endowments, which shape their learning and adaptation in infancy and fundamentally affect their ability to grow up into the adolescent and mature stages. By and large, one can identify three broad types of endowments that may differentially impact upon start-ups' survival chances and that have been discussed in the literature.

In the first place, start-ups differ according to the founders' pre-entry organizational experience and background. This may affect new firms' performance in two ways. On the one hand, one may presume that the founder of a new firm will transfer and install routines and blueprints from their previous company to the new venture. Through this transmission mechanism, 'offspring' start-ups inherit and are endowed with the organizational traits of their 'parent' companies (Klepper 2001). As a consequence of this hereditary process, one may expect start-ups founded by employees working for successful incumbents to be less liable to exit from the industry than start-ups founded by employees with no previous experience or with experience in a different producing environment (Klepper and Sleeper 2005). On the other hand, the pre-entry background may also crucially shape the technological and market opportunities available to the entrepreneur. In particular, one may expect spin-offs (i.e. start-ups founded by former employees of industry incumbents or backed by incumbents) to enjoy a decisive knowledge advantage over other types of start-ups, such as diversifying or *de novo* entrants (Agarwal *et al.* 2004). Once again, the mechanism at work is heritage from parents to offsprings. To the extent that employees of successful incumbents may exploit and transfer knowledge created through their employers' investments to start their own ventures, spin-offs will have a competitive advantage over other types of entrants. By working for a successful incumbent, would-be entrepreneurs are more likely to have access to and discover information on potential opportunities regarding unexploited technologies, new or better ways to serve markets, customers' needs and demand (Shane 2000; Shane and Stuart 2002; Agarwal *et al.* 2004). More broadly, one can argue that the greater the *level* of technological and market know-how inherited from a start-up founder and the more closely *related* these resources to the needs of a particular industry, the greater is the likelihood that the new venture will grow and survive (Helfat and Lieberman 2002).

In addition to organizational capabilities and resources, another type of endowment that may impact upon the performance of new firms relates to the individual human capital possessed by the firm founder or by the team of founders. Although the intellectual assets forming an individual's human capital may partly derive from experience gained during previous employment, this type of endowment may be kept distinct from the hereditary process shortly outlined above. On the one hand, formal training and education of entrepreneurs has been generally found to be associated with lower rates of failure of start-ups (Brüderl *et al.* 1992; Gimeno *et al.* 1997). On the other hand, industry-specific skills and human capital may also help start-ups to overcome the liability of newness and positively affect the performance of start-up firms. In particular, start-up experience, i.e. experience gained through the previous creation of new organizations, and industry experience, i.e. experience accumulated working for previous organizations in the same industry in which the new firm is operating, have been pointed out in the literature as some of the most important determinants of new venture success (Delmar and Shane 2006; Gimeno *et al.* 1997). Beside generic and industry-specific human capital acquired through training and experience, some authors have also emphasized the crucial role of idiosyncratic intellectual assets embodied in individuals. Particularly in the field of biotechnology (Zucker *et al.* 1998) and in the case of start-ups exploiting academic inventions (Shane and Stuart 2002), it has been shown that the performance of new ventures is related to the presence in the founding team of prominent scientists, i.e. individuals with an outstanding record or holding a high academic rank.

Finally, start-ups may be also differentially endowed with resources and therefore have different survival prospects due to the relations they entertain with the external environment or with other organizations. This type of resources endowment differs from either the hereditary mechanism or the founders' human capital mentioned above as it derives from the location in a physical or social space. On the one hand, the vast literature on knowledge externalities points out the many benefits that may accrue to start-up ventures by virtue of their location in a vibrant and well-developed high-technology cluster. Such benefits encompass easier and faster access to innovative ideas and market opportunities, but also to human and financial resources that are vital to successfully founding a new organization (Sorenson 2003). Actually, it is still unclear whether location in a dense cluster does indeed improve the survival chances of new firms, or whether it merely facilitates market entry. Recently, some authors have questioned the prevailing view of a positive cluster effect, by arguing either that knowledge spillovers are essentially neutral with respect to new firm survival (Klepper 2002) or that the increasing costs associated to dense clusters may more than offset any benefit from locating in them (Stuart and Sorenson 2003). Beside external effects arising from location in a physical space, start-up companies may acquire resources necessary for survival and growth through ties (formal and informal) to other organizations. In particular, it has been argued that establishing relationships with *prominent* actors works as a signal that affects the perceptions of the quality of a new venture, thereby enhancing the likelihood of its survival (Podolny 1993; Stuart *et al.* 1999).

In this chapter, we focus mostly on the first two sets of determinants of new firm survival, i.e. the organizational background and human capital of firm founders. The next section explains in more detail the methodology adopted.

3 Methodology

We aim at studying the performance of new innovative firms in the laser technology and in the field of data-switching networks, where performance is measured as longevity of firms. We are then interested in the evolution of a particular subset of industry actors: firms that are both innovative and recently established. This raises some methodological issues regarding the identification of the sample, since our primary source for identifying the object of our analysis is patent data. Our approach is therefore somewhat complementary to the one followed by most of the studies mentioned in the previous section as the entire firm population is not present in patent data. As a consequence, in what follows we first describe the steps through which we have identified our sample and provide a brief motivation for choosing the laser and data-switching networks technologies. We shall then present some descriptive statistics on the information available for each firm and briefly discuss how this may be related to the probability of a firm exiting from the market.

3.1 Identification of new innovators

The first step in our data collection was aimed at identifying the object of our analysis, i.e. new firms that innovate (i.e. apply for a patent) within a relatively short time after their foundation. Using the EP–Cespri database, which contains data on patent applications to the European Patent Office (EPO) from 1978, we collected data on all firms that patented for the first time in either the laser technology field or in the field of data-switching networks between 1990 and 2005. The two technology fields correspond to well-defined International Patent

Classification (IPC) codes, respectively H01S and H04L12. From the sample thus generated, we then extracted all companies located either in the United States or in one of the following European countries: Germany, the UK, France, Italy, Sweden and Denmark.

Starting from this sample, we collected further information regarding the history of each firm and its legal status. In fact, many of the companies that start patenting each year are not truly new ventures, but most often they are divisions/subsidiaries of long established companies, joint ventures, firms resulting from mergers of already existing companies, and so on. To sort out innovative entrants from other types of patenting organizations, we went through all the first-time patenting firms and searched for their establishment date and their year of exit from the market (if any) using both international company databases (e.g. AMADEUS), national company databases (AIDA, DIANE, FAME, etc.) and Internet-based searches. Since our focus is on *new* and *innovative* firms, we restricted our sample according to the following two criteria. First, we used all the above-mentioned sources to identify firms belonging to categories that were not relevant to the object of the research, namely categories that did not respond to the definition of *new* firms (i.e. firms that are truly new ventures). All firms that turned out to be subsidiaries/divisions of established firms or clearly owned by other institutions (e.g. universities) have been excluded from the analysis. Second, we assumed that truly innovative new entrants are more likely to have applied for their first patent shortly after their foundation; hence we restricted our sample to firms established no earlier than five years before their first patent application at the EPO.

The resultant sample includes 49 laser and 184 data-switching networks firms, all of which turn out to be established between 1990 and 2003.[1] A large share of these firms is located in the United States (31 laser and 129 data-switching networks firms, respectively). This raises a major concern as US firms might first apply for a patent at the USPTO and apply at the EPO only at a later stage. This implies that for all the US firms in our sample we could not be sure that their first patent at the EPO was their first patent overall. This could generate two sources of bias. First, we might treat as new entrants in either the laser or data-switching networks technologies firms already active in other technologies, which diversify in the two technologies of interest. Second, since we aim at studying the impact of the technological position of the innovative firm at the time of entry in the market as proxied by characteristics of the first patent (as we shall explain later), we would then mistakenly proxy a firm's initial technological position using a patent farther in time from the beginning of activity. This could be misleading also if the firm's true first patent were in one of the two relevant technologies. For these reasons we searched the USPTO database for all the US firms in our sample and excluded all the firms that were found to have applied for a patent at the USPTO before their first patent application at the EPO (based on this patent priority date). Firms with no previous patent applications at the USPTO or with a first patent application at the USPTO equivalent to the first EPO application are instead included in our final sample. We also keep US firms for which we find at least one patent application at the USPTO in the same year as the priority year of their first patent application at the EPO, even if this other application is not in the two technologies of interest and keep track of this information, which might signal that the new firm enters the market with a wider knowledge base.

After excluding from the sample all US firms that did not satisfy our selection criteria, we are left with our final sample, which includes 39 laser firms (21 located in the United States) and 152 data-switching networks firms (97 located in the United States). For each of these firms, using again Internet-based searches, we finally identified at least one founder.

3.2 Lasers and data-switching networks

Lasers represent a subfield of the 'optics' technological domain. In terms of the International Patent Classification code developed by WIPO, it is identified by the subclass H01S, whose definition is 'device using stimulated emission'. According to WIPO, this subclass comprises 'devices for the generation or amplification, by using stimulated emission, of coherent electromagnetic waves or other forms of wave energy and such functions as modulating, demodulating, controlling or stabilizing such waves'. Being a general-purpose technology (lasers are applied to a wide range of industries) deeply intertwined with underlying scientific domains, the laser industry provides valuable insights to the investigation of the science–technology linkage.[2]

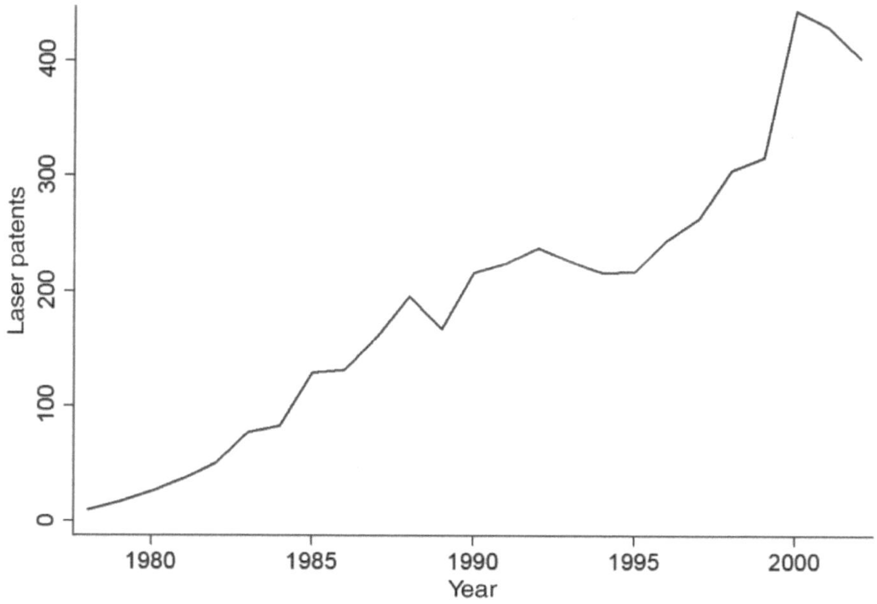

Figure 6.1 Number of patent applications at the EPO in laser technology.

The second technology we focused on is that of data-switching networks, which corresponds to the IPC code H04L12. Compared to lasers, data-switching networks represent a narrower technological domain, included within the wider field of transmission of digital information encompassing devices for 'transmission of signals having been supplied in digital form and includes data transmission, telegraphic communication, or methods or arrangement for monitoring'. Primary application for switches are ICT infrastructures: it is therefore not surprising the sharp increase in patenting which took place in the last 15 years, during the surge of the Internet era (see Figure 6.2).[3]

Both technologies have also been characterized by a large and increasing share of new (technological) entrants throughout the 1990s (Figure 6.3), which confirms they are highly dynamic technologies, where new entrants have a relevant role.

3.3 Patents and firm survival

The data collection steps described in the previous section allowed us to identify our sample and to obtain information on a firm's survival, based on its foundation year and lifespan.[4] Most of the firms in our sample are established in the second half of the 1990s (Figure 6.4): 40 (21 per cent) of them die during the period of observation, on average at 5.5 years of age. Almost all the firms

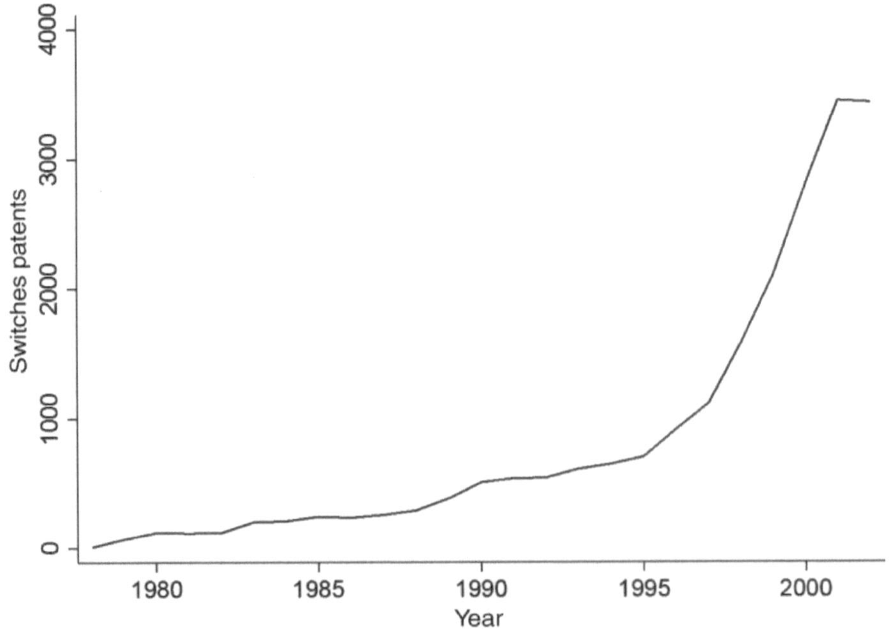

Figure 6.2 Number of patent applications at the EPO in the data-switching networks field.

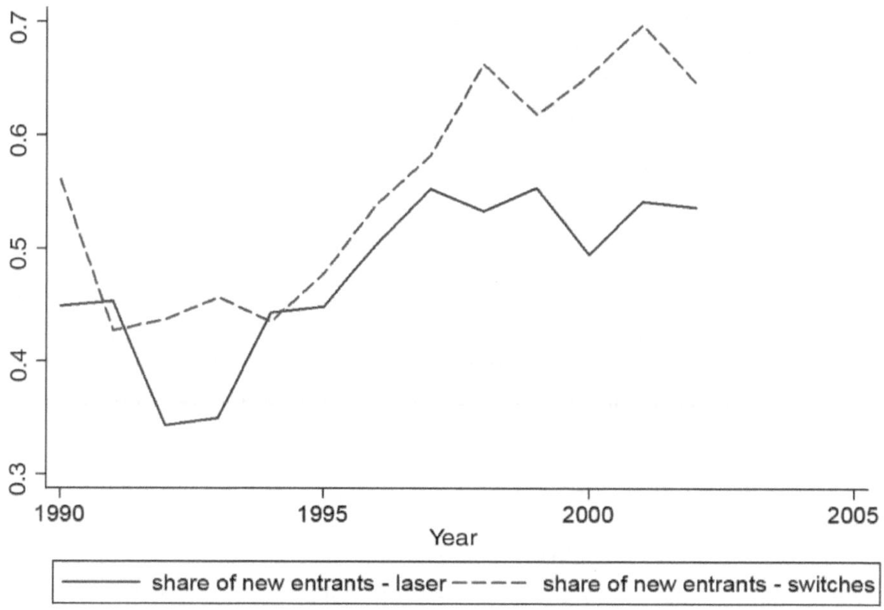

Figure 6.3 Share of new patenting firms over total number of applicants at the EPO.

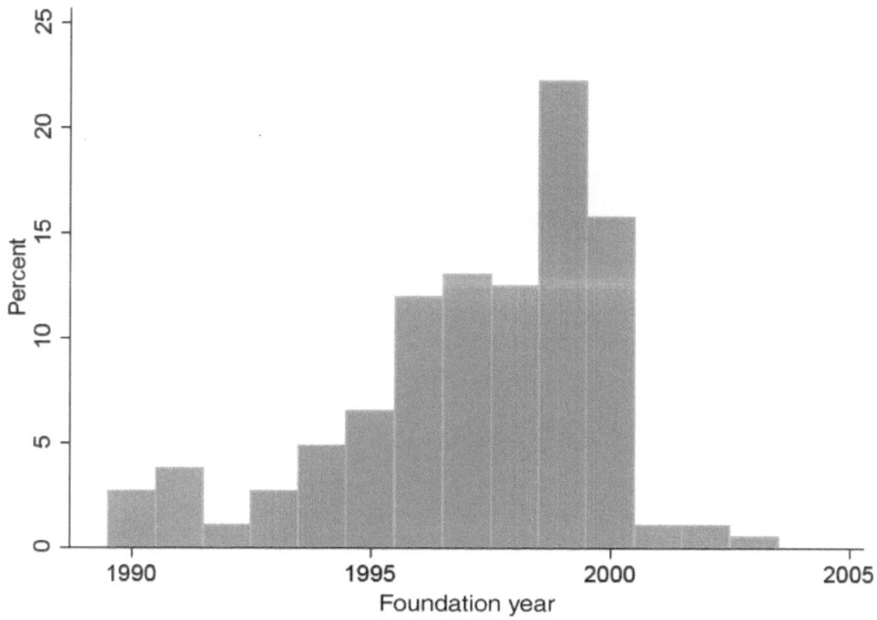

Figure 6.4 Distribution of firms' foundation years.

surviving at the end of the sample period are aged five or above, while most of the firms exiting do it within five or six years from their foundation (Figure 6.5).

To the information on firm survival we then added all the information on the firm's first patent application available from the EP–Cespri database. This gives us a measure of innovative performance, as measured by the lag between foundation and first patent application. Further, it includes information on firm-specific characteristics, based on their first patent application. If close in time to the firm's foundation, the first patent may serve as a proxy for the firm's knowledge capital at birth and hence possibly contribute to explain new innovators' performance.

Figure 6.6 reports the 'patent lag', calculated as the difference between the first patent priority year (i.e. the closest to the actual innovation) and the year of foundation. The figure confirms that most of our firms (70 per cent) do patent within one year from their foundation, some of them even before.[5] Although we explicitly exclude firms that patent over five years after their foundation, it is noteworthy that so many firms apply for a patent at the very beginning of this timespan. That implies that such firms do come to the market with clearly defined competences.

We therefore use information on first patent citation patterns and inventors to obtain variables that we take as representative of a firm's characteristics at birth.[6] For all the firms in our sample we gathered data on inventors' experience, forward and backward citation counts of their first patent, science–technology linkages and the founder's experience as an inventor.

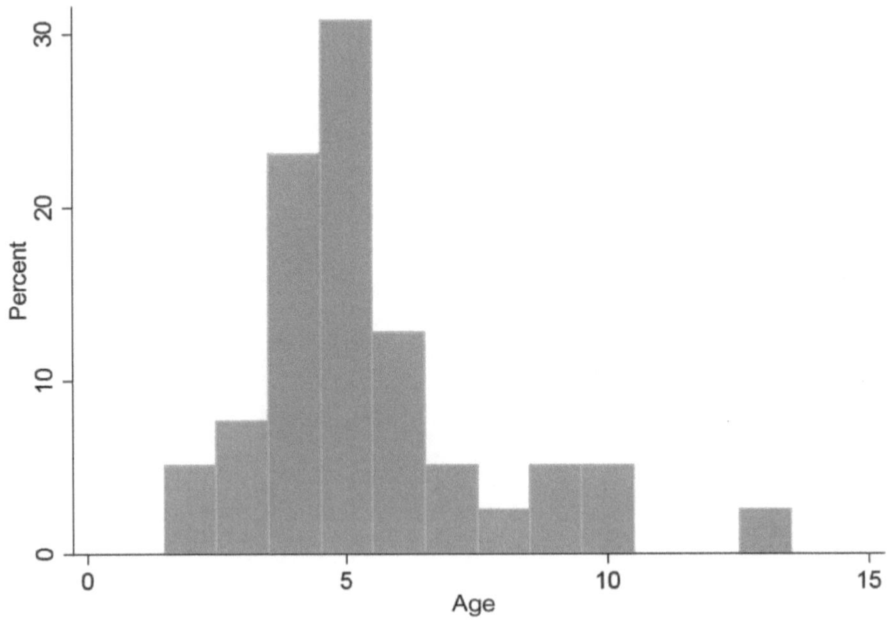

Figure 6.5 Age distribution at exit.

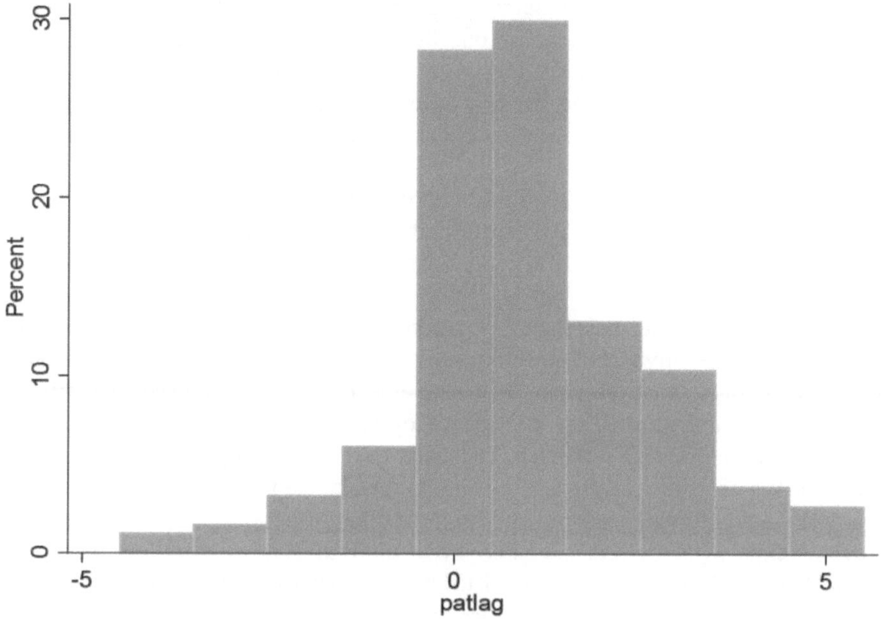

Figure 6.6 Patent lag distribution.

The first information we collected was aimed at assessing how many firms were directly established by inventors. For every firm in the sample, we identified its first patent in our database and extracted the names of inventors listed. Then we checked whether they were also founders of the firms through a comprehensive Internet search. Where we could not find any data, we contacted inventors directly by email or telephone interview. This methodology allowed us to obtain information for the whole sample. For 98 firms (53 per cent) founders were also listed as inventors in the first patent application. Most of these firms (65 per cent) are resident in the United States.

We used information about the inventors of the first patent to build two additional indicators. First, we checked if inventors listed in the first patent were also authors of scientific papers. The underlying hypothesis is that firms listing scientific authors among their inventors could benefit from a stronger link to scientific knowledge. Cutting-edge scientific knowledge is likely to be highly contextual, hence it probably flows more efficiently through people than through other codified media. As a consequence, we hypothesize that firms employing author–inventors can exploit a privileged channel to access new scientific knowledge, which influences their performance. We are particularly interested in authors of scientific papers that have potential impacts on innovation in the selected fields. For this reason, we adopted the following identification procedure. We first extracted all the citations to scientific papers made by all patents in the two selected fields (H04L12 and H01S) and we extracted their authors' names from the ISI Web of Science database. We then matched inventors' names listed in the first patent of our firms with the list of authors' names obtained from ISI and considered an inventor also an author of papers relevant for innovation in the field when there was an exact match of surname and two or more initials. All matches were then checked manually to minimize the risk of incurring false-positive cases.

Applying this methodology we find that only 8 per cent of our switches firms employ an inventor–author. By contrast, the percentage is significantly higher (57 per cent) for laser firms. This is consistent with the laser sector being largely based on scientific research.

The second indicator based on first patent inventors is related to their experience. We consider inventors' previous experience as a potentially relevant variable influencing firm performance. Using our EP–Cespri database, we try to assess the previous innovative experience of all the inventors listed in the first patents. This means that for each first patent's inventor, we establish whether that was his/her first patent application or he/she had previously been listed among the inventors of some other patent. In doing this, we check for inventors' previous experience both at the EPO and at the USPTO.[7] Slightly less than half of our firms (44 per cent) are found to employ experienced inventors, with an average tenure of around four-and-a-half years.

We finally collected information based on citations included in the document of the first patent of our firms and on the citations the same patent received within three years from its application.

With reference to backward citations, we distinguish between citations to other patents and citations to scientific (non-patent) literature (NPL). NPL citations may again signal that the innovation (and hence the firm's knowledge base) is highly science-based and hence points to firms' ability to exploit scientific knowledge. Firms with deeper links to scientific communities could enjoy a direct access to a broad and useful set of knowledge, which in turn might increase the probability of a firm's survival. We find that 55 per cent of our firms include citations into their first patent document, with an average of 1.6 citations per patent. Twenty per cent of our firms cite scientific literature and NPL citations represent, on average, 23 per cent of all citations.

Finally, we collected data on the number of citations received by the first patent of our firms within three years of its application,[8] excluding self-citations. It is widely recognized that the number of citations received might signal the relevance of the firms' technological contribution to the state of the art and hence proxy for the quality of the patent: 28 per cent of our firms have their first patent cited in the three years following application and receive, on average, 2.4 citations, although the distribution of citations received is highly skewed, with 68 per cent of cited firms receiving at most two citations.

4 Estimation results

In our setting, survival analysis models a firm's risk ('hazard') to exit from the market conditional on a vector of explanatory variables. There are different specifications of survival models. In proportional hazard models the systematic hazard is written as:

$$h(t|\mathbf{x}_j) = h_0(t)\exp(\mathbf{x}_j\beta_x) \quad (6.1)$$

In the Cox model $h_0(t)$, the baseline hazard, is left unparameterized and estimates of $\boldsymbol{\beta}_x$ are obtained by conditioning on failure times. By contrast, in the parametric approach a functional form for $h_0(t)$ is specified. Analogous studies on industry evolution have employed the parametric Gompertz specification, which assumes that:

$$h_0(t) = \exp(a)\exp(\gamma\, t) \quad (6.2)$$

The baseline hazard is thus assumed to increase or decrease exponentially with time. The reason for employing this specification in these studies is that for firm survival one would expect to find $\gamma < 0$, reflecting a 'liability of newness'.

Here we use the parametric exponential specification. The exponential specification assumes that the baseline hazard is constant over time, hence would not be able to capture the 'liability of newness' phenomenon, but seems to be more appropriate for our data, as we explain below.[9]

Before moving to the estimation results, two issues are worth discussing. First, the relatively small size of our sample and, in particular, the small number of non-censored exits (40) limits the scope and power of quantitative analysis and certainly affects the estimation results. Second, we need to deal with a potential sample selection problem due to the way we have identified the firms in our sample. We aim at having in our sample all new firms that apply for a patent at the EPO in the selected technologies within five years of their foundation and start from patent applications at the EPO over the period 1990–2005.[10] However, given the five-year maximum lag we have chosen and the possibility of a negative lag we might miss firms born after 2000, because they have not patented yet, or firms born in 1990 that have patented one year earlier (the most common case when the patent application anticipates firm foundation). For this reason, we limit our sample to firms founded in the period 1991–2000 and exclude from the sample firms that were founded in 1990 or later than 2000.

Table 6.1 reports estimation results of the survival model where firms are considered at risk from the year they apply for their first patent, i.e. for each firm the first patent priority year represents the beginning of the spell. This choice allows us to refer all firms' characteristics that can potentially influence their survival to the time the spell begins. All specifications include the age of firm at the time of the first patent application (*age*), which, with the exception of the first specification, is always found negative and significant. This is in line with a stylized fact of the industry life cycle: exit hazard declines as firms mature.

Table 6.1 Survival of new innovators, 1992–2000 (exponential specification)

	(1)	*(2)*	*(3)*	*(4)*	*(5)*	*(6)*	*(7)*	*(8)*
Cohort 1	−0.43							
	(0.48)							
Cum. no. entrants		0.77***	0.76***	0.78***	0.74***	0.73***	0.78***	0.71***
		(0.27)	(0.28)	(0.28)	(0.29)	(0.27)	(0.27)	(0.26)
Experienced inventor			−0.05					
			(0.34)					
Tenure				0.01				
				(0.04)				
Inventor = author					−0.16			
					(0.56)			
Founder = inventor						−0.41		
						(0.33)		
Npl citations							0.17	
							(0.41)	
Cited within three years								−0.91**
								(0.45)
Age	−0.15	−0.28*	−0.28*	−0.28*	−0.27*	−0.27*	−0.28*	−0.32*
	(0.15)	(0.16)	(0.16)	(0.17)	(0.16)	(0.16)	(0.16)	(0.16)
Constant	−3.08***	−7.56***	−7.49***	−7.70***	−7.39***	−7.18***	−7.66***	−6.96***
	(0.33)	(1.63)	(1.70)	(1.70)	(1.73)	(1.66)	(1.65)	(1.60)
Log-likelihood	−113.34	−109.10	−109.09	−109.06	−109.06	−108.35	−109.02	−106.64
Prob>chi^2	0.48	0.01	0.03	0.03	0.03	0.01	0.03	0.00
Observations	174	174	174	174	174	174	174	174

Notes

Standard errors in parentheses. All regressions include control for first patent application happening before foundation. Table reports coefficients: hazard ratios can be obtained by taking the exponential of the estimated coefficients.

* significant at 10%; ** significant at 5%; *** significant at 1%.

Model (1) investigates performance differences between different cohorts, distinguishing between firms entering between 1991 and 1995 (cohort 1) and entrants in the period 1996–2000 (cohort 2). The included cohort does not show a hazard significantly different from the control group. This confirms the absence of advantages in moving early found in recent studies on the US and German laser industry.

Model (2) includes a technology control to account for the dynamics of innovation in the field and reflects the technology conditions at the time of patenting and the attractiveness of the field itself. The one reported is the cumulative number of new technological entrants in the field at $t-1$, which is found positive and significant: as the number of new firms grows, the set of technological opportunities already exploited increases, thus increasing the hazard of exit. We also experimented with alternative measures (e.g. number of patents or number of firms in the field), which gave qualitatively similar results.

Next a series of models have been estimated to identify the potential effects of inheritance and human capital, ability to absorb scientific knowledge and founder's technological know-how. The inheritance variable, usually related to founders in the literature, is here based on inventors' experience. As explained in Section 3, we have a dummy equal to 1 if at least one of the members of the first patent inventors' team is experienced, that is he/she has already developed an innovation previously (i.e. has already patented before). This variable is included in Model (3), while Model (4) includes the tenure of the experienced inventor (the longest if more inventors of the same patent are experienced), measured starting from the first patent of the inventor we could find in the EPO or USPTO database. None of the two ever results significant. The same happens for the two variables that we take as indicators of the new firm's ability to absorb and mould scientific knowledge: a dummy for at least one member of the inventors' team being also an author of scientific papers relevant for the field (i.e. cited by some patent in the field), included in Model (5), and a dummy for the presence of NPL citations in the first patent document, included in Model (7). Finally, Model (6) includes a dummy for the founder being also among the inventors of the first patent, which does not seem to affect firm's survival.

The only variable that we find to significantly reduce the hazard of exit is the indicator variable for the presence of forward citations to the firm's first patent in the three years following application. This may signal that the quality or radicalness of the innovation representing the firm's technological entry into a sector is a key determinant of firm's future performance measured through longevity.

As a robustness check, all hazard models presented in Table 6.1 have been re-estimated using the semi-parametric Cox specification, which does not specify the temporal development of the baseline hazard, but only assumes that it changes in the same way for all firms. This yielded almost identical results (see Table 6.2) and an inspection of the resulting estimated cumulative baseline hazard confirms it is increasing at a constant rate (hence the appropriateness of the exponential distribution – see Figure 6.7).

Finally, as explained earlier, exit in our sample does imply the end of firm's existence, since firms involved in mergers, acquisitions and name changes have

Table 6.2 Survival of new innovators: exponential vs. Cox specification

	(1)	*(2)*
Cited within three years	0.40** (0.18)	0.38** (0.17)
Age	0.73* (0.12)	0.71** (0.12)
Cum. no. entrants	2.03*** (0.54)	2.18*** (0.64)
Log-likelihood	−106.64	−172.85
Prob>chi^2	0.00	0.00
Observations	174	174

Notes
(1) Exponential specification (2) Cox specification. Hazard ratios reported.
Standard errors in parentheses.
* significant at 10%; ** significant at 5%; *** significant at 1%.

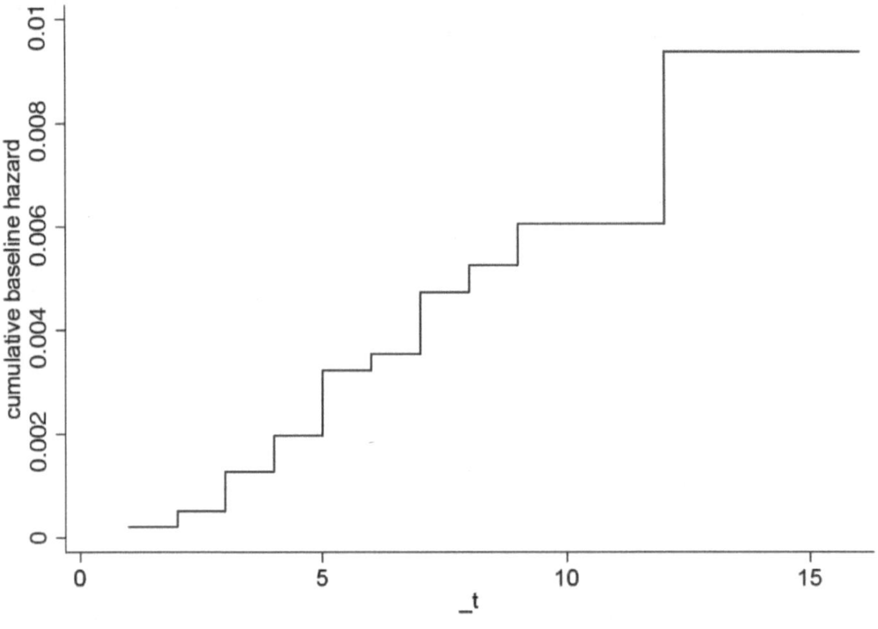

Figure 6.7 The cumulative baseline hazard.

been considered as continuing entities. This assumption might be questionable, particularly for acquisitions, since this event does not give systematic information on firm's performance. We then alternatively treated acquisitions as censored exits in the survival analysis, but this did not change the results.[11]

5 Conclusions

This chapter breaks new ground in the use of patents data as a new data source for research on industry evolution. As such, our approach is complementary to the one mainly followed by the established literature in the area. This approach provides new opportunities: it allows obtaining information not otherwise employed in the literature as, for example, the time to the first patent, information on inventors and a measure of the technological position of the firm.

We focus on a few potentially relevant determinants of new innovative firms' performance (survival) and identify our firms from their first patent application.

While our results do not confirm the relevance of inheritance and human capital, ability to absorb scientific knowledge and founder's technological know-how, we interestingly find that the quality of the innovation representing the firm's technological entry positively affects survival. A way to interpret these results is that the firms in our sample, being all innovative firms, are all well-endowed with technological knowledge and that only those able to introduce a truly 'successful' innovation enjoy a survival advantage.

The 'success' of the innovation could be technological in that the firm introduces a more radical innovation, thus opening a new technological trajectory, then followed by other firms. However, it could be argued that firms will likely choose to move along the new trajectory if it is promising in terms of economic returns, i.e. if it has opened new market opportunities. As such, firms with 'successful' innovations at entry could also be better positioned to learn about market opportunities. Our variable based on forward citations could then capture a firm's ability to exploit knowledge about market opportunities and customer needs. This may itself be related to the founder's previous market-related experience, which we cannot capture with our patent-based measures. Such interpretation would be in line with findings in Buenstorf (2007) indicating that technological capabilities were less important as determinants of firm success in the German laser industry than knowledge about market opportunities and customer needs.

Although we believe the point discussed above is particularly interesting, we should also emphasize that our results are certainly affected by some data limitations. In particular, the two technologies we have selected have different characteristics. As a consequence, they should be studied separately. At present, however, this is not possible because we do not have enough observations in the laser technologies, while excluding laser firms from the sample to study only data-switching networks firms will further reduce the number of uncensored exits. Another possible critique to our approach is that it is difficult to compare our findings to those from studies based on traditional sources because we cannot be sure that firms patenting in the same IPC class do compete in the same market. However, we claim that an appropriate choice of technology (IPC class) can easily overcome this problem. A further possible critique is that we focus our analysis only on innovative entrants, rather than all entrants. However, we believe this is an advantage of our approach as we concentrate on firms with growth potential, where it is more interesting to study the determinants of success.

Acknowledgements

We would like to thank Lorenzo Novella and Riccardo Valboni for excellent research assistance.

Notes

1 The final sample excludes firms for which it was not possible to find relevant information. These were 27 laser firms and 47 data-switching networks firms.
2 The high degree of science intensity of this field emerges also from analysis of citations to scientific literature contained in patent documents. While patents in optics cite on average two scientific papers, the average citation rate to scientific literature is 3.3 for laser patents. This citation rate is only slightly lower than the citation rate in biotechnology, which is equal to 3.8.
3 The growth rate of patent applications at the EPO up to 2003 has been 18 per cent in the laser technology and 56 per cent in the data-switching networks technology.
4 Firms that change name, merge with or are acquired by other firms are treated as continuing firms. Data on ownership changes, mergers, acquisition and name change were compiled primarily using the same sources of information discussed in the previous section.
5 As the figure reveals, there are two firms patenting four years before foundation and three firms patenting three years before. Although one, maybe two, (negative) year lags might be more common, such relatively long lags raise the suspicion that we have not correctly identified the firms. For this reason, we decided to eliminate these firms from the subsequent analysis.
6 Note that we only consider the firm's first patent in the two selected technologies, even though firms may apply in the same year for patents classified in other technologies. Indeed, 22 per cent of our firms simultaneously enter into multiple technologies (two, on average). Most of these (78 per cent) are US firms. This might indicate firms that do not enter in a narrow technological trajectory, but rather do it with a broad knowledge base.
7 For this purpose, we use both the information on the inventor's name and on his/her address (both reported in the patent document). This is easy within the EP–Cespri database, where inventors have been assigned a code. However, we also had to follow up on inventors, bridging from the EPO database to the USPTO. For this purpose we used an automated query process using the inventor's name and the city of residence. When these fields appeared identical we considered the inventors as the same person.
8 The reason of the three-year citation lag is twofold. First, the last year for which we have an application in our sample is 2002 and, second, our EP–Cespri database can be considered to include all (at least most) patent applications dated up to 2005.
9 All hazard model tested below have been re-estimated using the Gompertz specification, which yields very similar results.
10 The EP–Cespri database contains information on EPO patent applications up to date, however applications can be included only after the patent document is published, which requires, on average, 18 months from the date of application. For this reason, the database can be considered to include (almost) all patent applications at the EPO up to 2005.
11 Note, however, that this does not solve the above-mentioned problem of the low number of uncensored exits.

References

Brüderl, J., Preisendörfer, P. and Ziegler, R. (1992) 'Survival chances of newly founded business organizations', *American Sociological Review*, vol. 57 (2): 227–42.

Buenstorf, G. (2007) 'Evolution on the shoulders of giants: entrepreneurship and firm survival in the German laser industry', *Review of Industrial Organization*, vol. 30 (3): 179–202.

Delmar, F. and Shane, S. (2006) 'Does experience matter? The effect of founding team experience on the survival and sales of newly founded ventures', *Strategic Organization*, vol. 4 (3): 215–47.

Gimeno, J., Folta, T.B., Cooper, A.C. and Woo, C.Y. (1997) 'Survival of the fittest? Entrepreneurial human capital and the persistence of underperforming firms', *Administrative Science Quarterly*, vol. 42 (4): 750–83.

Helfat, C. and Lieberman, M. (2002) 'The birth of capabilities: market entry and the importance of pre-history', *Industrial and Corporate Change*, vol. 11 (4): 725–60.

Klepper, S. (2001) 'Employee startups in high-tech industries', *Industrial and Corporate Change*, vol. 10 (3): 639–74.

Klepper, S. (2002) 'Firm survival and the evolution of oligopoly', *RAND Journal of Economics*, vol. 33 (1): 37–61.

Klepper, S. and Sleeper, S.D. (2005) 'Entry by spinoffs', *Management Science*, vol. 51 (8): 1291–306.

Podolny, J.M. (1993) 'A status-based model of market competition', *The American Journal of Sociology*, vol. 98 (4): 829–72.

Shane, S. (2000) 'Prior knowledge and the discovery of entrepreneurial opportunities', *Organization Science*, vol. 11 (4): 448–69.

Shane, S. and Stuart, T (2002) 'Organizational endowments and the performance of university start–ups', *Management Science*, vol. 48 (1): 154–70.

Sorenson, O. (2003) 'Social networks and industrial geography', *Journal of Evolutionary Economics*, vol. 13 (5): 513–27.

Stuart, T.E. and Sorenson, O. (2003) 'The geography of opportunity: spatial heterogeneity in founding rates and the performance of biotechnology firms', *Research Policy*, vol. 32 (2): 229–53.

Stuart, T.E., Hoang, H. and Hybels, R. (1999) 'Interorganizational endorsements and the performance of entrepreneurial ventures', *Administrative Science Quarterly*, vol. 44 (2): 315–49.

Zucker, L.G., Darby, M.R. and Brewer, M.B. (1998) 'Intellectual human capital and the birth of US biotechnology enterprises', *American Economic Review*, vol. 88 (1): 290–306.

7 Brokerage roles in academic patenting

An analysis of Italian inventors

Francesco Lissoni[1]

1 Introduction

Recent research on university–industry technology transfer has dedicated great attention to the theme of university patenting. In the past few years, European economists and other social scientists have identified a large number of so-called 'academic inventors', namely academic scientists who appear as designated inventors of patents whose assignee is not necessarily a university or a public research organization, but also the scientist himself or, most often, a business company (Geuna and Nesta 2006; Verspagen 2006).

Most patents signed by academic inventors are co-authored by other inventors, who may be either employees of the assignee or graduate students brought in by the academic inventors themselves.

The relationship between academic inventors and their co-inventors is of great interest. Do academic inventors play a peripheral or central role in knowledge diffusion? Do academic inventors entertain stable collaborations with their co-inventors, or is the interaction limited to short-lived projects (those that have produced the observed patents)? Does the answer change with the professional status of the co-inventors (who may be either academic scientists, industrial researchers, or students)?

Fieldwork by Murray (2004) suggests that academic scientists involved in entrepreneurial ventures bring useful social capital to the latter, which consists both of their personal networks of current and former students and of the 'cosmopolitan network of colleagues and co-authors'. Nicolaou and Birley (2003) provide a theoretical treatment of academic inventors' entrepreneurial networks.

Extensive data collected by Meyer *et al.* (2003) for Finland suggest that academic inventors in that country are more closely connected to large firms rather than start-up companies. This suggests that framing the analysis of academic inventors' social network within the analysis of entrepreneurship may be reductive.

At a quantitative level, Balconi *et al.* (2004) examined the overall 'networks of inventors' in Italy and found that academic inventors tend to occupy rather central positions, which suggests that their social capital is considerable also within the community of industrial researchers.

Here we combine the same kind of relational analysis on inventors' data with the results of a short questionnaire submitted to a subset of Italian academic inventors, and with data on their scientific publication record and CVs. Our objective is to explore in greater depth the personal relationships that academic inventors entertain with co-inventors with a different background and/or professional status. In order to do so, we adapt Gould and Fernandez's (1989) definitions of brokerage roles to our data, and illustrate some of our results with selected biographical information.

In Section 2 we provide a brief summary of the literature on academic inventors, and of the technical literature on the notion of 'brokerage' as applied to social network analysis. In Section 3 we describe our data and methodology. In Section 4 we present our quantitative results and biographical examples. Section 5 concludes.

2 Background literature

2.1 Academic inventorship

Research on university patenting in Europe has been very intense over the past decade. Besides the studies surveyed by Geuna and Nesta (2006) and Verspagen (2006), as recently as 2007–8 new research has been published on Germany (Czarnitzki *et al.* 2007), Norway (Iversen *et al.* 2007), Italy, France and Sweden (Lissoni *et al.* 2008). All the evidence so far produced suggests that European universities, despite not having as large patent portfolios as their US counterparts, do contribute remarkably to their countries' patenting record via their scientists' inventive efforts (Lissoni *et al.* 2008).

Most patents covering inventions by academic scientists, however, do not belong to universities but to business companies. Some of them derive from contract research or consulting, while others are the result of research partnerships between companies and universities, which share the IPRs over the research results. The institutional specificities of the European universities, in fact, are such that the latter have either little interest or little ability in bargaining hard both with their own staff and with companies in order to reclaim or to share the IPRs over academic inventions, when compared to the US public and (above all) private academic institutions.[2]

Present enquiries focus now on the technological characteristics, ownership pattern and the value of what have now become known as 'academic patents', that is patents whose inventors comprise at least an academic scientist, irrespective of the assignee. Crespi *et al.* (2006) find that inventors' own estimates of the economic value of their patents are not affected by the ownership of the patent, that is they find no evidence that, by retaining the property of their scientists' inventions, universities would do a better job than companies in turning patents into valuable assets.

Other lines of enquiry focus not so much on the patents, but on the inventors. The latter can be interviewed, their CVs or publications can be retrieved, and analyses can be performed on the determinants of their productivity, and on their mobility in space or across organizations. Recent investigations, in particular, have found that academic inventors are very productive scientists, whose productivity further increases after patenting (see Chapter 4, this volume).

Finally, patent data reclassified by inventors are a key source of relational data, which can cast light on processes of innovation diffusion and, when academic inventors are involved, university–industry links. Patents authored by two or more co-inventors, in fact, are instances of cooperation and knowledge exchange. Co-inventors of one or more patents can be safely presumed to know each other, and to have exchanged some information and, possibly, knowledge. One can build networks of inventors where the latter are represented nodes, and patents are ties between them. Balconi *et al.* (2004) build the network of Italian inventors from data on patent applications registered at the European Patent Office, from 1978 to 1999, and find that academic inventors tend to occupy rather central positions in the network. Lissoni and Sanditov (2007) find that the same results hold for France and Sweden.

Network analyses based purely on patent data, however, cannot say much on the qualitative aspects of the relationships between academic inventors and their co-inventors, for at least three reasons. First, we may wish to know more on the quality of the relationships established between inventors. Are they stable or occasional? Do they provide both parties with chances of scientific collaborations, or are they a mere vehicle for information exchanges (such as those on funding opportunities, jobs, etc.)?

Second, the identity of non-academic co-inventors of academic patents is not always clear; they may either be industrial researchers (most typically, members of the patent assignee's R&D staff), graduate students or former academic colleagues who are now retired or have left the university.

Last, we would like to know more about the nature of the relationship between the academic inventors and the patent assignee and its R&D staff. Do academic inventors provide a bridge between university and industry (that is, between cliques of academic scientists and industrial researchers)? Or do they also provide a bridge between different companies (that is, between industrial researchers)? Are academic inventors instrumental in introducing their students to business R&D staff members?

2.2 Brokerage

In order to answer these questions, it is necessary to collect information not only on the academic inventors and co-inventors, but also on the relationship between them (more on this in Section 3). A useful instrument to guide this data collection and the following analysis is the notion of 'brokerage', as derived from economic sociology and, in particular, from applications of social network analysis.

Such notion builds upon that of 'in-betweenness' across nodes in a network (Freeman 1979). In any network (either social or physical), any node j is said to stand in between two other nodes i and z if the shortest path between the i and z

passes through it. Measures of in-betweenness for node *j* will be therefore based upon counts of how many shortest paths between any pair of nodes in the network pass through it.

If we consider only *j*'s ego-network (i.e. the subset of the overall network that comprises only the nodes in direct contact with j), *j* will result to be in-between its ego-network nodes to the extent that the latter are no connected to each other by any direct tie, that is they reach one another only through *j*.

The peculiar role of in-between nodes is most evident in directed graphs, that is graphs where ties between nodes do not necessarily run both ways, but may also run in one direction only (for example, from *i* to *j* and from *j* to *z*). In social network analysis, directed graphs are used to portray information flows in a community of agents: in-between agents are the necessary intermediaries of information flows between many agents who could not otherwise reach each other (in our example, *i* can send information to *z* only through *j*). In-between agents can take advantage of their position in many ways, which have been the subject of much sociological investigation (Burt 2001).

In-between agents are even more relevant when the community of agents they belong to are fragmented across different affiliation groups, as happens when citizens of the same town belong to different religious or ethnic groups or (as in our case) researchers in a given technological field belong to different types of organizations (university and industry). In this case, they may find themselves in-between not only two other agents, but two entirely different groups of people, whose chances to communicate in the absence of the in-between nodes are very few. In our example, if *i* and *z* belong to different affiliation group, their link through *j* (who may belong either to the same group of *i* or *z*, or possibly to a third one) is especially important, since their different affiliation may make it especially hard to build alternative links.

Terms such as 'brokers' and 'gatekeepers' have for long been used in sociology and organization theory to identify actors who are in-between other actors from different affiliation groups. Within social network analysis, Gould and Fernandez (1989) have proposed a set of measures of various brokerage positions any node can take in its ego-network, which are now available on most software packages for social network analysis, such as *Ucinet* and *Pajek* (both used for this chapter).

- Node *j* is said to act as a broker between *i* and *z* whenever the latter belong to the same affiliation group, and this group is different from *j*'s.
- Node *j* acts as a gatekeeper when the information flow runs from *i* to *z*, and *j* and *z* belong to the same group, which is a different one from *i*'s (*j* is said to act as a representative of *z* whenever the information runs the opposite way).
- Node *j* acts as a coordinator when all three nodes belong to the same group, and as a liaison when each of them belongs to a different group.

The number of each type of brokerage positions taken by node *j* are then divided by the expected number of position the same node would take if its ego-network were a random one, so that standardized measures are obtained, which can be used to compare nodes with ego-networks of different size and group composition.

A number of papers have been produced, which makes use of Gould and Fernandez's methodology, and find that actors in brokerage positions enjoy strategic advantages in a number of activities, ranging from the political to the entrepreneurial (Taube 2004).

In what follows, we adapt Gould and Fernandez's measures to data on academic inventors. However, we do not make use of those measures to draw conclusion on the structural property of the networks academic inventors belong to, but only on the most salient individual characteristics associated to being in a brokerage position.

3 Data and methodology

Data used in this chapter results from the integration of three different sources: the KEINS database on academic inventors; the information gathered from a sample of Italian academic inventors through a questionnaire; and some extra information on these inventors' publication record, obtained through the ISI–Web of Science database. In Section 3.1 we say more on these sources, while in Section 3.2 we present the brokerage measures employed in the rest of the chapter.

3.1 Data

The KEINS database is the result of a collective effort of many researchers (Lissoni *et al.* 2006). It contains information on academic patents and inventors in France, Italy and Sweden. It originates from the EP–INV database produced by CESPRI–Università Bocconi, which contains all EPO applications published since 1978, reclassified by applicant and inventor; and from three lists of university professors of all ranks (from assistant to full professors), one for each of the above-mentioned countries (PROFLISTs). Academic inventors have been identified by matching names and surnames of inventors in the EP–INV database with those in the PROFLISTs, and by checking by email and phone the identity of the matches, in order to exclude homonyms.

Each country section contains information on individual characteristics of the scientists (such as age, affiliation, academic rank, discipline), although the extent and quality of such information may vary across countries. Information on Italian academic inventors originates from the complete list (provided by the Italian Ministry of Education) of professors and researchers who, in 2000 and 2004, held a position in a scientific or technical discipline in an Italian university (including medical and engineering schools).

In the KEINS database, professors' disciplines are defined according both to national classification and to a harmonized one. All of them are quite detailed and allow for some compression into broader categories, which we will refer to as 'fields'.[3] In this chapter we focus on Italy and on the four fields with the highest share of academic inventors over the total number of professors in that country, namely chemical engineering (it includes technology of materials, such as macromolecular compounds), biology, pharmacology, and electronics and telecommunications. We also limit our analysis to professors who were already on duty in 2000, for a total of 301 academic inventors, who count for around 10 per cent of Italian professors active in the selected disciplines in 2000 (Table 7.1).

As shown in Table 7.2, most patents signed by the selected academic inventors are owned by business companies, rather than 'open science' institutions (such as universities and public research organizations); a minority of patents are assigned to individuals, most often the professors themselves, or a relative. The ownership patterns vary across the professors' disciplines, with biology standing out as the one that records the highest percentage of patents assigned to open science institutions and individuals.

Publication data were collected from the 1975–2003 online version of ISI's *Science Citation Index* for all the 308 academic inventors in the selected fields. A more detailed description of these data can be found in Breschi *et al.* (2007).

Finally, we contacted the academic inventors in order to ask them questions related to the various aspects of the research behind the observed patents. The questions most relevant for this chapter were those related to the identity of and the relationship with the co-inventors:

1 Whether they actually knew the co-inventors listed along with them on the patent document (all answers were positive).
2 Whether those co-inventors were academic colleagues, industrial researchers or students, at the time when the patent was signed.
3 Whether they had gone on collaborating with co-inventors at a later time.
4 Whether they planned to do so in the future.
5 And, in case of negative answers to both questions three and four, whether they were at least still in touch and exchanged information, albeit occasionally.

We successfully interviewed 156 academic inventors (51 per cent of the total), who provided information on 741 of their 771 co-inventors (interviewees who had worked with many co-inventors sometimes refused to provide information on all of them, since this would have prolonged the interview too much). However, due to the fact that several co-inventors signed patents both with

Table 7.1 Italian university professors in 2000, selected fields

Field	*Professors, active in 2000*	*Of which: academic inventors, no. and (%)*
Chemical engineering and materials technology	355	66 (18.5)
Pharmacology and pharmaceutical sciences	613	84 (13.7)
Biology	1,359	78 (5.7)
Electronics and telecommunications	630	73 (11.6)
Total	2,957	301 (10.2)

Source: Breschi *et al.* (2007).

Table 7.2 Ownership of academic inventors' patents[§] by type of applicant and field; number and (%) of patents

Field	*Business companies*	*'Open science' institutions*[a]	*Individuals*[b]	*Others (n.e.c.)*	*All applicant types*
Chemical engineering and materials technology	125 (78.1)	18 (11.3)	15 (9.4)	2 (1.3)	160 (100)
Pharmacology	192 (85.0)	24 (10.6)	10 (4.4)	–	226 (100)
Biology	91 (54.5)	43 (25.7)	30 (18.0)	3 (1.8)	167 (100)
Electronics and telecommunications	199 (81.9)	28 (11.5)	13 (5.3)	3 (1.2)	243 (100)
All fields	607 (76.3)	109 (14.2)	68 (8.5)	8 (1.0)	796 (100)

Source: EP-INV-DOC database.

Notes

a Universities, public labs and government agencies; both Italian and foreign.

b Same applicant's and the inventors' names.

§ Patents owned by more than one applicant were counted more than once.

academic inventors that agreed to be interviewed and with academic inventors who did not, we have information also on the co-inventors of 29 of the latter, for a total of 185 academic inventors.

3.2 Methodology: brokerage positions from patent data

Once the group affiliation of co-inventors is established, one can move to calculate the number and type of brokerage positions of academic inventors. However, several difficulties stand in the way of a straightforward application of Gould and Fernandez's methodology. Both derive from the fact that networks of inventors derive from archival data, rather than questionnaires.[4]

First, a questionnaire aimed at gathering network information can phrase questions in such a way that the resulting ties between actors are directed (for example: 'who do you rely upon for information among your friends?'; information requests may go from *j* to *z*, and not be reciprocated). This is not the case with patents: when *j* and *z* are found to be co-inventors of the same patent, there is no way we can tell who was asking for information from who, and whether he was reciprocated. No directed graph can be built.

Second, when *n* inventors have worked on the same patent, they immediately form a clique of size *n*. If $n=3$ this is a triangle whose vertices (nodes) are all connected to each other: no node is in between the other two. This means that counting how many times an academic inventor stands in-between two co-inventors (as required by Gould and Fernandez's method) makes sense only for academic inventors with at least two patents.

Last, standardized brokerage measures, based upon the ratio between the observed instances of in-betweenness and the hypothetical instances for a random ego-network, do not make sense. In fact, academic inventors' ego-networks can never be random, because they are the sum of several complete subgraphs, one for each patent signed by the academic inventor.

The first problem can be confronted simply by adapting Gould and Fernandez's methodology to our data, which implies giving up some of the nuances of the original definitions of brokerage positions. Figure 7.1 reports the definitions we will use in this chapter.

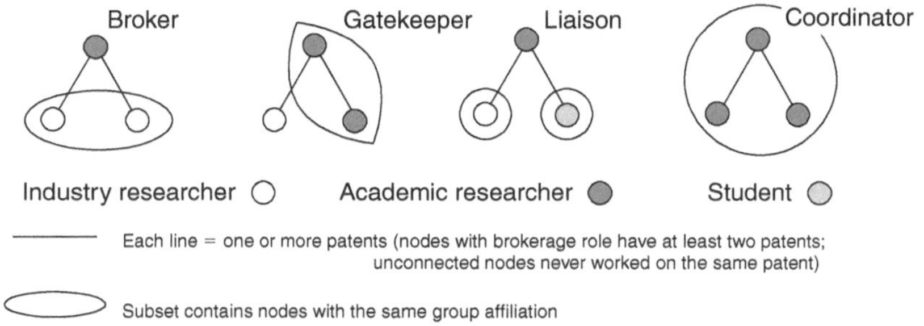

Figure 7.1 Brokerage positions: definitions.

Academic inventors find themselves in the position of brokers whenever they stand in-between two industrial researchers (or, more rarely, students). They act as gatekeepers whenever they stand in-between an industrial researcher (or a student) and an academic researcher;[5] and as liaisons when they stand in-between an industrial researcher and a student. Finally, they act as coordinators whenever they stand in-between two members of their own affiliation group, that is two academic researchers.

The second methodological problem mentioned above forces us to restrict our analysis only to academic inventors with at least two patents. They are a minority in our sample, only 74 out of 185 (see Table 7.3).

As for the last problem, we have found no other solution, so far, than considering the absolute number of brokerage positions taken by each academic inventor, but controlling for the number of patents signed by the inventor any time we use his brokerage score as an independent variable in a regression.

Figure 7.2 provides some summary statistics on the distribution of brokerage scores over the 74 academic inventors considered. Most academic inventors do not play any brokerage role at all, that is they are never in-between any two co-inventors; this is particularly true for 'liaison' scores, since students are a rare presence in the set of co-inventors.

Academic inventors' age and seniority also contribute greatly to explain academic inventors' brokerage scores. From Table 7.4 we see that it is only full professors who achieve the highest scores. However, age and seniority are also correlated to the number of patents held by each academic inventor, and possibly with other variables highly correlated to the observed brokerage scores. Therefore, in the next section, we move on towards a more systematic exploration of the characteristics of academic inventors with high brokerage scores.

Table 7.3 Number of academic inventors, by number of patents signed

Number of patents per inventor	*Number of inventors*
1	111
2	22
3	15
4	9
5	11
6–10	9
11–20	6
21–30	2
All inventors	185

Source: KEINS database.

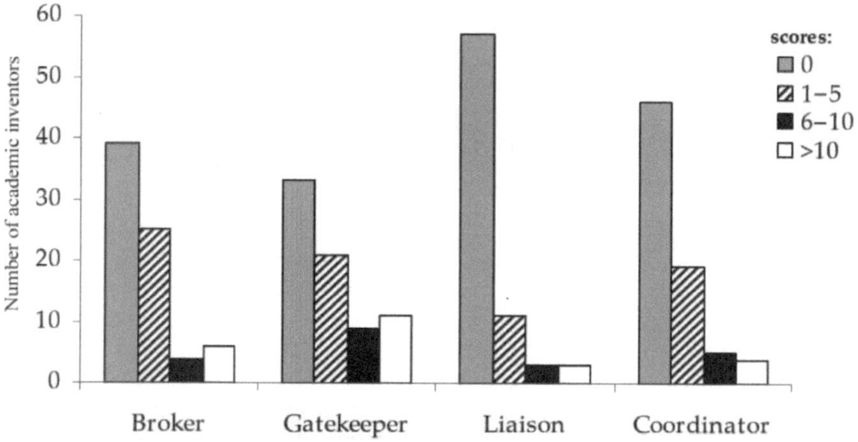

Figure 7.2 Distribution of brokerage scores.

4 Results

4.1 Who are the brokers?

In this section we examine how many of the 74 academic inventors in our database with more than one patent hold brokerage positions. On the basis of their biographical information, and of their publication record, we also produce a profile of those with the highest brokerage scores.

In order to do so, we first run a regression of various measures of brokerage over a number of characteristics of the academic inventors, such as their observed scientific productivity, their age, their academic rank (full, associate or assistant professor) and disciplinary affiliation. Productivity is measured as the number of articles published on academic journals listed on the ISI–Web of Science database between the academic inventor's twenty-fourth birthday (a conventional date for the start of his career) and 2003 (when the publication data were collected). Alternatively, a 'late-productivity' measure is similarly produced, which spans only over the last ten years of the academic inventor's career.

When running the regression, we control for the number of patents signed by the academic inventors which, by construction, are positively correlated to the brokerage scores. However, we distinguish between patents owned by business companies, open science institutions and individuals, as we expected only the former to be signed by co-inventors from the 'industrial researcher' affiliation group.

By no means the proposed regression can be interpreted as an explanatory one, that is no causation link can be assumed to run from the independent variables to the dependent ones. In order to 'explain' the brokerage position reached by any academic inventor, one would need to observe the accumulation of social

Table 7.4 Brokerage scores and academic seniority

Score	*Broker*		*Gatekeeper*		*Liaison*		*Coordinator*	
	Full	*Assoc. and assistant*	*Full*	*Assoc. and assistant*	*Full*	*Assoc. and assistant*	*Full*	*Assoc. and assistant*
0	45.5	63.3	40.9	50.0	77.3	76.7	54.5	73.3
1–5	31.8	36.7	20.5	40.0	11.4	20.0	25.0	26.7
6–10	9.1	–	15.9	6.7	4.5	3.3	11.4	–
>10	13.6	–	22.7	3.3	6.8	–	9.1	–

ties (with co-inventors) over time, and find a way to solve two problems of endogeneity in the regression. The main problem is that a fixed effect may explain a lot: highly productive professors have a higher probability of producing not only papers, but also patentable inventions; and those who have many patents have a high probability to record a high brokerage score.

However, one can interpret our regression coefficients as a set of partial correlation indexes, which help producing a portrait of who acts as a broker among academic inventors.

We can first look at Table 7.5, which provides descriptive statistics for the variables used in the regressions. It confirms that the distribution of the dependent variables is highly skewed, with the median always at zero (one for the gatekeeper score) and the mean well above one; the 'max' column indicates also the presence of a few outliers for the broker and gatekeeper score.

The distributions of independent variables concerning productivity and the number of patents signed are also highly skewed, with some academic inventors having produced no publications over the observed time span.

The 74 academic inventors considered are predominantly senior ones, with a mean age of 55; over half of them are full professors, less than one-fifth are assistant professors.

With respect to the original distribution of academic inventors across disciplines (see Table 7.2), biology is slightly under-represented, in favour of chemistry and pharmaceuticals.

Table 7.6 reports the results of the regressions. Since the proposed brokerage measures can be considered as count data (number of instances of in-betweenness), and have a highly skewed distribution, we made use of negative binomial regressions (we also tested whether zero-inflated negative binomial

Table 7.5 Descriptive statistics

Variables	*Obs*	*Mean*	*Median*	*Variance*	*Min*	*Max*
Broker	74	4.81	0	210.2	0	109
Gatekeeper	74	7.30	1	386.5	0	131
Liaison	74	1.61	0	23.5	0	31
Coordinator	74	2.08	0	26.6	0	25
Age	74	55.5	56	78.4	38	74
Productivity	74	2.72	1.98	6.26	0	14.68
Late productivity	74	3.81	2.88	13.8	0	20.38
Business-owned patents (no.)	74	4.64	3	30.0	0	28
University-owned patents (no.)	74	0.78	0	1.10	0	4
Individual-owned patents (no.)	74	0.35	0	1.57	0	7
Chemistry	74	0.28	–	–	0	1
Electronics and telecommunications	74	0.24	–	–	0	1
Pharma	74	0.31	–	–	0	1
Biology	74	0.16	–	–	0	1
Full professor	74	0.59	–	–	0	1
Associate professor	74	0.26	–	–	0	1
Assistant professor	74	0.15	–	–	0	1

Table 7.6 Brokerage scores; negative binomial regressions over academic inventors' characteristics

	Broker (1)	*Broker (2)*	*Gatekeeper (3)*	*Coordinator (4)*	*Liaison (5)*
Productivity	0.40*** (0.13)		0.36*** (0.13)	0.30** (0.13)	0.23 (0.26)
Late productivity		0.21** (0.09)			
Age	0.08** (0.03)	0.07** (0.04)	−0.01 (0.03)	−0.04 (0.04)	0.14* (0.08)
Business-owned patents (no.)	0.66*** (0.11)	0.68*** (0.11)	0.54*** (0.11)	0.44*** (0.12)	0.84*** (0.26)
Business-own pat.2	−0.02*** (0.004)	−0.02*** (0.00)	−0.01*** (0.00)	−0.01*** (0.00)	−0.02*** (0.01)
University-owned patents (no.)	0.002 (0.72)	0.10 (0.73)	1.66*** (0.63)	1.73** (0.68)	2.58* (1.40)
University-owned pat.2	−0.02 (0.25)	−0.05 (0.25)	−0.35* (0.22)	−0.29 (0.22)	−0.63 (0.42)
Individual-owned patents (no.)	−0.59** (0.29)	−0.44* (0.27)	−0.41* (0.24)	−0.34 (0.24)	−0.32 (0.50)
Chemistry	0.31 (0.68)	0.13 (0.66)	0.10 (0.62)	0.31 (0.70)	0.68 (1.27)
Electronics and telecommunications	0.95 (0.77)	0.63 (0.75)	−1.25* (0.71)	−1.42* (0.79)	1.59 (1.44)
Pharma	−0.32 (0.65)	−0.52 (0.63)	−0.90 (0.61)	−0.64 (0.69)	−2.74* (1.67)
Associate professor	0.20 (0.73)	0.11 (0.77)	−0.32 (0.62)	−0.53 (0.72)	1.58 (1.43)
Assistant professor	0.62 (0.84)	0.40 (0.89)	0.03 (0.78)	−1.77* (1.08)	1.38 (1.46)
_constant	−7.43*** (2.49)	−6.65** (2.57)	−1.12 (2.14)	0.59 (2.49)	−13.74** (5.92)
	*LR test of alpha = 0: $chi^2(01) = 352.78$****	*LR test of alpha = 0: $hi^2(01) = 363.14$****	*LR test of alpha = 0: $chi^2(01) = 412.3$****	*LR test of alpha = 0: $chi^2(01) = 103.3$****	*LR test of alpha = 0: $hi^2(01) = 139.0$****
	No. of obs = 74	*No. of obs = 74*	*No. of obs = 74*	*No. of obs = 74*	*No. of obs = 74*
	LogL = −126.6	*LogL = −128.8*	*LogL = −158.2*	*LogL = −97.5*	*LogL = −76.0*
	LR $chi^2(12) = 61.2$	*LR $chi^2(12) = 56.8$*	*LR $chi^2(12) = 54.0$*	*LR $chi^2(12) = 40.3$*	*LR $chi^2(12) = 25.5$*
	Prob > $chi^2 = 0.00$	*Prob > $chi^2 = 0.00$*	*Prob > $chi^2 = 0.00$*	*Prob > $chi^2 = 0.00$*	*Prob > $chi^2 = 0.01$*
	Pseudo $R^2 = 0.19$	*Pseudo $R^2 = 0.18$*	*Pseudo $R^2 = 0.15$*	*Pseudo $R^2 = 0.17$*	*Pseudo $R^2 = 0.14$*

Note
Standard errors in brackets.

regressions were more appropriate, but the Vuong test returned negative results; see Vuong 1989). Notice that the estimated coefficients cannot be directly interpreted as the marginal influence of the independent variable on the dependent one, as in OLS regression; instead, estimated coefficients indicate the difference between the log of the expected counts for a unit change of the independent variable; this is the same as the log of the ratio of the expected counts, which implies that by taking the exponential of the estimated coefficient one can measure the expected percentage increase of the dependent variable for a unit change of the independent one.

The regressions have been run for all of the 74 academic inventors susceptible to brokerage measurement; excluding outliers did not change much the estimated coefficients, nor their significance.

We will comment first on the estimated coefficient of control variables, and then explain how the number of patents, by type of ownership, affects the brokerage score, leaving the discussion of the effect of scientific productivity until last.

Age matters, but not much. It is not significant in regressions (3) and (4), and barely significant in regression (5). Regressions (1) and (2) indicate that being one year older increases an academic inventor's expected broker score by no more than 8 per cent. Seniority dummies are not significant. Dummies for disciplines are either insignificant or significant at no more than 90 per cent.

Estimated coefficients for the number of patents, classified by ownership type, depend directly on the technical constraints of the proposed brokerage measures, but they are also quite informative of the nature of contacts established by academic inventors through their patenting activity.

We first observe that individual patents are not associated with high brokerage scores. On the contrary, having signed one's own patent diminishes the brokerage score by up to 55 per cent (regression (1)). This is because individual patents are very often signed by one or very few academic inventors, most of them colleagues at the same university, which explains why the negative effect is particularly severe for the broker measure (which requires the academic inventor to stand in-between two industrial researchers).

University-owned patents do not seem to be associated with high broker scores, which again is a largely predictable result. University-owned patents are most often signed by colleagues in the same university, and occasionally by some students; at most, they can originate some brokerage position in-between students. On the contrary, one extra university-owned patent increases the expected coordinator score of 464 per cent. The effect on the gatekeeper score is lower, because it is non-linear, that is it depends on the total number of university-owned patents signed by the academic inventors, with a negative second derivative; moving from one patent to two, the score increases by 84 per cent; moving from two patents to three, we have a 8 per cent decrease (academic professors with three university patents only, that is with no additional business-owned patents, are very rare, however).

Business-owned patents affect all of our brokerage measures, in a non-linear way. Adding one more of these patents to the portfolio of an academic inventor with one patent only, increases the broker score by either 82 per cent or 90 per cent (depending on the regression: (1) or (2)), the gatekeeper score by 68 per cent, the coordinator score by 52 per cent, and the liaison score by 123 per cent. Even the academic inventor with the highest number of patents in our sample would see his brokerage score going up with one more patent: 6 per cent to 8 per cent for the broker score, 27 per cent for the gatekeeper score, 15 per cent for the coordinator score, and 27 per cent for the liaison score.

While the relationship between business-owned patents and broker and liaison scores is pretty obvious (both these scores derive from the academic inventor standing in between two co-inventors, one of which has to be an industrial researcher), it less so for the gatekeeper score, and even less for the coordinator score.

Academic inventors may be gatekeepers when standing in-between an academic colleague and either an industrial researcher or a student. While university-owned patents grant some chances for the academic inventors to stand in-between a colleague and a student, business-owned patents provide also the chance to stand in-between a colleague and an industrial researcher; in addition, mixed teams (of academic scientists and industrial researchers and/or students) are much more common with business-owned patents than with university-owned ones.

The existence of a positive effect of business-owned patents on the coordinator score is explained again by the high frequency of business-owned patents counting at least two academic co-inventors. If academic inventor *j* has worked on two business-owned patents, one with the academic colleague *i* and another with colleague *z*, he will end up acting as a coordinator between the two even in the absence of any university-owned patent in his portfolio.

Figure 7.3 may help clarifying some of these technicalities, and also provide nice examples of top brokers, gatekeepers and coordinators. It represents the ego-networks of selected academic inventors, who are represented by large black circles and labelled with a number, which is the code they have been given in the KEINS database; small circles are co-inventors and they are coloured according to their affiliation group; grey lines around subgraphs, each labelled with a different letter, indicate that the inventors/nodes inside the line are team members, that is they worked on the same patent(s).

Inventor *1622030* (centre-top of the figure) has not an especially high score, but is representative of what we may call a 'pure' coordinator, that is an academic inventor who stands in-between several of his colleagues, and has never worked with inventors from outside his affiliation group. He and the co-inventors circled by the 'A' line have worked on the same patent, as one can understand by noticing that every node is connected to all others. This means that the coordinator score of inventor *1622030* does not derive by his collaboration with the other inventors of patent 'A', since they are connected to each other independently of him. In fact, inventor *1622030*'s coordinator score derives from his work on a different patent, which he has produced by working with the two inventors represented by the nodes on the right hand side of his ego-network,

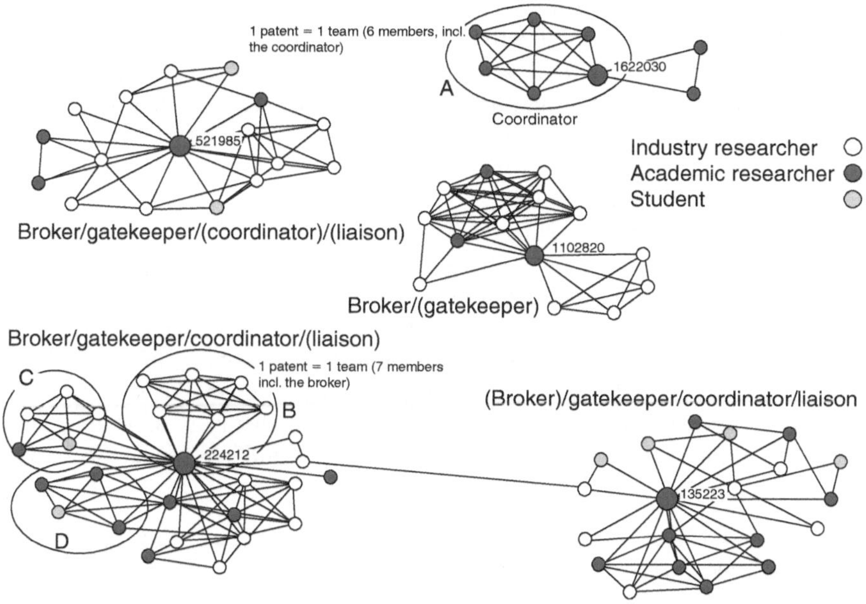

Figure 7.3 Top brokers, gatekeepers and coordinators.

outside the 'A' line. Inventor *1622030* acts as a coordinator between these two inventors, and those circled by 'A'.

Similarly, all of the brokerage scores of inventor *224212* (bottom-left corner of the figure) derive from his work on several different patents, such as those invented with co-inventors/nodes circled by the 'B', 'C' and 'D' lines (for the sake of tidiness, 'C' and 'D' have not been extended to include inventor *224212*, but the reader should be aware that he belongs to all teams). Notice that patent 'B' lists inventor *224212* as the only academic co-inventor, while 'C' and 'D' are more mixed (as much as the other patents, not represented by any circle line, which must have seen the participation of the co-inventors/nodes on the bottom-right hand side of *224212*'s ego-network). This heterogeneity explains why inventor *224212* records very high scores on all measures. A similar explanation applies to inventor *135223* (bottom-right corner of the figure), albeit this inventor's broker score is not particularly high, due to the limited number of industrial researchers in his ego-network.[6]

Finally, we notice that inventor *1102820* (centre of the figure) has zero liaison score, because his ego-network does not include any student, and also zero coordinator score, because the only two academic co-inventors in the ego-network are linked by a direct tie. A similar explanation applies to the relatively low liaison and coordinator scores of inventor *521985* (top-left corner of the figure), whose ego-network includes only two students and only three academic co-inventors, two of which are linked by a direct tie.

Going back to the regressions in Table 7.6, we notice that from all of them, with the exception of number (5), academic inventors with strong brokerage positions emerge as highly productive scientists, both when their overall publication record is considered and when the attention is restricted only to their recent years of activity.[7] Regression (1), for example, tells us that one more paper per year by the academic inventor increases his expected broker score by 49 per cent (from regression (2) we see that a similar increase in late productivity produces a 23 per cent increase).

These results cannot be explained by the positive association between scientific productivity and number of patents (found by the studies we reviewed in Section 2), since in all the regressions we control by the latter. Rather, it is suggestive of the possibility that highly productive scientists have, *ceteris paribus*, a more diverse set of collaborations, which includes actors from different affiliation groups, several of whom are connected only through the co-inventing activity performed along with the productive scientists themselves. This is confirmed by looking at the biographies of the top brokers and gatekeepers from our sample.

4.2 A close look at the top brokers and gatekeepers

In this section we briefly report some significant biographical information on the top brokers and gatekeepers in our sample, which we retrieved using standard search engines on the World Wide Web and from a close look at the identity of the patent assignees.[8]

Only six academic inventors have a brokerage score higher than ten. All of them are senior full professors, born between 1935 and 1956. For two of them, M.B. and F.P., who scored respectively 109 (top score) and 33, we could not find much biographical information of sorts, besides a short CV listing the affiliation (University of Rome, 'La Sapienza', the largest university in Italy, and the University of Modena and Reggio, respectively) and the discipline (pharmaceutical chemistry and materials technology, respectively). M.B.'s 15 patents belong to six different assignees, one of which is a large public research organization, while the others are medium-sized Italian pharmaceutical companies. Three out of F.P.'s 13 patents are assigned to General Electrics, all the others (but one) to companies within the ENI group.

The most junior of the remaining four academic inventors with high brokerage scores is G.P., a professor of Electronics at the University of Catania (Sicily), which he represents within *Plast_Ics*, an academic consortium supported both by the Ministry for University and Research and by STMicroelectronics, the Italian company with the highest number of academic patents. The consortium lies at the core of the most important high-tech district of southern Italy, the so-called 'Etna Valley', whose origin dates back to the decision of STMicroelectronics to locate some of its R&D facilities in Catania (Santangelo 2004). G.P.'s patents in our database are eight, three of which are for STMicroelectronics, the remaining ones for Italtel, once the manufacturing arm of Telecom Italia (when the latter held the monopoly over Italian telephone services), now a service company jointly shared by Telecom Italia and Cisco Systems.

STMicroelectronics also plays an important role in explaining the high brokerage score of R.Ca., a professor of Electronics at the University of Pavia. All 28 KEINS patents signed by R.Ca. belong to STMicroelectronics. In particular, R.Ca. is the founder of Microlab, a joint initiative of his university and STMicroelectronics that signed contracts also with Conexant Systems, Lucent Technologies and National Semiconductor, among others. More recently, R.Ca. played a key role in the decision of Marvell Semiconductor, a US multinational, to open an R&D centre in Pavia, and to locate its Italian headquarters there.

Also P.C., who heads the pharmacy department of the University of Parma, has been the promoter of a university consortium, Tefarco_Innova, for the production innovative pharmaceutical technologies. Before then, P.C. signed a number of patents for several pharmaceutical companies (ten different assignees for 24 patents).

Finally, R.Co. is the academic 'broker' with the longest experience in foreign universities, first in Cambridge, then in Heildeberg. A molecular biologist, he came back as full professor at his alma mater, the University of Naples, in 1980. In 1990 he also became head of IRBM–Angeletti, a laboratory staffed with over 150 researchers under the control of the pharmaceutical multinational Merck and located in Rome. Most recently (December 2007), he was shortlisted for the presidency of CNR (Centro Nazionale delle Ricerche), the largest public research organization in Italy. All of his six patents in our database are assigned to IRBM–Angeletti. He's the founder of an academic spin-off presently hosted within the premises of a consortium of private and public institutions in Naples.

With the exception of R.Co., all the academic inventors with a brokerage score over ten also have a comparable gatekeeper score. Among those who have a very high gatekeeper score but a broker score of less than ten, five have a gatekeeper score over ten.

The top gatekeeper scorer among them is L.N., a chemical engineer from the University of Naples with a long record as promoter of public and joint public–private research consortia, which ultimately launched into quite a remarkable political career, which culminated in his appointment as Minister for Reform and Innovation in the Public Administration, from 2006 to 2008. Almost all of his ten patents have different assignees, one which is L.N. himself (three are assigned to as many individual inventors: one to the Ministry of University and Research, one to the University of Brighton in the UK, and the remaining to Italian and foreign multinationals).

Another academic 'gatekeeper' of some relevance is M.M., who is presently deputy rector of the University of Urbino and was until recently the president of CIB, the Inter-University Consortium for Biotechnologies, to which 25 Italian universities are affiliated. He also is the assignee of one of his three patents in the KEINS database, alongside with his co-inventors; the other two belong to the European Union and to a local company, respectively.

Finally, G.G. also deserves some mention. He is full professor of telecommunications at the Tor Vergata University of Rome and a leading expert in radar technology, having represented Italy at a number of international organizations involved in air traffic control, such as ICAO (the International Civil Aviation Organization) and Eurocontrol (European Organisation for the Safety of Air Navigation). Four out of his 11 patents are assigned to his university, all the others, but one, to Selenia, now Selex–S.I., a manufacturer of professional electronic systems belonging to the Finmeccanica group.

These short biographical notes suggest that high brokerage and gatekeeping scores are not necessarily associated with in-between positions across companies, but only to in-between positions across co-inventors: several of our 'brokers' have patented just for one or two assignees, although they have worked with many different of the assignees' employees. All brokers and gatekeepers, however, appear also to have consulted for or cooperated with a variety of assignees, both from the business sector and from the public one (public research organizations, universities and consortia).

4.3 How good are brokers at keeping in touch with co-inventors?

Brokerage scores calculated so far amount to structural properties of the network of inventors, and of nodes therein. They do not tell us anything about the quantity and quality of information or knowledge which academic inventors may pass on or receive from co-inventors.

In this section we report the results of three questions asked to academic inventors about their relationships after the patent was applied for. In the first question it was asked whether the inventor and the co-inventor had worked together on any research project after the patent (the question was asked separately for each co-inventor). In case of a negative answer, a second question was asked, about whether the inventor foresaw any chance of future research cooperation with the co-inventor. In case of a further negative answer it was asked whether at least the inventor and the co-inventor were still in touch, for occasional exchanges of information.

Here we summarize answers to the first two questions in a 'cooperation' variable, which takes value one in case of a positive answer to at least one of the two questions, and zero otherwise. Another variable, 'contacts', takes a value of one in case of a positive answer to at least one of all the three questions, and zero otherwise.

Table 7.7 shows that instances of after-patent cooperation are much more frequent when the co-inventor is an academic, rather than an industrial researcher or a student: academic inventors have engaged or foresee to engage in more joint research effort with almost one out of two academic co-inventors, but only with one out of four industrial ones (and a bit less than one out of three students).[9]

As one may expect, academic inventors are still in touch with the vast majority (84 per cent) of their academic co-inventors. They are also in touch with the majority of their (former, at the time of the questionnaire submission)

Table 7.7 After-patent research cooperation and contacts, by affiliation group of the co-inventor (% of co-inventors)

Co-inventor's affiliation	*Research cooperation*§	*Contacts and info. exchanges*§
Academic	47 (135)	84 (128)
Industrial	25 (148)	45 (145)
Student	29 (24)	63 (24)

Note
§ Number of valid answers in brackets.

students (63 per cent), and a much lower percentage of industrial researchers (45 per cent).

A research question of immediate interest is the following: are academic inventors with high brokerage scores any better, or just more interested, at keeping in touch with their co-inventors? A positive answer would suggest that they may be pursuing the brokerage position in order to manage actively and take advantage of it, or at least that they take advantage of it once they reach it.

Logit regressions in Table 7.8 try to answer the question. Units of observations are inventor–co-inventor pairs, and the two dependent variables considered are the 'cooperation' and 'contacts' variables described above. The independent variables are pretty much the same as those summarized in Table 7.5. Once again, parameters do not lend themselves to an immediate interpretation as marginal effects of the regressors, which we report in the comments to the tables.[10]

Specification (1) includes as regressors the academic inventor's scientific productivity, number of patents and age (and related quadratic terms, if significant), as well as dummies for the academic inventor's rank and disciplinary affiliation, and for the co-inventor type (industrial researcher or student, with academics as the reference case). It also includes the academic inventor's brokerage, gatekeeping and coordination scores (liaison scores cannot be included, due to multicollinearity problems).

We notice that the number of patents signed by the academic inventor negatively and significantly affects the probability of the latter collaborating further on research with the co-inventor: obviously, academic inventors with many patents face increasing opportunity costs for maintaining strong ties with all co-inventors than with their colleagues with fewer patents.

We also notice that academic inventors are less likely to maintain research links with industrial researchers, as opposed to academics and students (the coefficient for students is negative but not significant). This result holds regardless of the academic inventor's rank and number of patents.

Finally, we notice that high brokerage and coordination scores negatively affect the probability of further research cooperation with both academic and industrial co-inventors; on the contrary, higher gatekeeping scores increase such a probability. This is tantamount to saying that academic inventors who are central in one of the two groups of co-inventors, but do not stand in between the groups (as gatekeepers do), entertain weaker relationships with their co-inventors. One

Table 7.8 After-patent research cooperation and contacts; logit regressions over brokerage scores and co-inventors' characteristics

	Cooperation – All co-inventors (1)	*Contacts – All co-inventors* (2)	*Contacts – Academic co-inventors only* (3)	*Contacts – Industrial co-inventors only* (4)
Productivity	0.02	0.23**	0.47*	0.16
	(0.07)	(0.09)	(0.26)	(0.12)
No. of patents	–0.44***	0.00	–0.17**	0.48**
	(0.13)	(0.04)	(0.09)	(0.20)
(No. of patents)^2	0.01**	–	–	–0.01**
	(0.00)			(0.01)
Age	–0.01	0.02	0.09	0.04
	(0.03)	(0.03)	(0.07)	(0.04)
Associate prof.	–1.39**	0.38	–0.72	2.13**
	(0.61)	(0.58)	(0.99)	(0.97)
Assistant prof.	–1.38*	–0.57	–1.06	0.24
	(0.71)	(0.66)	(1.13)	(1.00)
Chemistry	–2.41***	–3.18***	–1.25	–6.14***
	(0.54)	(0.73)	(0.81)	(1.71)
Electronics & telecom	0.73	–1.01	1.30	–3.81***
	(0.70)	(0.87)	(1.15)	(1.69)
Pharma	–3.09***	–2.38***	1.88	–6.15***
	(0.64)	(0.77)	(1.25)	(1.68)
Broker	–0.03	–0.07***	–0.04	–0.13***
	(0.02)	(0.02)	(0.04)	(0.04)
Gatekeeper	0.06**	0.08***	0.05	0.15***
	(0.03)	(0.03)	(0.05)	(0.05)
Coordinator	–0.13**	–0.22***	–0.15	–0.47***
	(0.06)	(0.06)	(0.11)	(0.15)
Industrial	–1.31***	–1.91***	–	–
	(0.35)	(0.33)		
Student	–0.90	–1.35**	–	–
	(0.58)	(0.56)		
Constant	4.46**	2.34	–2.95	0.65
	(2.08)	(2.11)	(4.60)	(3.09)
Obs	337	325	141	160
LogL	–147.6	–157.6	–44.6	–74.6
LR chi^2	134.68***	117.97***	54.01***	71.31***
Psuedo R^2	0.31	0.2724	0.3769	0.3233

possible explanation is that academic inventors with a high brokerage score and a low gatekeeping score may in fact be consultants, whose patents stem from several short-lived collaborations with industry. On the contrary, gatekeepers whose ties with industry arise alongside those with the rest of academia via public–private ventures and consortia maintain more stable relationships with all co-inventors. As for coordination scores, we know that they are much lower than brokerage and gatekeeping scores, which possibly indicates that academic inventors with high coordination and low gatekeeping scores (that is, pure coordinators) have fewer patents, which in turn arise from more occasional projects.

Finally, we note that discipline dummies are very important: a median academic inventor from chemistry or pharmaceutics has 50–64 per cent less probability of maintaining research ties with co-inventors than a similar academic inventor from biology or electronics and telecommunications. Marginal effects for assistant and associate professors, compared to full professors, are around −22 per cent, which is also the marginal effect for co-inventors being industrial ones.

Columns (2) to (4) regression exercises parallel those in column (1), but with reference to looser contacts, such as those for mere information exchanges. Here we note a different impact for scientific productivity and number of patents. The former has a significant coefficient in all specifications except (4), the latter on the contrary appears significant only in this specification. This suggests that highly productive academic inventors tend to maintain contacts with more co-inventors than less productive ones, but this holds for academic co-inventors and not for industrial ones. At the same time, having many patents is associated with keeping in touch with many more co-inventors from industry, but not from university (in specification (3) the sign of the coefficient for the number of patents is indeed negative and significant); however, such contacts are not for research, but for more generic exchanges of information.

The effect of academic rank on the probability of maintaining contacts with co-inventors is less marked than in regressions on research cooperation: full professors do not appear to behave differently from assistant professors, while associate professors seem more likely to maintain contacts than colleagues from both the other ranks, but the effect is significant only in specifications (2) and (4). Effects of disciplines are significant in all models except (3), which suggests they mainly refer to relationships with industrial co-inventors (that is, the probability of maintaining contacts with academic co-inventors does not vary across disciplines).

Coefficients in model (2) suggest that contacts and information exchanges, like research cooperation ties, are maintained with different probabilities according to the type of co-inventor. In this case, contacts with both industrial and student co-inventors are less likely to be maintained than those with academics.

The impact of brokerage, gatekeeping and coordination scores differ across types of co-inventors. In specifications (2), which refers to all co-inventors, and (4), which refers only to co-inventors from industry, all coefficients are significant. However, contacts with university co-inventors are maintained regardless of the network position of the academic inventor. Overall, this suggests that gatekeepers, as opposed to pure brokers or coordinators, do a better job of keeping in touch with industry, while relationships within the academy do not depend on the scientist's position within the network of inventors.

5 Conclusions

In this chapter we have adapted a set of brokerage and gatekeeping measures proposed by Gould and Fernandez (1989) that may be of help in clarifying the relationship between academic inventors and their co-inventors, whether from industry or from the academic or student body.

We find out that brokerage and gatekeeping positions are very few, and they are held by scientists with both a large number of patents and a strong publication record. Brokers and coordinators are not especially better than colleagues at further cooperating on research with co-inventors after the patent experience. Gatekeepers, do better, especially when it comes to keeping in contact with industrial researchers. More generally, network ties between academic and industrial researchers may be short-lived as far as knowledge exchanges are concerned, but may serve well other purposes.

The joint reading of our quantitative evidence and the top brokers' biographical notes suggests that the latter manage actively their relationships outside university. Some of them, especially those who have signed patents only for one or two different assignees, are likely to keep in touch mainly for research or research-funding purposes. Others, such as those academic inventors with many different assignees and/or many assignees such as public consortia and the likes, may nurture their personal links outside universities for more strategic purposes. The existing literature on university patenting has focused almost exclusively on academic inventors' monetary incentives (Lach and Shanckerman 2003; see also Stern 2004, for a more general treatment). Here we find that the social contacts gained through collaboration with industry may be part of the reward, as they help boosting the academic inventors' reputation and career both inside and outside the university.

Future research will have to test further this hypothesis through the collection of longitudinal data on academic scientists' careers, and more detailed questionnaire data on the exact contents of the information exchanges between academic inventors and their co-inventors, both during and after their collaboration on the patent. Longitudinal data will help testing whether social contacts with industrial co-inventors do help academic inventors throughout their career, while refined questionnaire data will also help building a directed graph, thus allowing more precise applications of the brokerage concepts and scores.

A related research line will possibly investigate benefits enjoyed by students of academic inventors with high brokerage scores. Data from the KEINS database for Italy have already been exploited to find out whether master's students supervised by academic inventors have easier access to the labour market, with positive results (Zinovyeva and Sylos Labini 2007). Inclusion of brokerage scores and extensions to other countries will help confirming the general validity of this result.

Notes

1 Università di Brescia and KITeS–Cespri–Università Bocconi.
2 US universities have been entitled with the IPRs over federally funded research since 1980, with the introduction of the Bayh–Dole Act, which has been imitated only recently by a few European countries. Some of the latter, in addition, held for long (or still hold now) the so-called 'professor's privilege', by which all inventions produced by university professors do not belong to their employers (as it is the norm with business companies and their R&D staff), but to the professors themselves, who dispose

of them freely. And even in countries where the 'privilege' never existed, university professors are civil servants under direct control of the state, whose disclosure duties towards their universities are rather unclear. More on this in Lissoni *et al.* (2008).

3 The major limitation of the KEINS database for Italy is that it includes only those professors and researchers who had passed a competitive examination for a tenured position (from now on, we will refer to them simply as 'professors'). Thus our data miss the large number of fixed-term appointees who, at the time, had been working in one or more universities for one or more years, as well as all the PhD students, post-doc fellows and technicians. In the current Italian system, assistant professor (called 'researcher') and associate professor positions, despite being only the first two steps of the academic career ladder, are not offered as fixed-term appointments, but as tenured ones. The main differences with the position of full professor lie in wage and administrative power and responsibilities.

4 For a more general treatment of the advantages and disadvantages of using archival data, as opposed to questionnaire results, as inputs for social network analysis, see Burt (1983).

5 In the absence of direction, it is impossible to distinguish between gatekeepers and representatives.

6 The ego-networks of inventors *224212* and *135223* share a common co-inventor/node, which explains the line linking them.

7 We report the results of the regression including 'late productivity' instead of 'productivity' only for the case with the broker score as the dependent variable. Results with different dependent variables are pretty much the same.

8 The academic inventors mentioned in this section checked and amended their biographical information, after a brief email exchange.

9 The number of co-inventors for which valid answers have been obtained is very low compared to the number of co-inventors whose affiliation could be retrieved. This is due to the fact that, even in the absence of valid answers to the questionnaire by one academic inventor, some information on his co-inventors' affiliation could be obtained by other academic inventors who also knew those co-inventors. It is also due to the fact that the questionnaire was rather long, and questions on after-patent cooperation were asked at the end of the interview, when many interviewees (especially those with many co-inventors) said they had run out of time and declined to answer any more questions.

10 They are available on request.

References

Balconi, M., Breschi, S. and Lissoni, F. (2004) 'Networks of inventors and the role of academia: an exploration of Italian patent data', *Research Policy*, vol. 33 (1): 127–45.

Breschi, S., Lissoni, F. and Montobbio, F. (2007) 'The scientific productivity of academic inventors: new evidence from Italian data', *Economics of Innovation and New Technology*, vol. 16 (2): 101–18.

Burt, R.S. (2001) 'Structural holes versus network closure as social capital', in Lin, N., Cook, K. and Burt, R.S. (eds) *Social Capital: Theory and Research*, New York: Aldine de Gruyter.

Crespi, G., Geuna, A. and Verspagen, B. (2006) 'University IPRs and knowledge transfer. Is the IPR ownership model more efficient?', mimeo, SPRU, University of Sussex and Ecis, Eindhoven University of Technology.

Czarnitzki, D., Glänzel, W. and Hussinger, K. (2007) 'Patent and publication activities of German professors: an empirical assessment of their co-activity', *Research Evaluation*, vol. 16 (4): 311–19.

Franzoni, C. and Lissoni, F. (2007) 'Academic entrepreneurship, patents, and spin-offs: critical issues and lessons for Europe', in Varga, A. (ed.) *Universities and Regional Economic Development*, Cheltenham: Edward Elgar.

Freeman, L.C. (1979) 'Centrality in social networks: conceptual clarification', *Social Networks*, vol. 1: 215–39.

Geuna, A. and Nesta, L.J.J. (2006) 'University patenting and its effects on academic research: the emerging European evidence', *Research Policy*, vol. 35 (6): 790–807.

Gould, R.V. and Fernandez, R.M. (1989) 'Structures of mediation: a formal approach to brokerage in transaction networks', *Sociological Methodology*, vol. 19: 89–126.

Iversen, E.J., Gulbrandsen, M. and Klitkou, A. (2007) 'A baseline for the impact of academic patenting legislation in Norway', *Scientometrics*, vol. 70 (2): 393–414.

Lach, S. and Shankerman, M. (2003) 'Incentives and invention in universities', NBER Working Paper no. 9727, National Bureau of Economic Research, Cambridge, MA.

Lissoni, F., Sanditov, B. and Tarasconi, G. (2006) 'The Keins database on academic inventors: methodology and contents', CESPRI Working Paper no. 181, Università Bocconi, Milano.

Lissoni, F., Llerena, P., McKelvey, M. and Sanditov, B. (2008) 'Academic patenting in Europe: new evidence from the KEINS database', *Research Evaluation*, vol. 16: 87–102.

Meyer, M., Sinilainen, T. and Utecht, J.T. (2003) 'Towards hybrid triple helix indicators: a study of university–related patents and a survey of academic inventors', *Scientometrics*, vol. 58 (2): 321–50.

Murray, F. (2004) 'The role of academic inventors in entrepreneurial firms: sharing the laboratory life', *Research Policy*, vol. 33 (4): 643–59.

Nicolaou, N. and Birley, S. (2003) 'Academic networks in a trichotomous categorisation of university spinouts', *Journal of Business Venturing*, vol. 18 (3): 333–59.

Santangelo, G.D. (2004) 'FDI and local capabilities in peripheral regions – the Etna Valley case', *Transnational Corporations*, vol. 13 (1): 73–106.

Stern, S. (2004) 'Do scientists pay to be scientists?', *Management Science*, vol. 50 (6): 835–53.

Taube, V.G. (2004) 'Measuring the social capital of brokerage roles', *Connections*, vol. 26 (1): 29–52.

Verspagen, B. (2006) 'University research, intellectual property rights and European innovation systems', *Journal of Economic Surveys*, vol. 20 (4): 607–32.

Vuong, Q.H. (1989) 'Likelihood ratio tests for model selection and non-nested hypotheses', *Econometrica*, vol. 57 (2): 307–33.

Zinovyeva, N. and Sylos Labini, M. (2007) 'The relationship between academic research, teaching quality and graduates' employment outcomes', paper presented at the EALE Annual Conference, Oslo, 20–22 September.

8 New innovators and knowledge-intensive entrepreneurship in European sectoral systems

A field analysis

Camilla Lenzi, Kate Bishop, Stefano Breschi, Guido Buenstorf, Patrick Llerena, Franco Malerba, Maria Luisa Mancusi and Maureen McKelvey

1 Introduction

This chapter is about the characteristics and factors of survival and success of knowledge-intensive entrepreneurship in some sectoral systems in Europe. Literature and evidence on the US case are rather rich and detailed but the European case is relatively less explored and documented. Also, in both cases, existing studies are concentrated on a limited number of sectors. They rarely take on a cross-sector cut. Differently, this chapter proposes a cross-sector and cross-country comparison at the European level with a specific focus on some key elements in the process leading to the birth, start-up and early evolution of a new firm.

In particular, this chapter reports and comments on the results of a survey conducted on a sample of companies defined as new innovators, i.e. recently founded firms able to innovate soon after entry into the market, and active in selected technologies characterized by strong innovative and competitive dynamics in the last 15 years: biotechnology, electronics and medical devices.

In the rest of the chapter, we first present the research streams that inspired this work and from which we draw the main dimensions to be analysed (Section 2). We next present the methodology used in this study (Section 3) and describe and comment the results of the questionnaire (Section 4). Finally, we present some concluding remarks (Section 5).

2 The main variables and dimensions considered

The literature on entrepreneurship and new ventures creation has largely investigated the determinants and factors of success of (innovative) start-ups and has tried to assess their relative importance, especially in technologically highly dynamic contexts in which innovative capabilities are crucial.

These factors include, among others, the human and intellectual capital of the founders and the personnel available to the new venture, the access to external sources of financing, the intellectual property rights (IPRs) portfolio, the participation of a network in collaboration with other companies (especially the founder's parent company), as well as other institutional actors (such as universities and research organizations).

The human capital of founders, namely their intellectual assets, is found to be a key element of a start-up success (Kogut and Zander 1996; Pfeffer 1994; Uzzi 1996). In parallel, founders' lack of experience in the field or founders' managerial incompetence are indicated as the main determinants of organizational failures (Boswell 1972; Dun and Bradstreet 1981). Consequently, personal talent, experience, skills and the parent firm inheritance and knowledge endowments an entrepreneur brings to the new venture are expected to influence new venture economic survival and success (Klepper and Sleeper 2005). More precisely, the higher a founder's intellectual assets the greater the chances of a new venture success in terms of both survival and economic performance (Bruderl *et al.* 1992; Uzzi 1996).

In knowledge-intensive and technologically sophisticated contexts, these considerations are particularly relevant. In fact, survival and successful adaptation require a high level of competences, skills and appropriate techniques (Nelson and Winter 1982). Part of this knowledge is theoretical – e.g. an engineer's understanding of the physical laws – part is the outcome of long personal experience, and part is obtained from the experience of others, through direct contact with their work (Winter 2006).

The resource-based view of the firm suggests that the role of financial capital should not be disregarded and that its interplay with intellectual capital sources is crucial to a firm's performance (Penrose 1959). In fact, during the early stages of development new companies need a great amount of financial resources in order to proceed with technological improvement and product marketing; they also need to acquire the critical tangible and intangible assets necessary to better and faster develop products and reach the market (Dierickx and Cool 1989). However, new ventures frequently lack legitimacy and are likely to experience unfavourable conditions of access to external resources, especially when they are engaged in risky lines of business (Lee *et al.* 2001). More precisely, new ventures face some disadvantages in comparison to incumbent firms, which have more legitimacy, a history of transactions and established relationships with the relevant actors (Dollinger 1985).

In particular, Venture Capital (VC) funding is found to strongly influence high-tech and innovative firms' performance. Hellmann and Puri (2000) indicate that VC-backed companies perform more radical product and process innovations and are faster in developing new products, thus reducing the time to market and the acquisition of competitive advantages. Similarly, Kortum and Lerner (2000) point out that VC-backed companies produce more valuable patents than other companies. VCs have indeed become one of the most important sources of funds also because of the consulting role and the provision of technical and managerial know-how (Barry *et al.* 1990; Kaplan and Stromberg 2000). Especially the monitoring and mentoring activities of VCs are critical for the success of many new ventures run by inexperienced entrepreneurs and complement the pure capital provision (Bottazzi and Da Rin 2002).

Moreover, the literature on market for technologies emphasizes the importance of IPRs in order to found a new innovative firm. In highly dynamic technological contexts, firms frequently are created around a patent or a set of patents that thus represent critical assets for the new ventures. In fact, when IPRs are strong and well enforced, it is possible that new firms enter the market by specializing in the development and sale of specific technologies. This strategy ultimately allows avoiding the costs and risks of large-scale investments necessary to move to the production and marketing phase (Arora and Merges 2004; Gans *et al.* 2002). Moreover, patents may provide a further source of assessment and an overview on the technological competences and economic value that might favour the establishment of partnerships and the access to external funding (Giuri *et al.* 2007).

Furthermore, empirical evidence indicates the increasing necessity of individual firms to engage in formal and informal agreements with other actors (i.e. networks) in order to collect and eventually share the relevant resources to production and innovation activities (Burt 1993).

Specifically in knowledge-based sectors, these networks are growing in terms of number of actors involved and of resources exchanged, thus generating more and more complex structures and new configuration forms. Indeed, the increasing complexity of knowledge-intensive sectors (and their relative knowledge base) implies an increasing dispersion and distribution of resources, knowledge and competencies among different actors. Thus, it is more and more important to set up agreements in order to collect and exchange resources and, ultimately, to coordinate the innovative activities and produce more and more sophisticated and knowledge- and technology-intensive products and services (Powell and Grodal 2005).

Finally, the highly systemic nature of innovation and technology and the different and distributed nature of the resources that entrepreneurs need to gather to launch a new venture, suggest that a systemic view can be very fruitfully adapted and exploited to study entrepreneurship phenomena and their relation to innovation. In this chapter in particular, we retain two systemic perspectives.

First, a sectoral system perspective. The empirical observation actually indicates that the rate and type of innovation and the organization of innovative activities (Malerba 2002) and the number of firms entering a new technical field (Breschi *et al.* 2000) greatly differ across sectors. Accordingly, also entrepreneurial ventures carrying out innovative activities might considerably differ across sectors in terms of the differential endowments, the key characteristics, resources and linkages they are able to activate and exploit in the early stages.

Second, a national system perspective. The empirical observation indicates that the rate and type of innovation and the organization of innovative activities greatly differ also across countries as well as entrepreneurial activities, which are a crucial function of national innovation systems (Nelson 1993).

All these streams of literature inspired our analysis and allowed us to identify the main dimensions explored in the questionnaire, i.e. the intellectual and human capital of the founders, the financial resources, the technological competencies and IPRs, the relationships with other firms and organizations.

This chapter, thus, focuses on these four dimensions and their relative relevance to the creation of a new firm in a cross-country and cross-sector perspective. More specifically, the analysis is conducted on a group of European new and innovative firms in technological contexts that experienced strong innovative and competitive dynamics in the last 15 years. New and innovative firms, i.e. new innovators, are defined as firms that innovate within a relatively short time after their establishment and are both knowledge-intensive (i.e. active in science-based and science-driven sectors) and technology-based (i.e. patenting in sophisticated and dynamic technological contexts). The next sections detail the methodology implemented and the results of the questionnaire.

3 Methodology

The data collection procedure included two different and succeeding phases. The former was primarily aimed at identifying the object of our analysis (i.e. new innovative firms) while the latter aimed at gathering additional data through the administration of a questionnaire addressed to the identified new innovative firms.

The identification of new innovator firms is based upon the combination of multiple criteria. In particular, we selected a group of firms applying for their first patent from 1990 onwards in specific technological classes and in a very short time after their establishment (we chose a five-year maximum lag between entry into the market and first patent application). Therefore, the selection criteria adopted lead to a sample of very young firms, mostly founded between 1995 and 2000; accordingly, our focus is on the start-up phase and the early evolution of the identified firms.

In more detail, we have identified a group of highly dynamic and growing technological classes according to the International Patent Classification (IPC) such as biotech, catalysis, lasers, mobile communication, stores, robots, surfate technologies, pharma-surgery, pharma-filters and prostheses, pharma-measuring and identification, pharma-electrotherapy, switch, image processing.

Next, we extracted from the EP–Cespri (now KITeS)[1] database the names of all the applicants from Italy, France, Germany, Great Britain, Denmark and Sweden that applied for their first patent at the European Patent Office (EPO) in each of the classes reported above after 1990.

The identification of new innovative firms could not be done *ex ante* because the EP–Cespri (KITeS) database contains no information on the origin and the type of firm that applies for a patent. We excluded from the analysis divisions/subsidiaries of established companies, joint ventures, universities, public research centres, hospitals, individual assignees, non-profit organizations, associations and foundations. For the remaining cases (i.e. truly private and independent companies), we searched all the available sources, such as Google, international company databases (e.g. AMADEUS), national company databases (e.g. AIDA,

DIANE, FAME, etc.), and collected the date of entry and exit, if any, in order to establish the time lag between entry into the market (year of foundation) and technological entry (year of the first patent). We defined the subset of new innovative firms as those firms patenting in the first five years after foundation.

Table 8.1 shows for each country the percentage of firms patenting in the first five years after foundation in the selected technological classes. These data show different patterns of patenting/innovating according to the country. Scandinavian countries exhibit the highest rate of firms innovating in the first years after entry into the market. Germany, Great Britain and France have similar rates, about 60 per cent. This means that the bulk of innovative firms innovate soon after entry: technological entry and market entry are very close in time. On the contrary, in Italy only 17 per cent of firms innovate soon after entry.

We next searched for at least one of the founders of new innovators and, most importantly, a contact for him/her (possibly an email address or a phone number), in order to be able to send them a questionnaire. The questionnaire was aimed at collecting further information on their experiences and characteristics. Founders were chosen as the target of the questionnaire because of their deep involvement and knowledge of the founding process and the start-up phase.

We addressed a questionnaire to all the founders of the selected companies. We were able to identify at least one of the founders of the selected firms in 344 cases out of 421 and to collect 376 contacts (including email addresses, company and personal telephone numbers).

The questionnaire was administrated between October and December 2006 and included a follow-up round of emails. We decided to implement an Internet questionnaire and a dedicated website was reserved and prepared in order to host it.

We made contact with the founders based on the first patent the firm applied for at the EPO and submitted the questionnaire by email. The email text described the research project and provided the respondents with the link to the questionnaire webpage and a personal and unique username and password in order to access and fill it in.

The response rate was 22 per cent, but there were considerable differences across countries.[2]

Table 8.1 Distribution and share of firms patenting in the first five years after entry into the market by country

Country	*Number of firms*	*Firms patenting in the first five years*	*Percentage*
Germany	232	138	59.5
Denmark	25	16	64.0
France	168	95	56.6
UK	119	72	60.5
Italy	88	15	17.1
Sweden	111	85	76.6
Total	743	421	56.7

4 Results

The final sample is composed of 46 firms from Germany, 16 from Denmark and Sweden, ten from France, 21 from the UK and six from Italy. They operate in different sectors: biotech (28); laser (ten); mobile communication (nine); stores (two); pharma-surgery (nine); pharma-filters prostheses (six); pharma-measuring identification (15); pharma-electrotherapy (four); switch (ten); image processing (six).

Given the reduced size of the sample in a few countries, we decided to group UK and Scandinavian firms (Northern Europe) as well as French and Italian ones (Latin Europe). Germany is considered alone.

Similarly, technological classes are grouped in three macro-sectors: biotech; electronics (which includes laser, mobile communication, stores, switch and image processing firms); and medical devices (which includes pharma-surgery, pharma-filters prostheses, pharma-measuring identification and pharma-electrotherapy firms). Table 8.2 shows the distribution and share of firms in the three technological sectors and geographical regions.

The analysis is developed according to questionnaire's structure. Two main areas of investigation are identified. The first is about firms and focuses on the origin and the assets exploited in the early stages (i.e. financial capital, IPRs and labour assets), and on the early evolution and the most important factors of success at present (i.e. networks, markets, products and customers). The second is about founders and examines their educational and professional backgrounds, as well as their entrepreneurial attitude and activity.

4.1 The firm

The dynamics of knowledge-intensive entrepreneurship

The firms in our sample are young and small.

In terms of firm age, 67 per cent of the firms are between five and ten years old; this suggests that most of them have been founded in the second half of the 1990s. One-third of the sample is composed of firms founded before 1990 or after 2000.

In terms of firm size, we have identified six different groups on the basis of their number of employees: 1–9 (46 per cent), 10–19 (17 per cent), 20–9 (13 per cent), 30–9 (10 per cent), 40–9 (5 per cent), >50 (8 per cent). Most of firms have

Table 8.2 Distribution and share of firms by sector and region

Regions	*Biotech*	*Electronics*	*Medical devices*	*Total*
Germany	17 (60.7%)	13 (135.1%)	16 (47.1%)	46
Northern Europe	7 (25%)	19 (51.4%)	11 (32.4%)	37
Latin Europe	4 (14.3%)	5 (13.5%)	7 (20.6%)	16
Total	28	37	34	99

ten employees at maximum; this result is primarily driven by firms' size in the medical devices sector (where 63.64 per cent of firms belong to the first category).

Empirical evidence shows that a great proportion of newly founded firms exit from the market in a rather short time and that the rate of exit does not decrease over time (Geroski 1995). Accordingly, some firms included in the sample are no longer active in the market or have changed their organizational and structural profile.

The questionnaire provides data about three different types of 'exits': merger and acquisition (M&A), name change and bankruptcy. M&A are more frequent in the biotech and electronic sectors, while medical devices firms present a relatively higher percentage of bankruptcies. In particular, 50 per cent of M&A and 50 per cent of bankruptcies are, respectively, in the electronic and medical devices sectors. Only ten firms went bankrupt (12.4 per cent of the firms): seven from Germany, two from Northern Europe, and only one from Latin Europe. This shows a prevalence of German firms among those that exited the market. M&As are more frequent in Northern Europe.

The role of finance

Easy and adequate access to funds is crucial to new firms' foundation, viability and performance (Cooper *et al.* 1994; Montgomery *et al.* 2005). Almost 70 per cent of the firms in the sample received financial support from an external organization in the start-up phase. However, there are clear differences between Northern and Southern regions; while in the former the percentage of firms that received financial support is 86 per cent, in the latter it is only 44 per cent. This reveals that the financial system provides a weaker support to entrepreneurship in these countries.

The main financial sources used at the time of foundation and during the early stages of the new company are VC (considered of high importance) and national government (considered of medium/high importance). Actually, the empirical evidence suggests that VC is key since it has a multiplicative effect with technological capabilities (Lee *et al.* 2001). On the other hand, the EU and banks have a marginal role in funding new companies.

The comparison across geographical areas signals the different degrees of development of financial markets of European countries. For instance, in the Northern region there is no entrepreneur who considers banks as highly important while VC is more frequently indicated as a highly important source of financial support. This may suggest a higher degree of development of financial markets. Differently, in the Latin region only 19 per cent of the respondents consider VC as highly important (along with local and regional authorities and national government) while 68 per cent consider it not relevant. In Germany, the use of the VC has not reached yet the levels of the Northern countries while local and regional authorities, national government and banks still represent very important sources of funding. This seems to suggest that financial markets are more developed than in the Latin region but less than in the Northern one (Table 8.3).

Interestingly, 25 per cent of entrepreneurs from Italy and France consider the support of the EU of medium importance. This could highlight a relatively higher value of EU policies in the Southern regions of Europe.

The three sectors differ in terms of financial sources used. Figure 8.1 shows that VC is fundamental in all sectors but especially in biotech; more than 60 per cent of biotech firms consider VC as an important source of funds compared to 44 per cent in electronics and 38 per cent in medical devices. Differently, banks are particularly relevant in the medical devices sector; 38 per cent of medical devices firms consider banks as highly important compared to 11 per cent in biotechnology and 14 per cent in electronics.

Intellectual Property Rights (IPRs)

IPRs can represent a very important asset to establish a new innovative company. In technologically highly dynamic contexts, firms are frequently established around a patent or a group of patents that consequently are crucial asset of the newly established company (Gans *et al.* 2002; Arora and Merges 2004). Patents actually represent, together with human capital, a measure of the new firm technological capabilities. Thus, patents can also provide a first assessment and outlook on a firm's value and technological standing of relevance to engage in commercial and research partnerships and to obtain external funds (Giuri *et al.* 2007). The questionnaire investigates this aspect by looking at the importance of the initial patent and licensing.[3]

The initial patent and licensing are considered of high importance to the founding of a new firm respectively by 43 per cent and 22 per cent of the founders interviewed. Northern Europe has the highest percentage of respondents (57

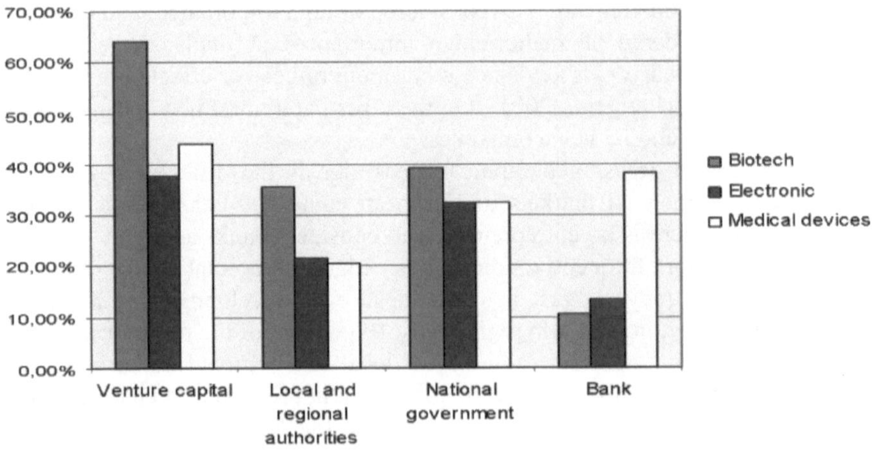

Figure 8.1 Importance of financial sources by sector (high and medium importance).

Table 8.3 Importance of financial sources by region (%)

Financial sources	*High importance*			*Not relevant*		
	Germany	*Northern Region*	*Latin Region*	*Germany*	*Northern Region*	*Latin Region*
Venture capital	30.4	56.8	18.8	47.8	32.4	68.8
Local and regional authorities	17.4	5.4	18.8	39.1	62.2	75.0
National government	15.2	10.8	18.8	43.5	43.2	56.3
EU Bank	13.0	0.0	6.3	43.5	78.4	50.0

per cent) who consider the initial patent as a very important asset to found a new company compared to Germany (35 per cent) and Latin Europe (38 per cent). Consistently, the percentage of entrepreneurs that consider patents not relevant is rather little. Also licensing is characterized by a similar pattern.

This is of relevance as firms in the sample are actually selected on the capacity to rapidly obtain the first patent. Also, this evidence confirms that in Northern Europe firms are more familiar with the use of IPRs in the form of patent protection as compared to Southern Europe ones, which is consistent with recent data from the European Innovation Scoreboard 2007.

The initial patent is considered of high relevance by almost half of the respondents and medium–highly relevant by 63 per cent of them. Patent protection is actually effective and largely exploited in all the sectors considered, but still considerable differences among them persist. The initial patent is especially important in biotech where it is considered of high or medium importance by 82 per cent of respondents as compared to 64 per cent in medical devices and 46 per cent in electronics. Licensing is characterized by similar differences across sectors. This suggests that the importance of IPRs varies according to sectoral specificities and the stage of development of market for technologies (Cohen *et al.* 2000).

Finally, biotech and electronic firms tend to innovate earlier than medical devices firms. In fact, most of them apply for the first patent one year after their foundation, while medical devices firms tend to patent from two years after their foundation.

Labour assets

Skilled and specialized labour and star scientists are a key resource in order to found a firm and can be considered as a proxy of its technological capabilities as well.

Access to specialized labour is widely perceived as one of the most important assets in all technological areas and only 3 per cent of respondents affirm that it is not relevant. Star scientists are relatively less important: only 31 per cent of respondents consider them as highly important and 21 per cent say that they did not play any role in founding a firm.

There are, however, sizeable differences across sectors: for instance, 75 per cent of biotech entrepreneurs consider star scientists of medium–high importance as compared to 24 per cent in electronics.

Also, a lower percentage of entrepreneurs from Northern Europe (62 per cent) consider the access to specialized labour as a highly important asset as compared to Germany (80 per cent) and Latin Europe (75 per cent) and a higher percentage of German respondents consider star scientists as not relevant as compared to the other two regions (respectively 30 per cent vs. 13 per cent in both Northern and Southern Europe).

Networks

The network of relationships in which a firm is embedded may have a strong impact on its activity (Thorton and Flynne 2003). Knowledge is increasingly complex and resources to set up and to run production and innovation activities are multiplying. As a consequence, they are increasingly fragmented and dispersed among different actors, thus requiring agreements and relationships to exchange knowledge and resources. Three different types of networks are explored: networks based on alliances and partnerships among companies, formal and informal links with universities and contractual agreements with the previous employer, if any.

On average, alliances and partnerships are considered as highly important (32 per cent of firms). This confirms the increasing trend in the number of partnerships recorded since the 1980s, especially in high-tech sectors (Hagerdoon 2002). This appears more important in Latin Europe (44 per cent) than in Germany (30 per cent) and Northern Europe (28 per cent). Instead, formal and informal links with universities are considered as highly important by 31 per cent of respondents in Northern and Latin Europe, while this percentage decreases to 24 per cent in Germany. However, more than 60 per cent of French and Italian firms consider them to be irrelevant: 31 per cent of companies have links with universities and consider them as very important but more than 60 per cent have no relationships with academic centres. This is consistent with several indicators available from the OECD Scoreboard or the European Innovation Scoreboard, such as patent citations to scientific literature, partnership with public research centres and university patents, that report greater difficulties in technology transfer in these countries, especially in Italy.

Collaborations with previous employers are maintained by 34.4 per cent of the firms, but relatively more frequently in Northern Europe (45 per cent) than in Germany (25 per cent) and Latin Europe (20 per cent). Founders previously employed at universities are more likely to maintain relationships and this occurs in 64 per cent of the examined cases, compared to 8 per cent of founders previously employed in the private sector. Consistently, the main area of relationships between the new firm and their founders' previous employers is R&D while it is significantly lower in other areas of activities (i.e. production, marketing and commercialization/distribution) where it ranges from 9.5 per cent to 12.7 per cent. This highlights that academic entrepreneurs consider the maintenance of linkages to the scientific community of relevance to ensure greater innovative performance (Cockburn and Henderson 1998).

Finally, at the sectoral level, there is a relevant percentage of entrepreneurs (50 per cent) who maintain relationships with their previous employers only in the biotech sector; in electronics and medical devices this percentage is about 30 per cent. This is consistent with the previous finding. Indeed, 75 per cent of founders in biotech come from academies and these are precisely those cases where relationships with the previous employer are more frequently maintained.

Markets, products and customers

The last part of the analysis is about markets, products and customers. Table 8.4 illustrates the relevance of products, markets and customers as sources of competitive advantages for firms. Uniqueness of the product/technology/knowledge is considered as the most important factor for the success of a firm. Marketing skills and customer services are also very important. This ranking does not differ across geographical areas.

However, there is a clear difference between biotech on the one side and electronics and medical devices on the other. In fact, while uniqueness of the product/technology/knowledge is important for nearly all firms in all sectors, marketing skills and especially customer services are more relevant in electronics and medical devices, thus suggesting the need of demand-related competencies to successfully compete in these sectors (Bresnahan and Malerba 1999; Gelijns and Rosenberg 1999).

A similar pattern emerges according to the type and the importance of the knowledge acquired during the founder's previous employment. In fact, while knowledge about product and technology is important in all sectors, consistent with the selection criteria of the sample, knowledge about demand is relatively less relevant and it is considered as highly important by 65 per cent of respondents. However, this also suggests that demand too plays some role in affecting the performance of start-ups. In fact, in a lot of sectors start-ups link with some key users with which they interact closely during the early years of existence (Von Hippel 2005). In addition, start-ups often focus on some market segments and niches that provide some shelter from competition from incumbent firms (Thompson and Klepper 2003). Actually, some interesting differences among sectors emerge in this regard (Figure 8.2). Knowledge about customers is more important for medical devices and electronic firms, where products are frequently conceived and designed to meet users' needs and to accommodate their requirements (Von Hippel 2005), but far less in biotechnology where products are frequently far from the commercialization stages. However, the relatively lower importance of managerial and organizational competencies and knowledge can also be affected by the initial choice to focus on technology-based sectors.

Table 8.4 Factors of success (%)

Factors of success	*Relevance*			*Not relevant*
	High	*Medium*	*Low*	
Uniqueness of the product/technology/knowledge	85.2	12.4	0.0	2.5
Marketing skills	46.9	32.1	16.1	4.9
Customer service	34.6	32.1	21.0	12.4

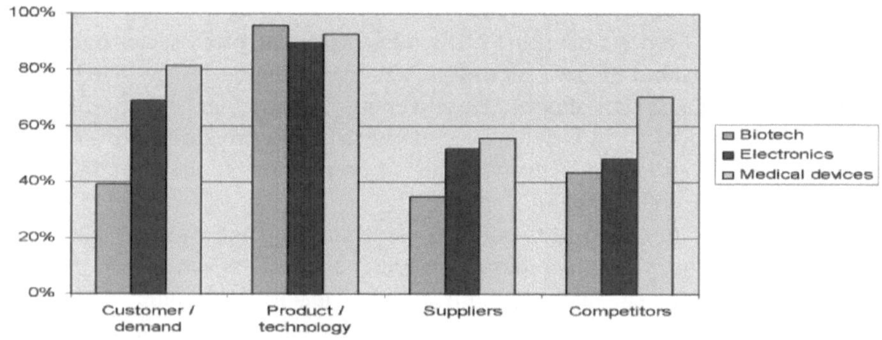

Figure 8.2 Type and importance of the knowledge acquired by sector (high and medium importance).

4.2 The founder

The second part of the questionnaire is about the founders of the new firms, their educational attainment (i.e. academic qualification), their professional background and knowledge and technological inheritance from their previous employment, and their entrepreneurial capacity (i.e. further entrepreneurial experiences, that is serial entrepreneurship).

Table 8.5 reports on founders' academic qualification in the sample and in each of the sectors considered. In general, founders exhibit a very high educational profile in all sectors. Not surprisingly, however, there is a relatively higher percentage of biotech firms' founders with a university PhD, consistently with the relatively greater importance of scientific knowledge base in this sector (Malerba 2002). This also confirms the importance of founders' intellectual capital both as advanced knowledge endowments and considerable individual competences, which both are key to the identification and exploitation of rich and profitable technological and market opportunities (Bates 1990; Klepper and Sleeper 2005; Shane 2000).

While in Germany and in Northern Europe most of the founders have a university PhD, in France and Italy they have a university master. This suggests a slightly lower educational profile of entrepreneurs in the two Latin Countries,

Table 8.5 Academic qualification by sector (%)

Academic qualification	*All sectors*	*Biotech*	*Electronic*	*Medical devices*
University PhD or equivalent	51.95	75	37.9	50.0
University Master's or equivalent	32.47	15	44.8	32.1
University BA, BS or equivalent	6.49	5	6.9	7.1
High school	5.19	–	6.9	7.1

consistently with OECD statistics on tertiary education and post-lauream training.

Also, Figure 8.3 reports the share of founders by country and sector of previous employment. About 85 per cent of founders were previously employed, generally either in the private sector or at a university. In particular, in Latin Europe most of the founders were previously employed in the private sector (70 per cent). Differently, in Germany and Northern Europe most of the entrepreneurs come from the university system.

These findings, thus, suggests some consideration on the linkages between university and industry. The number of academic spin-offs is actually an indicator of technology transfer and of a university's capacity to commercialize the results of the scientific research. This is again consistent with several indicators available from the OECD Scoreboard or the European Innovation Scoreboard, such as patent citations to scientific literature, partnership with public research centres, university patents, that reports greater difficulties in technology transfer in these countries, especially in Italy.

There are also considerable differences across sectors. In biotechnology, 75 per cent of founders come from university as compared to 62 per cent and 52 per cent respectively in electronics and medical devices. This further stresses that biotechnology is strongly much more linked to the scientific knowledge base as compared to the other sectors that rely on a more diversified knowledge base.

The main motivations to found a firm change according to the different sector of previous employment (Table 8.6). Higher income, independence and disagreement with previous employer are more important in the private sector than in university and education. Instead, commercial exploitation of own ideas is indicated as highly/medium important by all entrepreneurs previously employed at universities and by 84.6 per cent of those previously employed in the private sector.

Finally, founders might have had other entrepreneurial experiences. The average percentage of entrepreneurs who founded other firms (i.e. serial entrepreneurs) is lower than 50 per cent but it is lower in Latin Europe (40 per cent) than

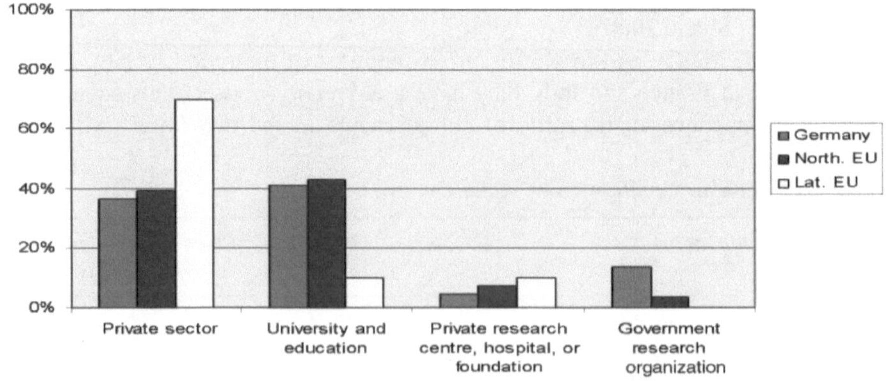

Figure 8.3 Sector of previous employment by region.

Table 8.6 Motivations to found a firm by sector of previous employment (high and medium importance) (%)

Motivations	*Private sector*	*University and education*
Higher income	57.7	45.5
Independence (being your own boss)	92.3	63.6
Commercial exploitation of your own ideas	84.6	100.0
Disagreement with previous employer about the innovations to pursue	34.6	4.6
Disagreement with previous employer about general strategies/management issues	38.5	9.1

Note
We consider only private sector and universities as previous employers because they are the most frequent categories in the sample.

in Northern Europe (55 per cent) and Germany (50 per cent). Only in the Latin region most of the entrepreneurs founded other companies exclusively before the company which was the object of the questionnaire. However, it is worth mentioning that serial entrepreneurs, differently from those that set up a single venture and operate it as a career, might develop specific competencies in founding new ventures (for instance because of privileged linkages with the financial community) but not necessarily in managing and running a firm.

5 Conclusions

This chapter was about the characteristics and factors of survival and success of knowledge-intensive entrepreneurship in some key sectoral systems in Europe.

Literature and evidence on the US case are rather rich and detailed but the European case is relatively less explored and documented. Also, in both cases, existing studies are concentrated on a limited number of sectors and rarely takes on a cross-sector cut. Differently, this chapter proposed one of the first cross-sector and cross-country comparisons at the European level with a specific focus on some key elements in the process leading to the birth, start-up and early evolution of a new firm.

The literature on entrepreneurship and new ventures creation has largely investigated the determinants and factors of success of (innovative) start-ups and has tried to assess their relative importance, especially in technologically highly dynamic contexts in which innovative capabilities are crucial.

These factors include, among others, the human and intellectual capital of the founders and the personnel available to the new venture, the access to external sources of financing, the IPR portfolio, the participation in a network of collaboration with other companies, especially the founder's parent company, as well as other institutional actors such as universities and research organizations.

Accordingly, this chapter aimed at assessing the relative importance of such factors by looking at a sample of European new and innovative firms, defined as those firms that innovate within a relatively short time (i.e. five years) after their establishment. In particular, we focused on the experience of firms that are both knowledge-intensive (i.e. active in science-based and science-driven sectors) and technology-based (i.e. patenting in sophisticated and dynamic technological contexts).

The results of the long process of sample identification, data collection, questionnaire elaboration and administration highlight a number of interesting features.

The importance of the intellectual capital of the founder available to the new venture is confirmed (Kogut and Zander 1996; Pfeffer 1994; Uzzi 1996) as well as the importance of scientific and technological knowledge endowment and capabilities accumulated and developed in previous professional and educational experience (Bates 1990; Bruderl *et al.* 1992; Klepper and Sleeper 2005; Uzzi 1996); these indeed affect and shape the initial conditions of the new firm creation process. Also the availability of human and financial capital (in particular, access to external sources of financing) plays a crucial role to start a new venture (Barry *et al.* 1990; Dierickx and Cool 1989; Hellmann and Puri 2000; Kaplan and Stromberg 2000; Sahlman 1990) as well as the exploitation of IPRs (Arora and Merges 2004; Gans *et al.* 2002) and the capacity to enter into a network of relationships with relevant actors possessing complementary knowledge and assets such as other firms, the founder's parent organization, universities, research centres (Etkowitz *et al.* 2000; Mowery and Shane 2002; Powell and Grodal 2005).

More interestingly, findings indicate remarkable differences across countries and technological fields, thus suggesting that both national innovation systems and sectoral innovation systems (Malerba 2002) and their interplay play a key role in sustaining entrepreneurship in technologically sophisticated contexts. In fact, the analysis highlights that the initial decision of entrepreneurship as well as the early evolution of new innovative firms is strongly shaped by the combination of both country and sectoral specific elements.

Differences across geographical areas are primarily related to two main aspects, i.e. the degree of development of and the ease of access to financial markets and the different functions performed by the university system and its effectiveness, and point to the existence of structural components and characteristics of different national innovation systems.

On the one side, there is evidence about different degrees of development of and the ease of access to financial markets. The relevant literature clearly indicates and documents their key role to spur and support entrepreneurship. This has significant policy implications on a single country's capacity to develop and sustain innovative capabilities in fast-changing sectors.

On the other side, there is evidence about the different functions performed by the university system and its effectiveness. This emerges according to a series of dimensions of analysis such as the educational profile of founders, the frequency of relationships and collaborations with the private sector and also the rate of spin-off from university. These findings confirm the difficulties in technology transfer from university to industry and in the commercial exploitation of scientific research mainly for Latin European countries.

Differences across sectors and technological fields are primarily related to two main aspects, i.e. the factors exploited in the establishment of the new firm and the competences exploited in order to be successful on the market, and point to the existence of structural components and characteristics of different sectoral systems of innovation.

On the one side, there are differences according to the importance and exploitation of IPRs and the strength and the areas of the relationships with the previous employer. In particular, IPRs are more intensively exploited in biotechnology as compared to the other sectors as well as the relationships with the previous employer, especially as far as research activities are concerned. On the other side, there are differences about the key competences exploited and the knowledge endowment and inheritance from the previous employer. While uniqueness of the product/technology/knowledge is important in all sectors, marketing skills and especially customer services are more relevant in electronics and medical devices. Similarly, while knowledge about product and technology is important in all sectors, consistently with the selection criteria of the sample, knowledge about demand and customers (though relatively less relevant) is more important for medical devices and electronics firms where products are frequently conceived and designed to meet users' needs and to accommodate their requirements, but far less in biotechnology where products are frequently far from the commercialization stages. Both sets of factors affect the mode of entry into entrepreneurship and the likelihood of survival on the market.

Notes

1 The EP–Cespri database collects patent applications registered at the EPO since 1978. Cespri is now KITeS.

2 Response rates were as follows: Germany: 27.4 per cent; Denmark and Sweden: 13.7 per cent; France: 16.7 per cent; Great Britain: 22.3 per cent; Italy: 40.9 per cent.

3 Licensing is considered as acquisition of one or more licences in order to start an entrepreneurial activity.

References

Arora, A. and Merges, R.P. (2004) 'Specialised supply firms, property rights and firms boundaries', *Industrial and Corporate Change*, vol. 13 (3): 451–75.

Barry, C., Muscarella, C., Peavy, J. and Vetsuypens, M. (1990) 'The role of venture capitalists in the creation of public companies', *Journal of Financial Economics*, vol. 27 (2): 447–71.

Bates, T. (1990) 'Entrepreneur human capital inputs and small business longevity', *The Review of Economics and Statistics*, vol. 72 (4): 551–9.

Boswell, J. (1972) *The Rise and Decline of Small Firms*, London: Allen and Unwin.

Bottazzi, L. and Da Rin, M. (2002) 'Venture capital in Europe and the financing of innovative companies', *Economic Policy*, vol. 34 (17): 229–70.

Breschi, S., Malerba, F. and Orsenigo, L. (2000) 'Technological regimes and sectoral patterns of innovation', *Economic Journal*, vol. 110 (463): 388–410.

Bresnahan, T. and Malerba, F. (1999) 'Industrial dynamics and the evolution of firms' and nations' competitive capabilities in the world computer industry', in Mowery, D.C. and Nelson, D.D. (eds) *The Sources of Industrial Leadership: Studies of Seven Industries*, Cambridge: Cambridge University Press.

Bruderl, J., Preisendorfer, P. and Ziegler, R. (1992) 'Survival chances of newly founded business organizations', *American Sociological Review*, vol. 57 (2): 227–42.

Burt, R. (1993) 'The network entrepreneur', in Swedberg, R. (ed.) *Explorations in Economic Sociology*, New York: Russle Sage Foundation.

Cockburn, I.M. and Henderson, R.M. (1998) 'Absorptive capacity, co-authoring behaviour, and the organisation of research in drug discovery', *Journal of Industrial Economics*, vol. 46 (2): 157–82.

Cohen, W.M., Nelson, R.R. and Walsh, J.P. (2000) 'Protecting their intellectual assets: appropriability conditions and why US manufacturing firms patent (or do not)', NBER Working Paper no. 7552.

Cooper, A.C., Gimeno-Gascon, F.J. and Woo, C.Y. (1994) 'Initial human and financial capital as predictors of new venture performance', *Journal of Business Venturing*, vol. 9 (5): 371–95.

Dierickx, I. and Cool, K. (1989) 'Asset stock accumulation and sustainability of competitive advantage', *Management Science*, vol. 35 (12): 1504–11.

Dollinger, M.J. (1985) 'Environmental contacts and financial performances of the small firm', *Journal of Small Business Management*, vol. 23 (1): 24–30.

Dun and Bradstreet, Inc. (1981) *The Failure Record*, New York: Dun and Bradstreet, Inc.

Etzkowitz, H., Webster, A., Gebhardt, C. and Terra, B. (2000) 'The future of the university and the university of the future: evolution of ivory tower to entrepreneurial paradigm', *Research Policy*, vol. 29 (2): 313–30.

European Commission (2007) 'European Innovation Scoreboard 2007 – A comparative analysis of innovation performance', PRO–INNO Europe Paper no. 6, available online at: www.proinno-europe.eu/admin/uploaded_documents/European_Innovation_Scoreboard_2007.pdf.

Gans, J.S., Hsu, D.H. and Stern, S. (2002) When does start-up innovation spur the gale for creative destruction?', *Rand Journal of Economics*, vol. 33 (4): 571–86.

Gelijns, A.C. and Rosenberg, N. (1999) 'Diagnostic devices: an analysis of comparative advantages', in Mowery, D.C and Nelson, D.D. (eds) *The Sources of Industrial Leadership: Studies of Seven Industries*, Cambridge: Cambridge University Press.

Geroski, P.A. (1995) 'What do we know about entry?', *International Journal of Industrial Organization*, vol. 13 (4): 421–40.

Giuri, P., Mariani, M., Brusoni, S., Crespi, G., Francoz, D., Gambardella, A., Garcia-Fontes, W., Geuna, A., Gonzales, R., Harhoff, D., Hoisl, K., Lebas, C., Luzzi, A., Magazzini, L., Nesta, L., Nomaler, O., Palomeras, N., Patel, P., Romanelli, M. and Verspagen, B. (2007) 'Inventors and invention processes in Europe: evidence from the Pat–Val survey', *Research Policy*, vol. 36 (8): 1107–27.

Hagerdoon, J. (2002) 'Inter-firm R&D partnerships: an overview of major trends and patterns since 1960', *Research Policy*, vol. 31 (4): 477–92.

Hellman, T. and Puri, M. (2000) 'The interaction between product market and financing strategy: the role of venture capital', *Review of Financial Studies*, vol. 13 (4): 959–84.

Kaplan, S. and Stromberg, P. (2000) 'Financial contracting theory meets the real world: an empirical analysis of venture capital contracts', CEPR Discussion Paper no. 2421.

Klepper, S. and Sleeper, S.D. (2005) 'Entry by spinoffs', *Management Science*, vol. 51 (8): 1291–306.

Klepper, S. and Thompson, P. (2005) 'Spin-off entry in high-tech industries: motives and consequences', mimeo.

Kogut, B. and Zander, U. (1996) 'What firms do? Coordination, identity and learning', *Organization Science*, vol. 7 (5): 502–18.

Kortum, S. and Lerner, J. (2000) 'Assessing the contribution of venture capital on Innovation', *Rand Journal of Economics*, vol. 31 (4): 674–92.

Lee, C., Lee, K. and Pennings, J.M. (2001) 'Internal capabilities, external networks, and performance: a study on technology-based ventures', *Strategic Management Journal*, vol. 22 (6): 615–40.

Malerba, F. (2002) 'Sectoral systems of innovation and production', *Research Policy*, vol. 31 (2): 247–64.

Montgomery, M., Johnson, T. and Fraisal, S. (2005) 'What kind of capital do you need to start a business: financial or human?', *The Quarterly Review of Economics and Finance*, vol. 45 (1): 103–22.

Mowery, D.C. and Shane, S. (2002) 'Introduction to the special issue on university entrepreneurship and technology transfer', *Management Science*, vol. 48 (1): v–ix.

Nelson, R.R. (1993) *National Innovation Systems*, Cambridge, MA: Oxford University Press.

Nelson, R.R. and Winter, S.G. (1982) *An Evolutionary Theory of Economic Change*, Cambridge, MA: Harvard University Press.

Penrose, E.T. (1959) *The Theory of the Growth of the Firm*, New York: ME Sharpe.

Pfeffer, J. (1994) *Competitive Advantage Through People*, Boston, MA: Harvard Business School Press.

Powell, W.W. and Grodal, S. (2005) 'Networks of innovators', in Fagerberg, I., Mowery, D. and Nelson, R.R. (eds) *The Oxford Handbook of Innovation*, Oxford: Oxford University Press.

Sahlman, W. (1990) 'The structure and governance of venture-capital organizations', *Journal of Financial Economics*, vol. 27 (2): 473–521.

Shane, S. (2000) 'Prior knowledge and the discovery of entrepreneurial opportunities', *Organization Science*, vol. 11 (4): 448–69.

Thompson, P. and Klepper, S. (2003) 'Submarkets and the evolution of market structure', Florida International University, Department of Economics, Working Paper no. 0303.

Thorton, P.H. and Flynne, K.H. (2003) 'Entrepreneurship, networks and geographies', in Acs, Z. and Audretsch, D. (eds) *Handbook of Entrepreneurship Research*, Dordrecht: Kluwer Academic Publishers.

Uzzi, B. (1996) 'The sources and consequences of embeddedness for the economic performance of organizations: the network effect', *American Sociological Review*, vol. 61 (4): 674–98.

Von Hippel, E. (2005) *Democratizing Innovation*, Boston: MIT Press.

Winter, S.G. (2006) 'Toward a neo-schumpeterian theory of the firm', *Industrial and Corporate Change*, vol. 15 (1): 125–41.

9 Knowledge-intensive entrepreneurship in Central and Eastern Europe

Results of a firm-level survey

Slavo Radosevic,[1] *Maja Savic*[1] *and Richard Woodward*[2]

1 Introduction

A 2005 article in *Business Week* asked whether the countries of Central and Eastern Europe (CEE) could be considered a 'rising powerhouse' in high-tech industries. It noted, among other things, that Poland's Warsaw University was ranked first in the world in top coder events, that software companies are springing up in Bulgaria, and that Romania has become an important centre of engineering and R&D activities for a number of industries, including automobile manufacturing. It then concludes: 'As the race for top talent heats up globally, it turns out that Central Europe houses one of the planet's richest creative pools.'

This chapter represents an attempt, based on evidence from a survey of 304 firms in six countries of the region, to separate myth from reality and identify both the strengths and weaknesses of Knowledge-Based Entrepreneurship (KBE) in the CEE countries in the light of the factors that affect the performance of firms. These factors can be broadly divided into those referring to the entrepreneur, to the firm, and to the environment within which firms and entrepreneurs operate.

Our study of KBE was carried out in Hungary, Lithuania, the Czech Republic, Croatia, Poland and Romania. For the purposes of this chapter we assume that knowledge-based entrepreneurship refers to the generation, utilization and transmission of knowledge that has been generated within the fields of science, technology and innovation in anticipation of commercial application. In particular the emphasis is on the development of new technologies and the introduction of new products and processes (including new services) and the actors (new firms, universities and sources of external finance). We adopt Grant's (1996) definition of a knowledge-based venture, whereby knowledge is the firm's most valuable asset. Grant (1996) proposes that the ability of a firm to integrate knowledge held by individuals within the organization creates its competitive advantage. When employees are mobile, the organizational capability depends more on the integration mechanisms than on the specialist knowledge that individuals posses (Grant 1996). Apart from the importance of transmission of knowledge within the organization we emphasize the importance of external environment and in particular the role of networks or innovation systems. We also acknowledge that some characteristics of entrepreneurs as individuals can significantly contribute to our understanding of the topic.

In Section 2 we will describe the sample of CEE KBE firms studied. In Section 3 we look more closely at the entrepreneurs behind those firms and identify the markets and customers they are serving. Section 4 analyses success factors and barriers in the firms' growth and describes categories of firms defined on the basis of those success factors and barriers. Section 5 explores the role of networks in firms' growth and again categorizes the firms based on the types of networks in which they are involved. We conclude with a summary of our findings.

2 Description of the sample

During 2007 we administered an extensive questionnaire to the owners and/or chief executive officers in a sample of 304 firms in Hungary (50), Lithuania (52), Czech Republic (50), Croatia (50), Poland (62) and Romania (40). The two-page questionnaire consisted of 18 questions divided into four categories: (a) information about the firm; (b) information about the entrepreneur; (c) information about the demand and sources of knowledge; and (d) information about growth, internationalization and networks of the firm. The data refer to 2006, except the financial information, which relates to 2001 and 2005.

The firms were selected, first, on the basis of industrial classification, using a list, prepared by Clendenning, E. Wayne, & Associates (Clendenning, E. Wayne, & Associates 2000), of industries considered to be knowledge-based. Since the exclusive use of industry classification as a selection criteria would likely result in some companies being inappropriately categorized as knowledge-based, we have additionally employed a set of auxiliary criteria for final selection of firms. These included, for example, whether the firm invests in R&D and whether it employs highly skilled personnel (with MScs or PhDs).[3]

The total number of employees in the firms in the sample is 52,883, which represents between 0.47 per cent and 11.9 per cent of overall industry employment (Table 9.1). In this respect, the sample can be considered on average fairly representative.

2.1 Industry breakdown

We have grouped our sample of firms into five broad sectors: pharmaceuticals, chemicals and plastics; manufacturing; software; media; and R&D (for the list of NACE industries included in each, see the Annex).

The industry structure of our sample is as follows: pharmaceuticals, chemicals and plastics (13 per cent), manufacturing (46 per cent), software (16 per cent), media (5 per cent) and R&D (18 per cent). There are significant cross-country differences[4] with respect to the sectoral composition of the sample. In the Hungarian and Lithuanian samples, R&D enterprises are represented heavily, while in the Czech, Polish and Romanian samples, manufacturing enterprises

Table 9.1 Description of the sample

Country	*Number of firms*		*Number of persons employed*		*Representativeness*	
	All firms	*Sample*	*All firms*	*Sample*	*No. of firms (% of total)*	*No. of persons employed (% of total)*
Hungary	45,586	50	422,111	5,660	0.11	1.34
Czech Republic	59,672	50	604,435	10,207	0.08	1.69
Lithuania	3,413	52	74,384	8,852	1.41	11.90
Croatia	NA	50	NA	4,111		
Poland	80,195	62	948,318	21,553	0.08	2.27
Romania	22,673	40	527,017	2,500	0.18	0.47
Total	211,539	304	2,576,265	52,883	0.14	2.05

Source: Eurostat, http://epp.eurostat.ec.europa.eu, except for Croatia as data are not available on Eurostat website.

are heavily represented. The Croatian sample has an unusually large group of media enterprises (Figure 9.1).

2.2 Ownership

The overwhelming majority of sampled companies are domestically owned. Percentage of domestic ownership ranges from 82 per cent in Poland to 96 per cent in the Czech Republic. The highest percentage of foreign-owned companies is found in Poland (18 per cent) and Hungary (16 per cent) (Figure 9.2). However, with respect to negligible differences across countries,[5] we can conclude that

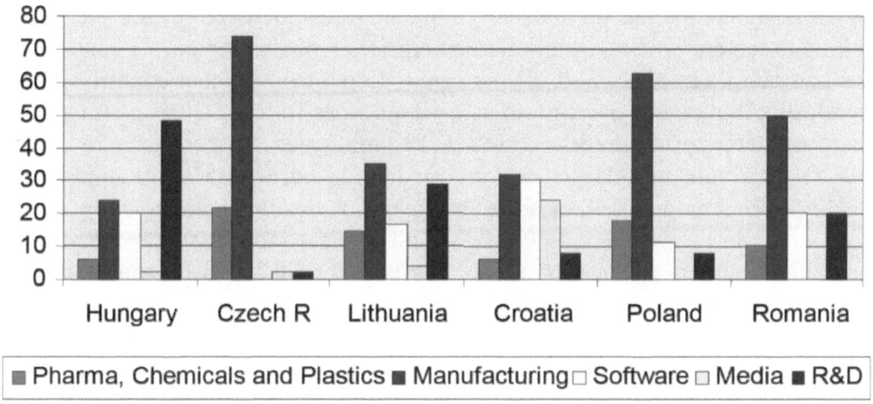

Figure 9.1 Industry breakdown.

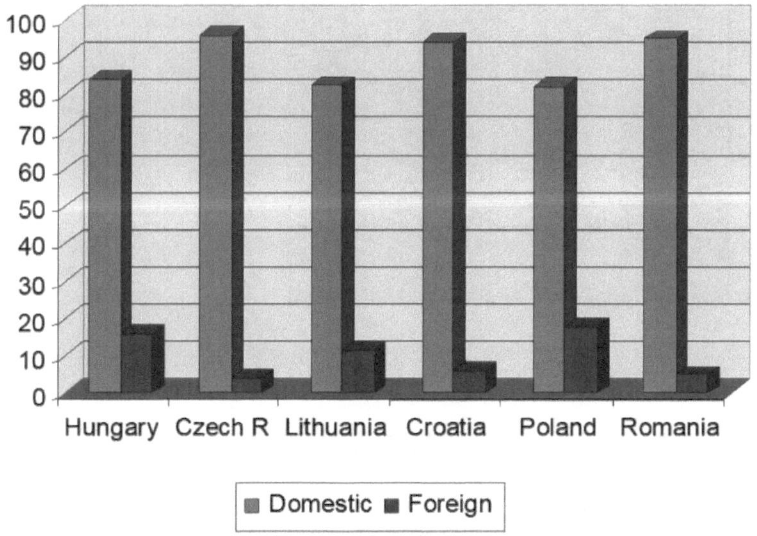

Figure 9.2 Ownership by country.

KBE firms in CEE countries are largely domestically owned. In other words, the process of internationalization (globalization) of KBE in these countries is still limited (for a similar conclusion see UNCTAD 2005). This is the same with respect to the industry breakdown, with the share of domestic ownership ranging from 98 per cent in IT to 84 per cent in the R&D sector.

2.3 Size

The vast majority (88 per cent) of the firms are small- or medium-sized enterprises (i.e. they have fewer than 250 employees), with 63 per cent of them having under 50 employees. Cross-country differences are not strong.[6] However, the Croatian sample has an unusually high share of microentreprises (i.e. firms with up to ten employees), with microentreprises representing 56 per cent of the Croatian sample as opposed to 14 per cent – 45 per cent in other countries.[7] The overall distribution for the sample as a whole is as follows: 30 per cent of the enterprises are microentreprises with 1–10 employees, 35 per cent are small, with 11–50 employees, 23 per cent are medium-sized, with 51–249 employees, and 12 per cent employ 250 or more employees. A detailed breakdown by country is shown in Figure 9.3. The Polish sample is biased towards a few very large companies (outliers), hence the mean distribution of the Polish sample is high compared to the other countries (Table 9.2).

There are no statistically significant differences across industries with respect to the number of employees.[8] Nevertheless, we can report that high proportions

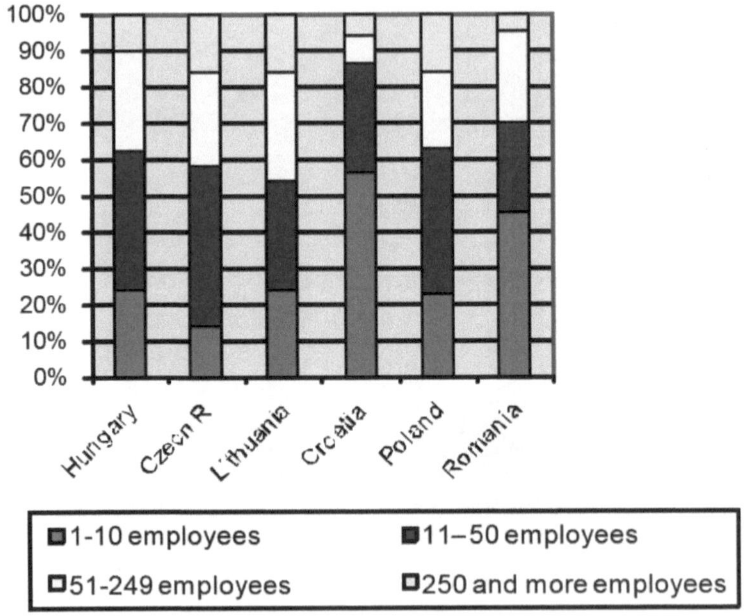

Figure 9.3 Size of KBEs.

Table 9.2 Size of KBEs by country

Country	*Average number of employees*	*Median number of employees*
Hungary	113	33
Czech Republic	204	43
Lithuania	188	45
Croatia	82	10
Poland	348	24
Romania	63	13
All countries	177	25

of microenterprises are found in software (40 per cent), media (56 per cent) and R&D (38 per cent), whereas the pharma, chemicals and plastics industry is biased by a few large outliers, making the mean size of companies in this industry much higher compared to other industries (Table 9.3).

2.4 Year of establishment

Most (64 per cent) of the firms were established during the initial transition period of the 1990s. Seventeen per cent of companies were established during the planned economy period (before 1990), and 20 per cent are very young companies established after 2000. Differences between countries are not statistically significant.[9] There are also no statistically significant differences between industries with respect to year of establishment.[10]

3 Knowledge-based entrepreneurs: their characteristics, motivations and markets

3.1 Education

Obviously, in a treatment of knowledge-based entrepreneurship, we are interested in the educational attainment and other human capital of the entrepreneur (defined in our research as the CEO). We begin our discussion of human resources, therefore, by presenting the educational structure of core staff and CEOs in our sample of firms.

Table 9.3 Size of KBEs by industry

Industry	*Average number of employees*	*Median number of employees*
Pharma/chemicals and plastics	359	55
Manufacturing	211	23
Software	46	15
Media	19	10
R&D	127	20
All industries	152	20

KBE enterprises are clearly a highly skilled segment of CEE firms. This is demonstrated by the fact that, in the sample as a whole, only 4 per cent of the CEOs have completed only secondary education; 31 per cent have a BA or the equivalent, 44 per cent a master's degree and 19 per cent a PhD. With 96 per cent possessing at least some higher education, the entrepreneurs in this group are much better educated than the founders of companies studied in a European Commission study of companies founded in ten CEE countries in the period 1995–2000. This study found rates ranging from 17.3 per cent to 42 per cent for the period as a whole (CEC 2002). The highest percentage (44 per cent) of PhDs was found in Hungary. Poland had relatively few PhDs (6 per cent), but a very high (85 per cent) percentage of master's degrees. Croatia had no PhDs and very few MAs; 90 per cent of the Croatian entrepreneurs had BA-equivalent degrees.

3.2 Entrepreneurs: where do they come from?

Next, we consider the employment of the entrepreneur prior to his or her founding of the firm. A common feature here is that a high percentage (68 per cent) of entrepreneurs come from the business sector (either private or state); one-quarter (25 per cent) come from the science and technology (S&T) sector (in which we include educational, research and healthcare institutions). The private sector is the previous employer of 61 per cent of the entrepreneurs in the sample (ranging from 57 per cent in Hungary to 67 per cent in Croatia and the Czech Republic); state-owned enterprises were the former employer of 7 per cent of them (from 2 per cent in the Czech Republic to 13 per cent in Poland). Research and healthcare institutions were the former employers of 25 per cent of our entrepreneurs (from 8 per cent in Croatia to 28 per cent in Hungary). The education sector employed 8 per cent of the sampled entrepreneurs (from 4 per cent in Poland to 19 per cent in Hungary). Finally, for 14 per cent of our entrepreneurs, work in the firm studied here was their first job (from 0 per cent in Romania to 33 per cent in Croatia).

We note two country-specific peculiarities. First, Hungary has a high share of entrepreneurs from the S&T sector (43 per cent), corresponding to the high percentage of PhDs in this group and the high share of R&D companies in the Hungarian sample. Croatia had a remarkably high percentage of entrepreneurs for whom this was their first job, while Romania's knowledge-based entrepreneurs have all worked before.

How important is the knowledge acquired by entrepreneurs during their previous employment for the firms in which they are currently working? We see some answers in Table 9.4, which shows the number of responses indicating high importance (six or seven points on a seven-point Likert scale) of this knowledge. As we see, the most important is knowledge of products and technology. There is varied, but much lower importance of knowledge on customers, competitors and suppliers. Thus, it seems that the creation of new firms involves the entrepreneurs repositioning themselves on the market, but not with respect to technology. This pattern is observed across all the countries studied.

Table 9.4 Previous employment of CEOs of KBEs

Sector	*Range*
Private sector (61%)	57% (H) – 67% (CR, CZ)
State-owned enterprises (7%)	2% (CZ) – 13% (PL)
Research (25%)	8% (CR) – 28% (H)
Education (8%)	4% (PL) – 19% (H)
First job (13%)	0% (ROM) – 33% (CR)

Table 9.5 High importance of knowledge acquired during previous employment (% of respondents previously employed elsewhere)

	Number of firms	*Percent*
Products/technology	164	62
Customer base/demand	100	38
Competitors	83	32
Suppliers	68	26
Other	20	69

3.3 Key rationale for establishing a company: market opportunity

We see in Figure 9.4 that the major rationale for establishing the company was market opportunity (commercial potential and financial opportunity). Technological opportunity was relatively less important than market opportunity, but still quite an important rationale in all countries except the Czech Republic and Poland. In these two countries the share of firms where technology was the major rationale was 22 per cent and 15 per cent respectively, which can be explained by the large share of manufacturing firms in both these samples. The Hungarian sample has the largest share of firms where technological opportunity is a major motivation, which reflects their high R&D intensity and orientation towards the public R&D sector.

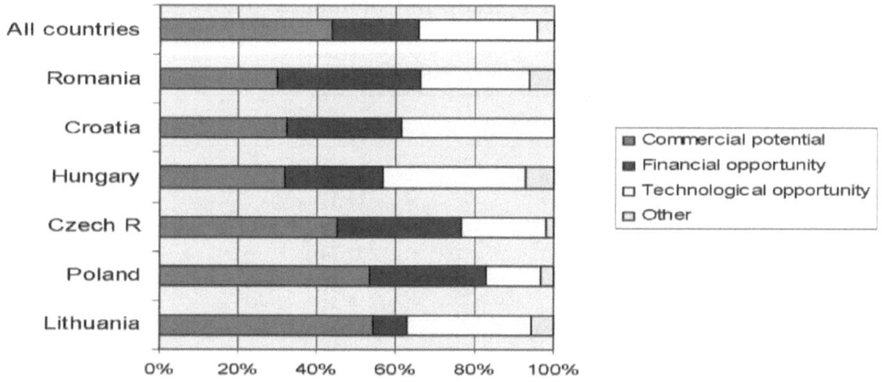

Figure 9.4 Key rationale to establish company.

It is interesting that a certain percentage of firms cite other motivations. Most of these are institutional opportunities related to privatization and their share is the highest in Hungary (10 per cent), Romania and Lithuania (both 8 per cent).

3.4 Market orientation: where are the customers and markets for CEE KBE firms?

The market orientation of sales of KBEs is quite different across countries. Table 9.6 shows the distribution of sales based on the share in total values (here, shares are highly influenced by the largest firms and do not reflect average values). Polish and Romanian firms are much more domestic-market oriented than firms in smaller countries, which are mainly oriented towards export. The Czech sample is quite balanced in this respect.

In Table 9.7 we present sales to specific types of customers as percentages of total sales of all surveyed companies in the relevant country. Table 9.7 shows that Lithuania and Croatia have the highest share of sales to foreign consumers, Hungary to foreign manufacturing firms, Romania to domestic manufacturing firms and Poland to domestic services companies. The Czech Republic again has a balanced orientation between sales to domestic and foreign manufacturers. However, we must note that the data in Table 9.7 may be strongly influenced by the largest firms.

4 Sources of knowledge, networks, barriers and firms' typologies

4.1 Beyond value chains: sources of knowledge for innovation

In this research we are particularly interested in investigating the importance of various types of relationships and networks for the firms studied and, in particular, how important networks are for knowledge transfer. In Table 9.8 we see the major sources of knowledge used for innovation. In-house knowledge, customers and suppliers are major sources. In Community Innovation Survey data

Table 9.6 Regional distribution of sales: foreign and domestic market*

Country	*Domestic*	*Foreign*	*Others*
Hungary	26	74	0
Czech Republic	49	51	0
Lithuania	24	76	0
Croatia	25	74	1
Poland	80	18	2
Romania	79	21	1
All countries	50	50	0

Note
* Expressed as % of total values.

Table 9.7 Distribution of sales by customer type (% of total sales in relevant country)

Country	*Domestic manufacturing*	*Domestic services*	*Foreign manufacturing*	*Foreign services*	*Domestic consumers*	*Foreign consumers*	*Domestic scientific/ research*	*Foreign scientific/ research*	*Domestic public procurements*	*Others*
Hungary	8	8	**64**	7	4	1	1	2	5	0
Czech Republic	**48**	0	**51**	0	0	0	0	0	0	0
Lithuania	8	6	4	8	4	**61**	0	3	6	0
Croatia	2	3	10	9	19	**55**	0	0	0	1
Poland	12	**51**	8	5	11	3	2	1	5	2
Romania	**44**	7	1	5	18	15	9	0	1	1
All	**41**	6	**45**	1	2	3	0	0	1	0

Note
Sum of row values = 100.

Table 9.8 Sources of knowledge as basis for innovation

	High and medium importance of sources of knowledge as a basis for innovation						
	All countries	*Hungary*	*Czech Republic*	*Lithuania*	*Croatia*	*Poland*	*Romania*
In-house	**99**	**98**	**100**	**98**	**100**	98	**98**
Customers	**84**	64	**84**	**96**	74	94	**95**
Suppliers	**76**	62	62	74	**88**	87	**85**
Fairs and exhibitions	**71**	50	58	82	**80**	83	73
Patents, journals	**69**	62	44	65	**84**	89	70
Research organizations	**68**	**82**	46	69	62	74	75
Other (Internet etc.)	**85**	100	100	100	100	100	10

Notes
% as a percentage of all firms that answered the relevant question.
High and medium importance = answers 7–3 on a scale 1–7.

(see Eurostat 2005) in-house and value chain partners (customers and suppliers) are also major sources of information for innovation. However, in the case of CEE KBEs the importance of other sources located outside the value chain is quite high. Fairs and exhibitions, patents, journals and research organizations have a very high share, ranging from 68 per cent to 71 per cent. It is also interesting that other sources of innovation (most often the Internet) are very often a major source of knowledge for innovation. Amongst enterprises mentioning the Internet as an additional source of knowledge, 85 per cent of them assign high to medium importance to this source.

The share of KBEs where in-house and value chain partners are of high and medium importance is high in all six countries. Only in Hungary do value chain partners play a somewhat less important role, which is again expected given that Hungarian KBEs are mainly public-sector oriented and R&D-intensive firms. In the case of Hungary, research organizations are the second most frequently cited important source of knowledge for innovation (85 per cent). The importance of other non-value chain partners (fairs and exhibitions, patents and journals) is important with varying degrees for specific sources in all six countries.

The key message from Table 9.8 is that the knowledge networks that underpin KBEs extend beyond value chains. In that respect, the KBEs in our sample do represent a distinctively different segment of firms in CEE countries.

To learn more about the patterns affecting the use of various sources of the knowledge utilized in innovation, we have undertaken factor analysis on the sources of knowledge (Table 9.9). Table 9.9 shows that sources of knowledge are grouped into three underlying or latent factors. We can distinguish firms where value chain partners are the main source of knowledge, as both variables – suppliers and customers – load highly on one factor with the majority of others being unrelated. We can also distinguish firms where formalized R&D knowledge (as contained in patents, journals and research organizations) is the major source, and those firms where in-house knowledge is the key knowledge resource in innovation. In both cases, factor loadings are very high and results quite robust. Fairs and exhibitions are the only source which loads relatively highly on two factors – value chains and formalized R&D knowledge. This may

Table 9.9 Results of factor analysis of sources of knowledge for innovation

	Rotated component matrix		
	Value chain	*Formalized R&D knowledge*	*In-house*
Suppliers	0.827	0.113	−0.067
Customers	0.813	−0.046	0.228
Fares/exhibitions	0.581	0.430	−0.190
Patents/journals	0.197	0.847	−0.044
Research organizations	−0.027	0.819	0.183
In-house	0.036	0.091	0.953

Note
Rotation Method: Varimax with Kaiser Normalization.

be expected given that knowledge for innovation very often resides in, and is easily accessible through, social and business networks. Fairs and exhibitions seem to be important for both meeting business partners and catching up with developments in the R&D area.

In summary, factor analysis has simplified the structure of the sources of knowledge for innovation and generated three factors which are robust proxies for three major sources of innovation: value chain partners, formalized R&D and in-house knowledge.

4.2 Success factors and typologies of KBEs

We asked firms about factors which lay behind their success. We have offered them a range of factors and asked them to assess their importance on a scale of one (completely unimportant) to seven (crucial). Based on these results we have undertaken factor analysis in order to simplify the structure of success factors and explore whether factors of success belong to specific underlying factors. Table 9.10 presents the results of factor analysis, which gives a very robust grouping of firms into three groups based on latent factors or variables shared across different types of firms. Factor analysis shows that, on the basis of their success factors, we can observe three groups of KBEs in CEE:

1 *Networkers*, or firms highly dependent on links with other scientific organizations, on EU Framework programmes, on government support and links with other firms.

Table 9.10 Types of firms with respect to success factors (results of factor analysis)

	Networker	*New technology based firms*	*Customer oriented organizational capabilities*
Links with scientific organizations	**0.754**	**0.313**	−0.071
EU Framework programmes and other EU support	**0.749**	0.086	0.095
Government support	**0.681**	0.239	0.172
Alliances/partnerships with other firms	**0.639**	−0.108	0.185
Links with previous employer	**0.489**	0.139	−0.179
Uniqueness of product/technology/knowledge	0.062	**0.813**	0.076
Patents and licences	**0.400**	**0.716**	**0.028**
People and training	**0.245**	**0.410**	**0.369**
Knowledge and customers' needs	−0.023	0.230	**0.663**
Management	0.396	0.014	**0.603**
Capability to offer expected services/products with low cost	0.014	−0.146	**0.601**
Quality	−0.115	0.376	**0.576**

Note
Rotation Method: Varimax with Kaiser Normalization.

2 *New Technology Based Firms*, or firms whose competitiveness is based on a unique technology, and on patents and licences.
3 Firms whose competitiveness is based on *customer-oriented organizational capabilities*. These firms are strong in knowledge of customer needs and in management, and are able to offer expected services/products at low cost.

The factor loadings on each of three factors are very high, and the three-factor solution is very neat and robust. The very interesting finding arising from this analysis is that CEE KBEs are not homogenous entities competing based on new technologies. In fact, new technology based firms are only one of three types of KBEs.

4.3 Barriers to growth

We asked KBEs about major barriers to their growth, in both domestic and foreign markets. Table 9.11 shows the frequencies with which various factors were identified as barriers of medium to high importance (3–7 on the seven–point Likert scale). The major barrier seems to be a limited domestic market (78 per cent), followed by lack of public support (75 per cent) and high costs of labour (71 per cent). Responses vary by country, but the limited market is among the top two answers everywhere except Croatia and Poland. For Croatia, Poland and Romania, the high costs of labour are also among the most frequent answers. Increased competition is among the major barriers in four of the six countries.

These answers suggest that KBEs are faced with limited domestic demand and face quite strong competition. In these market conditions, labour costs are of high concern.

Next, we explore whether barriers to firms' growth are systematically correlated and thus grouped into specific latent factors. The factor solution (Table 9.12) shows that high costs of labour, lack of skills and increased competition load on one factor. We interpret high labour costs and poor access to skills and knowledge as barriers that cause firms to feel weak with respect to their competitiveness. Hence, we take skills and labour as one important factor. Lack of access to finance and lack of public support are quite closely connected factors which load on one underlying factor together with limited domestic demand. It seems logical that limited finance and limited domestic demand are closely correlated in the case of KBEs, as these are businesses which have relatively high fixed costs and low marginal costs and which are thus faced with frequent cash-flow problems. Hence, we take lack of finance as a true firm-based barrier and limited domestic demand as an exogenous, structural constraint.

We have undertaken an identical analytical procedure for barriers on foreign markets. However, neither regression analysis nor factor analysis were able to provide either robust estimates or a factor solution.

Table 9.11 Barriers of high and medium importance on domestic market (% of firms)

	All countries	*Hungary*	*Czech Republic*	*Lithuania*	*Croatia*	*Poland*	*Romania*
Limited market	**78**	**78**	**70**	**74**	**80**	**79**	**88**
High cost of labour	**71**	68	68	60	**88**	**93**	**88**
Increased competition on market	54	59	**82**	59	**80**	**80**	**75**
Lack of access to finance	59	**78**	48	49	**84**	**81**	**75**
Lack of public support	**75**	52	50	43	**72**	**84**	**90**
Lack of skill and know-how	65	42	40	44	**72**	52	**88**
Other	57	100	0	50	100	67	10

Notes
% as a percentage of all firms that answered the relevant question.
Medium and high importance = answers 3–7 on a scale 1–7.

Table 9.12 Factor analysis results of barriers on domestic market

	Rotated component matrix	
	Skills and labour	*Finance*
Increased competition on market	**0.835**	0.076
Lack of skill and know-how	**0.724**	0.098
High cost of labour	**0.722**	0.412
Lack of access to finance	0.036	**0.851**
Limited domestic market	0.145	**0.686**
Lack of public support	0.418	**0.597**

Note
Extraction Method: Principal Component Analysis.

5 Role of networks in knowledge-based entrepreneurship

A point of departure of the KEINS project is that networks are central to entrepreneurship. In this section we explore different network dimensions of KBEs in our sample.

5.1 Role of various types of relationships

Table 9.13 shows the percentage of firms indicating the medium to high importance of relationships with different types of partners. The most important links are with value chain partners (domestic and foreign buyers and sellers). This pattern of networking is similar to that of 'normal' firms. However, what distinguishes KBEs is the relatively high importance of universities, research institutes as well as alliances, i.e. partnerships with other firms including licensing partners. This points to the importance of links that go beyond commodity/service flows as well as to the importance of institutions of national innovation systems (universities and research institutes) for KBEs. In this respect the pattern of networking is similar to that of importance of sources of knowledge for innovation (Table 9.8), which also goes beyond value chain partners. The lowest importance of international joint ventures points to weak equity links between local KBEs and foreign firms in CEE countries. This is consistent with the ownership structure of KBEs, which are dominantly domestically owned firms.

5.2 Types of firms based on intensity of links, country and industry patterns

We want to explore whether firms could be grouped in terms of different patterns of underlying network relationships. We undertake factor analysis using data on the intensity of network relationships. Table 9.14 shows the results of factor analysis for intensity of links.

Table 9.13 Relationships of medium to high importance (% of firms)

	All	*Hungary*	*Czech Republic*	*Lithuania*	*Croatia*	*Poland*	*Romania*
Domestic firm (buyers)	**86**	**76**	**92**	**75**	**94**	**94**	**88**
Domestic firm (suppliers)	**77**	**64**	**86**	55	**84**	**84**	**88**
Foreign firms (buyers)	**70**	**76**	**88**	**76**	**63**	57	54
Foreign firms (suppliers)	**67**	52	**70**	**78**	**65**	**61**	**72**
Public authorities	59	54	54	**67**	58	59	**64**
Vocational/higher education institute	56	**68**	36	53	**63**	**61**	54
Strategic alliances	57	**70**	48	44	46	**70**	**63**
Public/private research institute	52	**68**	28	39	44	**64**	**69**
Licensing	49	51	26	34	56	**64**	**62**
Consultants	46	34	38	37	56	52	58
International joint ventures	38	26	26	40	32	49	56

Notes

% as a percentage of all firms that answered the relevant question.

High and medium importance = answers 3–7 on a scale 1–7.

Table 9.14 Types of firms based on intensity of their links

	(Foreign) Network dependent	*Public research system dependent*	*Foreign value chain dependent*	*Domestic value chain dependent*
Strategic alliances	**0.798**	0.166	0.183	0.046
Licensing	**0.757**	0.170	0.019	−0.006
International joint ventures	**0.665**	0.066	0.330	0.074
Consultants	**0.546**	0.223	−0.112	0.369
Public authorities	0.468	0.453	0.039	0.297
Vocational higher education institute	0.096	**0.888**	0.091	0.032
Public/private research institute	0.281	**0.805**	−0.039	−0.029
Foreign firm suppliers	0.069	0.071	**0.844**	0.195
Foreign firm buyers	0.189	−0.021	**0.782**	−0.168
Domestic firm suppliers	−0.062	0.140	0.221	**0.839**
Domestic firm buyers	0.324	−0.155	−0.223	**0.681**

Notes
Extraction Method: Principal Component Analysis.
Rotation Method: Varimax with Kaiser Normalization.

Factor analysis shows that there are four types of firms based on the intensity of their links with external partners. This factor solution is quite robust with high loadings on underlying factors.

Foreign network dependent firms are those whose links with other firms, licensors, joint venture partners, consultants and public authorities are highly correlated. As Table 9.13 showed, the percentage of firms giving importance to these types of relationships is not very high in our sample when compared to value chain linkages. It seems that most of firms in this group are dependent on foreign partners other than buyers and suppliers (i.e. value chain partners). In addition, these firms are to some extent also dependent on public authorities but are not dependent on the public research system.

The second group of firms are those that are dependent on institutions of national innovation system like universities, research institutes and public authorities. Hence, it seems appropriate to define these firms as *public research system dependent firms*.

The third and fourth types of firms are those that are dependent on either foreign or domestic buyers or suppliers, i.e. *value chain dependent*.

The factor analysis solution has demonstrated the different network orientations of KBEs in CEE. It shows that although on the aggregate, KBEs are dependent on both value chain links as well as on national innovation system links (universities, research institutes), this dependence is highly differentiated. Some firms are indeed dependent on the public research system, and some on foreign networks, while some KBEs are very similar to 'ordinary' firms by being dependent mainly on value chain partners, domestic or foreign.

6 Conclusions

Our findings allow us to make the following conclusions about KBE in CEE countries. Knowledge-based entrepreneurs in this region usually start their careers in the business sector rather than the scientific sector, and start knowledge-based firms in order to take advantage of market (i.e. commercial and financial) opportunities (technological opportunities are frequently mentioned as a key rationale for establishing companies only in the Hungarian sample, where we see more entrepreneurs coming from the science sector). These entrepreneurs bring knowledge about products and technology from their previous employment and then develop new markets with their new firms. In that respect, KBE in CEECs can be considered as primarily a market repositioning activity.

KBEs in CEE are not 'gazelles' (i.e. fast-growing new technology based firms (NTBFs) which have the potential to reshape the industrial landscape). They consist of distinct types of companies, of which NTBFs are only one. The key factor in KBE firms' growth is most often firm specific capabilities which do not always involve R&D. Based on what the firms have identified as their success factors, we have identified three groupings of companies: new technology based firms, 'networkers' and companies whose success is based on 'customer-oriented organizational capabilities'.

The most common developmental barrier in domestic markets is the low level of demand on those markets. This is followed by high labour costs, increased competition and lack of public support. Firms fell into two groups with respect to the kinds of barriers that were most important for them. The first group is one where the main barriers concern skills shortages and high labour costs. For the second group, the major barriers are related to finance (lack of access to finance and of public support).

Innovation survey data for CEE countries show that in-house and value chain partners (buyers, suppliers) are key sources of knowledge for innovation. Unlike standard companies, which tend to limit their strategic interactions to value chain partners, the networks of KBEs are broader and more frequently involve innovation system actors (research institutes, universities), including professional networks (fairs and exhibitions). Indeed, here again we can identify a number of distinct types of firms (with respect to the sources of knowledge that are most important for their innovation processes): those where *value chain partners* are the key source of knowledge for innovation; those where *formalized R&D* like patents and journals and research organizations are the key source, and those where *in-house* or firm specific innovation activities are a key source of knowledge for innovation.

Another grouping of companies with respect to the intensity of their links with external organizations allows us to distinguish four types of firms: network dependent, public research system oriented, foreign and domestic value chain dependent firms. This shows that for different types of KBEs different networks are important. In general, these are either vertical (foreign and domestic value chains) or horizontal (links with the domestic public research system).

Acknowledgements

The authors would like to thank local experts that have collected or/and organized data collection: Andrea Szalavetz in Hungary, Anna Kaderabkova in the Czech Republic, Martina Hatlak in Croatia, Monika Kriaucioniene in Lithuania, Gheorgiu Radu in Romania, and the researchers at Biuro Badan Spolecznych in Poland. The usual disclaimers apply.

Annex: NACE classifications in industrial breakdown

Pharmaceuticals, chemicals and plastics

NACE Division 24 (manufacture of chemicals and chemical products)
NACE Group 24.4 (manufacture of pharmaceuticals, medicinal and chemicals and botanical products)
NACE Division 25 (manufacture of rubber and plastic products)

Manufacturing

NACE Division 29 (machinery and equipment)
NACE Division 30 (manufacture of office machinery and computers)
NACE Division 31 (electrical machinery)
NACE Division 32 (radio, TV and communication equipment)
NACE Division 33 (instrument engineering)
NACE Division 35 (other transport equipment)
NACE Group 35.3 (manufacture of aircraft and spacecraft)

Software

NACE Group 72.1 (hardware consultancy)
NACE Group 72.2 (software consultancy and supply)
NACE Group 72.21 (publishing of software)
NACE Group 72.22 (other software consultancy and supply)
NACE Group 72.3 (data processing)
NACE Group 72.4 (database activities)
NACE Group 72.5 (maintenance and repair)
NACE Group 72.6 (other computer related activities)

Media

NACE Group 64.2 (telecommunications)
NACE Group 92.1 (motion picture and video activities)
NACE Group 92.2 (radio and television activities)
NACE Group 92.3 (other entertainment activities)
NAICS 52.8 (internet service providers; NACE code not available)

R&D

NACE Group 73.1 (research and experimental development in natural sciences and engineering)
NACE Group 74.2 (architectural and engineering activities and related technical consultancy)
NACE Group 74.3 (technical testing and analysis)

Notes

1 School of Slavonic and East European Studies, University College London.
2 CASE – The Center for Social and Economic Research and University of Edinburgh.
3 As our classification is product-based, we were unable to include in our sample firms that are innovative in terms of knowledge-based or high-tech production processes but produce traditional products. Hence it is likely that some types of knowledge-based firms are omitted from our sample. However, any other approach to identification would be prohibitively costly.
4 On the basis of the Kruskal Wallis test, differences across countries with respect to sectoral structure are found to be significant (Chi-Square=60.502, Sig=0).
5 A Kruskal Wallis test reveals no significant differences between countries in our sample with respect to ownership structure (Chi-Square=9.696, Sig=0.084).
6 An ANOVA test shows no significant differences between countries with respect to firms' size (Sig=0.142; F=1.668).
7 The high share of micro enterprises in the Croatian sample is consistent with the high share of owners in Croatian sample for whom this is the first job. See below.
8 An ANOVA test of significance shows non-significant differences between industries with respect to size of KBEs (ANOVA F=1.908, Sig=0.109).
9 An ANOVA test of significance shows non-significant differences with respect to year of establishment across countries (ANOVA F=2.195, Sig=0.055).
10 An ANOVA test of significance shows that differences between industries regarding year of establishment are statistically non-significant (ANOVA F=1.777, Sig=0.133).

References

Business Week (2005) 'Rise of a powerhouse: how the young knowledge workers of Central Europe are pushing the region to a new level (International Cover Story)', *Business Week*, 12 December.

CEC (2002) *New Enterprises and Development of Enterprises in Central European Countries: Data 1995–2001*, Luxembourg: Commission of the European Communities, Office for Official Publications of the European Communities.

Clendenning, E. Wayne, & Associates (2000) 'Comparison and reconciliation of SIC and NAICS industry codes used to define knowledge-based industries (KBIs)', prepared for Industry Canada, Small Business Policy Branch.

Eurostat (2005) 'Patent classifications and technology areas', November (available at http://europa.eu.int/estatref/info/sdds/en/pat/pat_meth_hitech.pdf; last accessed 31 May 2007).

Grant, R.M. (1996) 'Toward a knowledge-based theory of the firm', *Strategic Management Journal*, vol. 17, Winter Special Issue: 109–22.

UNCTAD (2005) *World Investment Report: Transnational Corporations and the Internationalisation of R&D*, Geneva: UNCTAD.

10 Diversity of knowledge-intensive entrepreneurial firms

Struggling biotech firms despite opportunities

Johan Brink and Maureen McKelvey

1 Introduction

Scientific and technological knowledge is increasingly understood as one of the key variables to explain the competitiveness of national economies and of firms (Rosenberg and Birdzell 1986).[1] New scientific and technological knowledge helps creates opportunities for entrepreneurship, by renewing the knowledge base within existing industries as well as in emerging industries. Hence, as knowledge progresses, entrepreneurial opportunities are opened up for business ventures, which are vital to the renewal of the economy (Shane and Venkataraman 2000). Still, the development of technological opportunities does not automatically lead to the establishment and development of successful firms. Despite an opportunity-rich environment, many knowledge-intensive entrepreneurial firms often struggle for their survival. Hence, these struggling firms can be used to address one particularly relevant issue in knowledge-intensive entrepreneurship, namely the ways in which the development of scientific and technological opportunities affects the diversity of development within new firms.

This chapter addresses how and why these knowledge-intensive entrepreneurship firms frequently struggle for survival, while trying to benefit from knowledge as a dynamic and evolving asset. By choosing firms within one opportunity rich sector, biotechnology, we address whether and why differences in technological opportunities – and thereby in their knowledge assets – fundamentally affect the future firm development. We have chosen case studies in the biotechnology industry as much of the recent theoretical and empirical work on industrial dynamics and entrepreneurship has focused on biotechnology as an extreme example of how scientific knowledge is linked to commercialization (McKelvey and Orsenigo 2006).

2 Dissecting the concept of firms active in knowledge-intensive entrepreneurship

Our starting point is that much knowledge-based entrepreneurship results from technological opportunities and innovation that in turn result from both knowledge processes and economic processes (Fagerberg *et al.* 2005; McKelvey

1996). Knowledge-based entrepreneurship is thereby seen as a concept which captures several aspects of the modern stream of thought about the importance of especially technical and scientific knowledge for stimulating firms and economic growth.[2] Partly due to a combination of 'science and technology push' as well as 'market pull' which open up new opportunity space, these firms have incentives to develop innovations. This chapter will contribute to these debates, by showing why doing so can be difficult, however, for the firm. Section 3 presents a taxonomy, which was developed in order to specify the dimensions upon which knowledge matters, as an evolving asset within knowledge-intensive firms. This taxonomy relates the actors' ability to use and develop technological opportunities to the specific nature and dynamics of the underlying knowledge as useful in products and services. Four case studies explore the relationship between the theoretically based taxonomy and illustrations from specific biotechnology companies.

This proposed perspective about how to analyse the dynamics of knowledge-intensive entrepreneurship draws upon two bodies of theories, one around knowledge as defining the firm in itself and one around the differentiated nature of technological opportunities. Our combination of theories therefore enables us to propose a specific, conceptual matrix in order to analyse the development of knowledge-intensive entrepreneurship firms. Knowledge-intensive entrepreneurial firms are here defined as firms which use and develop new types of technological knowledge and artefacts. This definition stresses the use of technology for a business context, and not simply the development of science or technology per se. Or, in other words, this definition stresses the innovation aspect, as well as the invention aspect.[3] The broad theoretical starting point here is that knowledge-intensive entrepreneurship is an extreme form of the theory of firm literature, which assumes that firms can use knowledge as a primary asset to compete (Barney 1991; Grant 1996). Therefore, we expect that scientific and technological knowledge can be considered a particularly important asset, which helps shape the future development of these firms.

However, our interest is in linking these knowledge-intensive entrepreneurship firms to the economic context. At a broader political economic level, the concept of entrepreneurial activities and opportunities rests upon an evolutionary economic view of the economy as changing due to endogenous forces (which go beyond the price mechanism). These concepts stress the search for profit resulting from new knowledge about resources, markets and their inter-relationship. Hence, the notion of opportunities is central to understanding economic transformation in a capitalist economy, both at a micro and a macro level. The theoretical starting point in these theories is thus one which stresses 'opportunities open to new enterprise and investment' (Schumpeter 1942: 113). The concepts of technological opportunities have been developed as a key concept to capture the driving forces behind economic development in sectors. The original definition of the concept of technological opportunities refers to differences in the marginal rate of technological return on R&D investments such as the likelihood of innovating (Scherer 1965). Further theoretical development about the differences in technological opportunities in industrial sectors conceptualize why technology advances much faster in some sectors than in others (Klevorick *et al.* 1995; Nelson and Winter 1982). Hence, technological opportunities differ across sectors and industries.

This chapter goes further, by relating the concept of technological opportunities within specific sectors to the firms and their ability to compete. This enables us to analyse how and why knowledge-intensive entrepreneurial firms grow and develop over time, in relation to specifically defined and diverse knowledge bases. As such, we conceptualize that the knowledge-intensive firms must identify, act upon and realize innovative opportunities (Holmén *et al.* 2007). We therefore presume that going further in understanding the dynamics and developments of knowledge-intensive entrepreneurial firms per se requires also understanding how their knowledge competencies and business opportunities develops. This requires a long-term view of firm evolution and knowledge development, where the build-up of capabilities within the firm is strongly path dependent and contingent on past capabilities and accumulated knowledge (Nelson and Winter 1982). A significant proportion of the firm's competitive advantage is hence represented by the potential for further develop knowledge and technologies underlying new competitive products and services (Teece *et al.* 1997). Competitiveness at the firm level will come from both the degrees of novelty, as well as the breadth of application of knowledge to products and services.

Let us now turn to our theoretically based taxonomy, by developing the two main dimensions to analyse knowledge-intensive entrepreneurship, namely differences within the opportunities for continued knowledge development (such as driven by R&D, etc.) and differences within opportunities for the firm to re-use knowledge whereby the firm can use that technological knowledge to develop a narrower or wider range of goods and services. These elements, we argue, will help us analyse the networks and changes visible at the firm level, over time. Based on the above literature review and previous work, this subsection proposes a theoretically derived matrix to specify the differential opportunities available to a knowledge-intensive entrepreneurial firm in a sector (Brink and McKelvey 2006).

The matrix uses two dimensions of knowledge as an evolving asset for the knowledge-intensive entrepreneurial firm, namely technological opportunities and opportunities for the firm in terms of a range of products and services. Table 10.1

Table 10.1 The two proposed dimensions of technological opportunities for development and opportunities to use for knowledge resources within knowledge-intensive entrepreneurship firms combined into a 2 × 2 matrix

Technological opportunities	*Technological opportunities to use*	
	Specialized	*Broad*
Science	II Science-specialized	I Science-broad
Engineering	III Engineering-specialized	IV Engineering-broad

refines the concept of technological opportunities in relation to knowledge-intensive entrepreneurship firms, by using the two dimensions specified below.

The first dimension belongs to the realm of differences across technological opportunities, which can be said to be related to the likelihood of invention in the future. More specifically, the first dimension defines technological opportunities in terms of the dimensions of a more engineering vs. a more basic science style of technological knowledge development.

On the one hand, analysing aspects of technological change has long been a central theme within economics and management research, as well as the history of technology. One main proposition is that the evolution of technology traditionally depends upon a self-reinforcing paradigm, in which the historical development of such knowledge can be explained in terms of a given set of problems and problem-solving activities which are identified and addressed by practitioners (Dosi 1982). Vincenti (1990) further specifies how the breakdown of problem solving into subcomponents gradually enhances and broadens the underlying, required knowledge to solve the engineering problem. A similar progression of technological knowledge has been described by Nightingale (1998) as a process of pattern recognition and extrapolation into new, but related, problem situations–solutions, in which the successful outcomes reinforce the interpretation of the current paradigm.

On the other hand, this traditional engineering-style knowledge development can be contrasted with a science type of knowledge development. In this chapter, we are primarily interested in science type of development only in relation to problem-solving, with implications for the 'breadth' of the opportunity. Science and engineering are often seen as fundamentally different ways of knowledge development, and often linked to the radical–incremental discussions, and the history of science has long had a complex debate over these issues. Engineering practice (technology) can here be seen as activities predetermined of accumulated prevailing practices while science is seen as a venture expansion, unknown and 'un-expectable' (Klein and Rosenberg 1986).

These two styles of development are relevant for the knowledge-intensive entrepreneurship firms. A distinction can thus be made between technological development as composed of refinements and that as an effort by research to deliberately discovery novelty. Technologies are hence characterized by inherited and more or less formalized methods, and guided by well-structured problem solving. This results in new combinations and solutions, which are still well within the realm of the well understood. In other words, technologies are within the given trajectory and extrapolated from past solutions. Given the potential to break down such technologies into predetermined subcomponents and proven working solutions, complicated development activity can be roughly predefined and compartmented, allowing for simultaneous development rather than sequential development (Sanchez and Mahoney 1996). In contrast, a scientific type of technological opportunity and knowledge development expands new knowledge more radically. The introduction of discoveries thus occurs outside the current coherent and well-known areas of knowledge. In the same vein, a radical innovation may open up for later stage, more detailed understanding, and thereby the development of a coherent paradigm and the eventual emergence of a new technological trajectory. Such developments are, however, by no means given by the initial discovery per se (Kline and Rosenberg 1986). Therefore, engineering and scientific styles of development will be related to the differing ways of developing products and services within the firm.

The second dimension stresses the differences within opportunities for the firm to re-use knowledge whereby the firm can use that technological knowledge to develop a narrower or wider range of goods and services. Technological opportunities are thus here further specified, relative to the use of that knowledge asset within current and future products and services. The distinction is between the firm's technological knowledge as a specialized asset, or as a broad asset, defined in relation to production of a product and/or delivery of a service.

The conceptualization that the firm must combine their technological capabilities to produce future goods and services is an idea more prevalent within management research, than within economics. Accordingly successful firms frequently excel in the integration of individuals and specialized knowledge rather than specialized knowledge per se (Grant 1996). A more specific reasoning can be found on literature about complex product systems, especially the tradeoffs between the width and depth of knowledge base (Prencipe 2000). Moreover, research suggests that a majority of technology based firms have accumulated broad, distributed competencies rather than narrow, distinct technological competencies (Granstrand *et al.* 1997). For the firm, possession of such technological knowledge results in a broad set of potentials for gradual refinement of the technologies as well as a high number of potential combinations for future goods and services. Hence, this second dimension relates to the width of application in the market, to which knowledge assets can be applied, generalized and recombined rather than to the development of new knowledge per se.

Using these two dimensions, Table 10.1 thereby specifies a taxonomy with four categories of knowledge-intensive entrepreneurship firms. They are science-broad, science-specialized, engineering-specialized and engineering-broad knowledge-intensive entrepreneurship firms.

3 Research design and methodology

Our research design combines theoretical reasoning with explorative case studies of four knowledge-intensive entrepreneurship firms within biotechnology, which match the four theoretically derived quadrants identified above. The case methodology partly draws upon a grounded theory approach but it has more elements of explorative and theoretically relevant cases (Eisenhardt 1989; Yin 1989). More specifically, our case studies are used to strengthen and explain theoretically derived propositions, rather than derive and create new theory directly from empirical work or to illustrate theory. The combination of the theoretical review as well as the empirical cases should thus help us to develop and generate a more coherent and detailed set of hypotheses, as a result of the analysis.

3.1 Empirical material and indicators

Our case study research is primarily based upon historical developments for each firm, based on multiple sources. In terms of time period, we have studied each firm from their start-up date to 2005. Hence, a variety of complementary empirical methods are used, in order to develop a detailed understanding of the firm per se and in particular of the relationship between technological assets, opportunities and growth dynamics of these studied firms. A summary of the variables and empirical sources can be found in Table 10.2.

Our research design is built upon a three-stage process in which we initially focused upon choosing four firms, to match the matrix and its four quadrants, as specified in the next section.

First, we needed to find a way to specify and differentiate the scientific and knowledge assets of specific firms. We use the number of patents and publications as crude indicators of scientific and technological output for the firm. Further on, different patent classes are used as an indicator of the potential for combining or diversifying into related knowledge areas. The knowledge ego-networks are analysed, primarily based on formal points of collaboration through patents and research/publication networks.

Second, in order to interpret the relationship between knowledge characteristics and firm development we needed to find a way to specify and differentiate the innovative efforts, technological opportunities and successes facing the specific firms. For these cases, we used qualitative data, about the specifics of the firms. We were particularly interested in the relationship between new products and existing products.

Finally we were interested in how the firm was able to translate the evolution of knowledge/technologies into growth of the specific firm. Such an indicator should capture the growth of firms either by sales, size and employees or by geographical aspects such as expansion or contraction nationally and internationally. Quantitative indicators of growth are measured by figures on sales and employees from annual reports and changes thereof.[4] More qualitative descriptions did arise from the narratives of the development of these four firms. Given the lack of sales in entrepreneurial industries, financial indicators are difficult to use so we have used knowledge performance indicators (Murphy *et al.* 1996).

3.2 Relating cases and theory: biotech knowledge-intensive entrepreneurship firms and the matrix

The exploratory case study design was chosen to examine opportunities and firm development of specific knowledge-intensive entrepreneurship, over time. These four firms have all been started up as small, independent firms, and they are involved in what can be defined as the Swedish bioscience industry (McKelvey *et al.* 2003). The four case firms were therefore chosen according to case study logic of internal diversity, in that they can be described according to the conceptual differences proposed in Table 10.1 in Section 2. One firm was chosen per

Table 10.2 Major conceptual themes and their specific research design

	Method	*Indicator*	*Specific information*	*Source*
Technological competencies	Quantitative	Patent	Number, and classes of patents	Patent database
Scientific competencies	Quantitative	Publication	Magnitude and distribution of collaborations	Web of science
Innovativeness	Qualitative	Radical products	Estimated number	Annual reports
Technological integration	Qualitative	Number of technological fields of activity	Distribution of patents and publications	Areas of competencies
Firm development and growth	Quantitative	Employees' international activity	Number	

quadrant, as shown in Table 10.3. The research design thereby allows us to explore whether, and how, our initial theoretical ideas find any basis in the empirical material.

Quadrant I is science-broad, and hence it combines the dimension of scientific development with many opportunities for technological recombination in products and services. Therefore, the firm chosen here should represent a type of knowledge-intensive entrepreneurship firm where we expect both a broad knowledge base with continuous product development, but with occasional addition and amalgamation of radical new ideas. This quadrant is represented by the case of the 'Diagnostic'. This firm was chosen because they have managed the complex task to gradually improve and extend their initial products while at the same time introduce and integrate a radically innovative new product class.

Quadrant II is science-specialized and it combines the dimension of scientific development with the dimension of low opportunities for technological recombination and re-use of knowledge into other products and services. Therefore, the firm chosen here should represent a knowledge-intensive entrepreneurship firm where we expect strong focus on knowledge competencies, accumulated over time. This type of knowledge-intensive entrepreneurship should have a narrow and highly specialized knowledge/technological base and an innovative driven project, and the type is here represented by the case of 'Pharma'. They try to introduce new pharmaceutical active molecules based upon radical research about cell receptors.

Quadrant III is engineering-specialized, and it should represent knowledge-intensive entrepreneurship firms with a knowledge base defined both with an established technological practice but also the relative absence of any complex integration of detailed and differentiated knowledge in order to solve recurring problems. In the bioscience industry this type of firm has not gained as much attention but can exist, for example, as a local provider of services and production. The chosen case study is of firm 'Test' which has a rather narrow technological base. The chosen case firm performs clinical research activities in collaboration with small, research-intensive firms, within the pharmaceutical research chain.

Finally Quadrant IV is engineering-broad, and it should include a type of KBE firms which are active within a well defined and extensive technological area, but yet have a low rate of truly radical scientific development. The chosen case firm is 'Bioprocess' which has broad technological base and relatively large

Table 10.3 Four selected case firms according to the proposed differences within knowledge resources for knowledge-intensive entrepreneurship firms

Technological opportunities	*Technological opportunities to use*	
	Specialized	*Broad*
Science	II 'Pharma'	I 'Diagnostics'
Engineering	III 'Test'	IV 'Bioprocess'

opportunities for knowledge recombination. They have managed to jump from project to project, gradually increasing their understanding of their engineering capabilities with the upscaling of biotechnical processes.

Thus, these four cases of knowledge-intensive entrepreneurship firms studied and analysed here were chosen according to the logic and criteria related to similarity and diversity, suitable for performing exploratory and theoretically driven case studies.

4 The four knowledge-intensive entrepreneurial firms

These historical developments of these four knowledge-intensive entrepreneurship firms are presented in turn, following the four ideal types defined above.

4.1 'Diagnostics'

Based on an initial overview of the firm, 'Diagnostics' was chosen as a knowledge-intensive entrepreneurship firm in Quadrant I, science-broad type.

'Diagnostics' was founded 1984 in Gothenburg. The initial innovation was the outcome of a research project undertaken by three researchers at Gothenburg University. The research group had, during the previous two years in collaboration with the Sahlgrenska Hospital, developed and clinically tested a diagnostic application based on monoclonal antibodies for the testing of various cancer forms. The initial investment in 'Diagnostics' came from a Swedish investment fund started in 1979 by the Swedish government for the investment in new technology based firms. The major ownership was however controlled by a large Swedish industrial corporation from which the CEO was recruited. 'Diagnostics' immediately focuses on initiating commercial production and marketing of the first version of a specific cancer test kit.

In 1985, a newly graduated PhD researcher was recruited to lead the internal research and further product development. 'Diagnostics' then maintained continuously relationships with the original research group at the Gothenburg University for future product developments. This research relationship was, in 1986, complemented with a joint R&D project with a cancer institute in Beijing, China. In 1987 'Diagnostics' was sold to a larger Swedish pharmaceutical firm, which expanded the research effort towards various cancer diseases, aiming to incorporate the outcomes into its internal cancer therapeutics programme. In the beginning of the 1990s 'Diagnostics' grew to around 20 employees, of which about half were occupied with the diagnostic business unit. In 1992, the larger pharmaceutical firm restructured and concentrated all cancer research to Lund. The ten employees in the 'Diagnostics' business unit, however, made a management buy-out, and licensed the developed technologies, which still remain under the ownership of the large pharmaceutical firm.

During the 1990s, the firm focused upon continuously developing new diagnostic kits and commercially introducing them for various cancer forms. These products were sold to both hospitals and distributors in an increasing number of countries. Together with the cancer test kits, 'Diagnostics' developed a second business element where reagents, such as antibodies and antigens, were sold on long-term contracts to external firms, which develop diagnostic applications on their own. In the mid 1990s foreign sales offices were established in the United States and in 1997 a representation office was established in Beijing.

Around 2000, 'Diagnostics' grew to around 20 employees and engaged in a radical innovation strategy based upon technologies around DNA, mRNA. In addition a new type of product generation was introduced, which broadened the applications from cancer diseases towards brain disorders. In 2001, the patent held by the large pharmaceutical firm expired and 'Diagnostics' was no longer dependent upon external relations for its core technology. To further develop new products 'Diagnostics' entered relationships between a research institute in Gothenburg and deepened the collaboration with Gothenburg University, as well as entering a network of European firms and research groups supported by the European Union.

In the beginning of twenty-first century, 'Diagnostics' added sales representatives in both Russia and Italy. By 2004 'Diagnostics' had around 30 employees of which one-third were employed outside Sweden. Further on, 'Diagnostics' started with the development of a new factory in China; the construction, however, stopped in 2004.

This specific firm illustrates the science-broad type of knowledge-intensive entrepreneurship and it can be interpreted as follows. The case of 'Diagnostics' shows that knowledge arising under the development, production and usage of bioscience is potentially essential for further product generation. The patent structure shows high technological diversity, although the numbers of patents are low. The external relationships in knowledge generation, measured by the

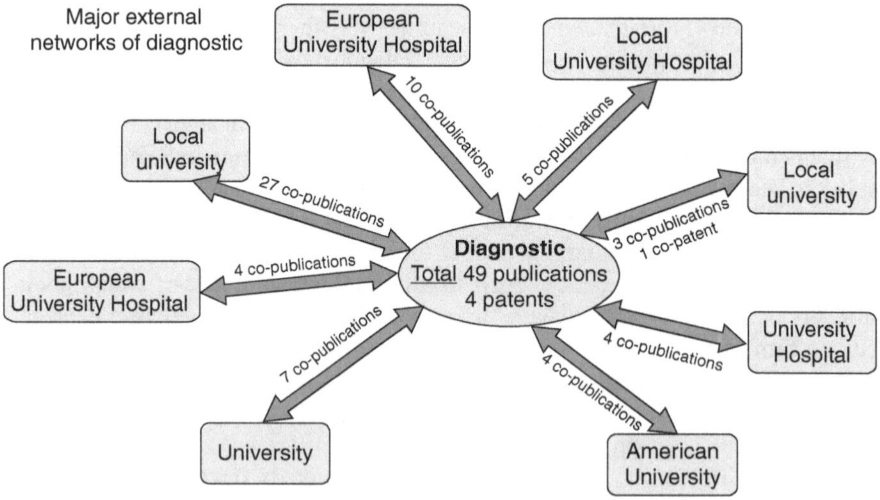

Figure 10.1 Knowledge network for 'Diagnostics'.

publication network of 'Diagnostics', show an increasing diversification with several independent relationships both between different universities as well as different users and applications of the developed technologies which were crucial for the development and integration of the radical innovation of DNA technology applications.

The technological opportunities for 'Diagnostics' are thus seen as 'widening' and the firm is able to transfer knowledge to new generations of technologies. Knowledge development arises within the relationship of applying and using the known knowledge/technology deepening the design space. This implies that learning occurs simultaneously as the technology is being applied. This could be exemplified by the dual strategy of 'Diagnostics' to both be a developer of actual test kits, and as an initial developer of reagents sold to other diagnostic application manufacturers.

Over time, however, the knowledge-intensive entrepreneurship has changed its strategy quite a lot. 'Diagnostics', however, has gone from a single-project research spin-off, to product and technology diversification and actually produces, markets and distributes the different projects. With the event of DNA technology 'Diagnostics' radically expanded their design space. This allowed them to fuse external knowledge gained through networking with the internal practices and evolved knowledge around the initial technologies.

The main results as relevant to specifying different categories of knowledge-intensive entrepreneurship firms are:

- 'Diagnostics' managed to gradually improve and diversify their product portfolio with related products and build up a organizational structure supporting their business.
- Further on, 'Diagnostics' managed to introduce a second radical innovation, which partially related to their existing competencies but partially involved a degree of newness. This innovation was dependent upon a large external research collaboration and effort to expand the knowledge base.

4.2 'Pharma'

Based on an initial overview of the firm, Pharma was chosen as a knowledge-intensive entrepreneurship firm in Quadrant II, science-specialized type.

'Pharma' was founded in 1998 as the outcome of a research project at Uppsala University, which had started at the beginning of the 1990s. 'Pharma' located within the Uppsala University science park, with six researchers with experience from the two major pharmaceutical firms in Sweden. Four different subtypes of receptors are thought be involved in a numerous disorders such as inflammation (especially skin), sexual dysfunction and hunger control and related disorders, e.g. obesity. The goal of the firm is to develop pharmaceuticals targeted towards a special type of cell receptors which had been characterized in the university research project. A first round of financing was obtained when a Swedish venture capital fund invested the equivalent of 1.5 million euro in

'Pharma'. Apart from knowledge around the target, e.g. the specific receptors, 'Pharma' acquired competencies in combinatorial chemistry and chemometrics, which is an information-intensive chemical method used in the search for pharmaceutically active entities. These competencies were acquired by recruiting a researcher originally from Umeå University where that technology had partially evolved.

In 2000, a second round of financing involved two additional major Swedish venture capital firms. This financing triggered an expansion in the number of employees to ten, and 'Pharma' relocated to new facilities outside Uppsala where the firm had better opportunities for continuous growth. Growth in the number of employees continued, and by the autumn of 2001 'Pharma' employed 23 persons. In addition between ten and 20 chemists worked regularly for the firm, through a contract research organization in Riga, Latvia.

During 2002 'Pharma' merged with a newly established research firm in Stockholm, in order to broaden its project portfolio. Around the time of the merger, a third round of financing was obtained, taking on additional investors. The prime project entered clinical trials, and a collaboration agreement for further development was signed with a major American pharmaceutical firm.

In the autumn of 2003, 'Pharma', now employing around 40 researchers, ran into financial troubles. The multiple and distributed venture capital owners demanded a restructuring. This reduced the number of researchers to 27, and the firm explicitly focused its efforts upon the prime project towards the specific receptor. However, in the second phase of clinical trials in the beginning of 2004 'Pharma' declared bankruptcy because the second largest owner no longer agreed to additional funding. 'Pharma' was liquidated and parts of the intellectual property rights were sold to another small research firm.

This case of 'Pharma' can be interpreted as a specific type of knowledge-intensive entrepreneurship firm, in terms of impact of technological opportunities and competencies in the development of the firm. In this case, the pharmaceutical research activity is science based, dependent upon clinical research as well as upon university and public (open) research as the primary new source of innovations. Moreover, knowledge about biological relations are generated in the scientific community rather than accumulated internally in the firm due to prior research projects. Hence, 'Pharma' received the initial idea from the Uppsala University researchers, and it maintained strong linkages with the initial inventors. Medical scientific knowledge was not enough, however. Additional research activities are also dependent upon new technologies, which allow problem-solving during the development process. For 'Pharma', they identified chemometrics and hired a researcher from Umeå University as well as contracted research at a research institute in Latvia. Critical for the (un-)sustainability of 'Pharma' were the networking capabilities where new ideas, financing and technologies were accessed externally during projects. They had intense relationships with a few partners. 'Pharma' combined relatively unique knowledge and idiosyncratic radical design concepts. When the project failed the firm was scattered, and the intellectual property rights partially transferred to an alternative organization.

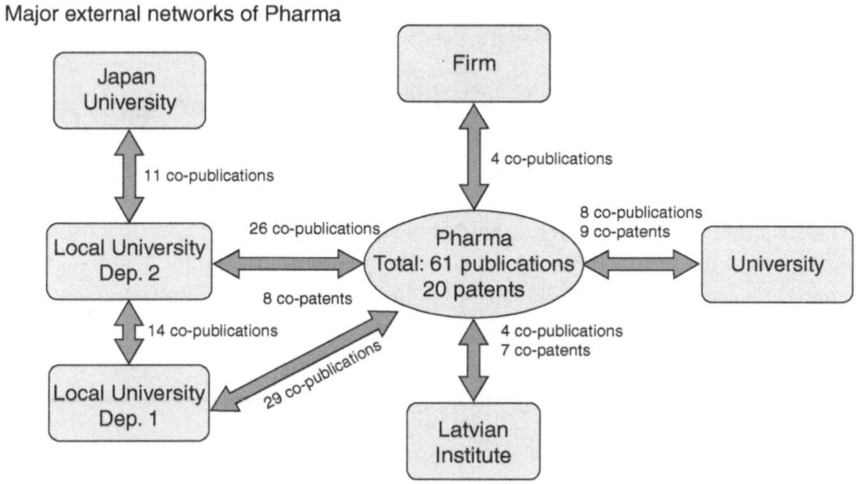

Figure 10.2 Knowledge network for 'Pharma'.

This analysis is also visible in the quantitative material. From the data of patent and publications, we can describe the technological opportunities as 'narrow'. We have argued that a narrow opportunity is reflected in a specific structure of relationships and network structure. As expected, 'Pharma's' ego-network is characterized by high intensity, strong ties, with a few external actors.

This specific firm illustrates the science-specialized type of knowledge-intensive entrepreneurship, and the reasons for this can be analysed through our concepts as follows. Biotechnology and pharmaceuticals represent a sector where competencies and products differ, because each new project requires significant effort. Scientific knowledge which emerges and competencies which are accumulated during the research process might seldom be leveraged in new, related projects. The design spaces are vague and can vary between projects. Of course, a larger firm may be able to leverage such knowledge, due to its deep and specific knowledge around a therapeutic area, but a smaller firm seldom has resources to cover this broader area. Hence, for each given research project within the biotechnology sector, the final product in terms of a pharmaceutical is often not a natural starting point for the future development of competencies and incremental innovations. Path dependency in terms of learning by usage of such a product is thus not a critical source of innovations nor new product generations. This causes difficulties for the knowledge-intensive entrepreneurship firm, attempting to develop the technological opportunities.

The main results and implications as relevant to specifying different categories of knowledge-intensive entrepreneurship firms are:

- Knowledge did not become embedded in any wider organizational architecture or organizational structure. This holds for both the receptors and the enabling technologies. Even if the biomedical design space concepts (drug targets) were proven to be doubtful, the firm and individuals did not leverage organizational capabilities in research technologies (chemometrics) into any new ventures.

- The main outputs of the firm are visible as codified knoweldge in patent applications and in the formed (related) chemical libraries. The main research results were published and left 'open' as contributions to public science. Possibly, another actor could develop the ideas in new directions.
- The top management team retained some of their competencies in a project-based organization. They formed an independent consultancy firm to leverage their complementary knowledge in project management, intellectual property rights and financial issues.

4.3 'Test'

Based on an initial overview of the firm, 'Test' was chosen as a knowledge-intensive entrepreneurship firm in the Quadrant engineering-specialized type.

In 1990 'Test' was founded in Lund as a small clinical trial company by an experienced manager from a Swedish major pharmaceutical firm. At the start 'Test' employed three people but this steadily grew to around 30 by the mid 1990s. During the initial five years 'Test' gradually developed relationships with the Swedish regulatory authority, the medical products agency, and gained competencies in regular approval procedures as the initial projects entered later stages. Simultaneously 'Test' established relationship with several university hospitals in Sweden for performing early stage (Phase I) clinical projects. By now 'Test' had internal experience in planning, performing, evaluating and filing for final pharmaceutical regulation approval.

By the end of the 1990s 'Test' employed over 100 persons and a geographical expansion occured. 'Test' established offices in the United States, Japan and Spain. Simultaneous to the geographical expansion, a collaborative agreement was entered into with a major US clinical research firm in 1999, providing access to joint clinical trials both in Europe and the United States. The Spanish office located in Madrid and Barcelona gradually became responsible for 'Test's' internal data handling such as statistical analysis, programming and biostatistics while the Swedish branch specialized into regulatory affairs.

After 2000 a new period emerged in the evolution of 'Test', which had expanded to over 120 employees at its main office in Sweden and around 60 in Spain. Growth in the number of employees stopped; instead 'Test' entered collaborative agreements with two Danish firms, one for distribution and handling of test substrates and one for extensive pre-clinical testing. In 2003 and once again in 2004 'Test' was forced to reduce the number of employees in Sweden as three major projects ended ('Test' does not have linkages patents/papers, so no diagram).

This specific firm illustrates the engineering-specialized type of knowledge-intensive entrepreneurship, and this case can be interpreted as follows. Build up, and replication of organizational capabilities with applications towards production and research services is an important category of industrial activity within the bioscience-based industry. Such capabilities could be targeted towards production and developing new technologies and pharmaceuticals but also providing services towards the research phase, such as various clinical practices. Critical for 'Test' was thus both to appropriate from the observed demand without losing competitiveness and to observe and avoid the emergence of radical discontinuous innovations, eliminating the original demand. For 'Test' this could be seen by gradually shifting parts of the activity outside Sweden as well as an immediate reduction in employees as the demand upon the delivered services fell.

However, 'Test' maintained the close links with different clinical agencies (university hospitals and regulative agencies) and this emphasizes the importance of such stable external relations (maintaining a reputation). Technologically and scientifically related opportunities have been hard to be perceived by 'Test'. No external relations towards technology development can be found, nor any patents. The only attempts to internally develop new technologies were sold off. However, 'Test' tried to diversify by entering into a relationship with two external firms in the clinical services industry. In 2005 'Test' acquired a competing firm. The motivation for the acquisition was to reduce the vulnerability of the firms to individual service agreements. Accordingly, no single customer relationship should extend above 15 per cent of the new firm's total turnover.

The main results as relevant to specifying different categories of KBE firms are:

- 'Test' had trouble with innovating: new ideas did not succeed. Neither knowledge integration and recombination, nor radical innovations have been a major source of additional competencies.
- The firm's competitive advantage has been based upon organization replication, in terms of expanding demand the firm has been able to scale up its activities, subsequently when demand dropped the firm reduced its size while maintaining its crucial competencies and relationships with both regulative agencies and public hospitals.

4.4 'Bioprocess'

Based on an initial overview of the firm, 'Bioprocess' was chosen as a knowledge-intensive entrepreneurship firm in the Quadrant engineering-broad type.

The history of 'Bioprocess' can be traced back to the beginnings of the 1960s where it was created as a business unit within a major Swedish food and agricultural company. By the early 1980s 'Bioprocess' was transformed into a joint venture owned by the initial food and agricultural company and by a major Swedish equipment manufacturer. The underlying idea was to develop and license biotechnical production technologies. In 1986 'Bioprocess' became an independent business unit. 'Bioprocess' acquired a patent from Russia, and a production unit for the biological production of L-threonine, an amino acid used in animal feed, was constructed in Lund. In the beginning of the 1990s the scale-up effort of L-threonine was finalized and the technology was sold to a major US agricultural corporation. 'Bioprocess' employed around 20 employees and changed its major strategy. The skills gained in scale-up and biological production acquired during the L-threonine project were now used towards the contract development and production of biologically produced active pharmaceutical ingredients. The production facility in Lund was upgraded to good manufacturing practice, GMP, and certified for pharmaceutical production. In 1994 a long-term contract was entered into with a major Swedish pharmaceutical diagnostic firm for the production of monoclonal antibodies.

In 1997, a Swedish dedicated biotechnology firm acquired 'Bioprocess'. Together they worked upon the scale-up and production of their biological product 'Feel–Good'. In the end of the 1990s 'Bioprocess' entered several agreements with European firms for the scale-up of biological processes of pharmaceutical production. The final production was performed both externally and internally, which forced 'Bioprocess' to expand its production facilities and to gradually employ over 60 people.

In 2001, the long-term contract with the major Swedish pharmaceutical diagnostic firm was extended. 'Bioprocess' now took additional responsibility and integrated purification of monoclonal antibodies to its initial production. In 2002 the scale-up and development of production methods for 'Feel–Good' were finished and 'Bioprocess' was sold to a Danish biotechnology firm. 'Bioprocess' however maintains its focus upon scale-up and commercialization of biotechnological production by the transfer of gained knowledge from project to project.

This specific firm illustrates the engineering-broad type of knowledge-intensive entrepreneurship, and this case can be interpreted as follows. 'Bioproc-

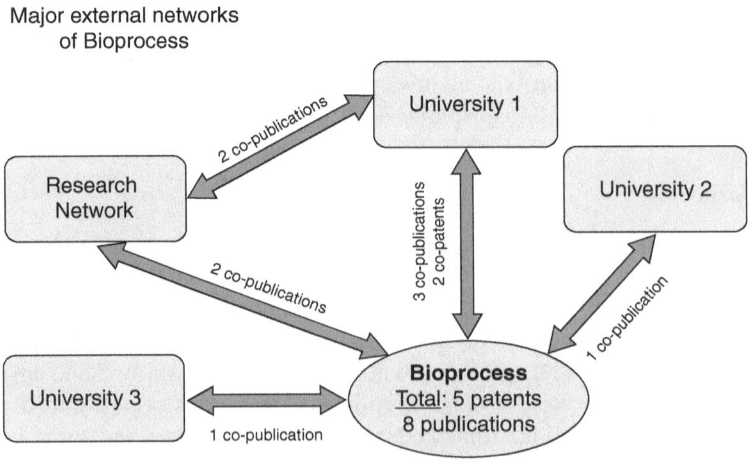

Figure 10.3 Knowledge network for 'Bioprocess'.

ess' shows that knowledge arising under the development, production and usage of bioscience is potentially essential for further product generations. Both the science and patent activity of 'Bioprocess' are distributed, as well as their external relationships. Knowledge development arises within the relationship of applying, using the knowledge/technology. Critical for 'Bioprocess' were the ability to maintain capabilities, access and integrate the different technological skills needed for designing and up-scaling biological production.

'Bioprocess' thus gained new knowledge within their ongoing activities and maintained relationships with universities and public research organizations to maintain their competitiveness within their design space. Internal research and radical innovations were given less emphasis. 'Bioprocess' tried to combine each development project, with leverage upon its knowledge base, with a stable standardized market for scale-up and outsourcing of production but has not yet mastered a standard business organization, as each business relation has a high degree of idiosyncrasy. Naturally the projects are unique, but the firm tries to leverage its knowledge and is now trying to 'standardize' towards more repetitive 'contract service' projects.

The main results as relevant to specifying different categories of knowledge-intensive entrepreneurship firms are:

- 'Bioprocess' has been involved in a number of different business models, yet their core competency of scale-up of biotechnical process has remained stable.
- To continuously upgrade the evolving knowledge base 'Bioprocess' has participated in a diverse knowledge community as well as learned from each individual project. Yet 'Bioprocess' has not made any effort to radically alter the scale-up processes, but has rather increased the internal distribution of knowledge to provide a 'complete' service in the individual projects.

5 Analysing the four cases of knowledge-intensive entrepreneurial firms

Our theoretically sampled case studies were designed to explore ideas about the dimensions by which knowledge-intensive entrepreneurship firms differ in terms of technological opportunities and competencies as well as within the resulting firm dynamics.

At the start, all firms acted upon an initial entrepreneurial opportunity based upon scientific and technological knowledge. In the four cases, the different entrepreneurs were active in other related activities, before starting the specific knowledge-intensive entrepreneurship firms studied here. As such, these firms were all spin-offs, occurring specifically within the bioscience sector either from an already existing firm or from a research university. Despite that similarity, these firms developed rather differently, due to a combination of factors, and so the development of these knowledge-intensive entrepreneurship firms differs tremendously, even if they all represent an opportunity-rich sector such as biotechnology. Table 10.4 summarizes the major empirical indicators for the four firms, in terms of the indicators defined in the research design section.

Hence, these indicators do suggest that the firms differ in terms of technological competencies, scientific competences, and firm growth and development.

These differences in firm development can be explored, in relation to the taxonomy presented in Section 2. The first dimension of a more engineering vs. a more basic science style of technological knowledge development has one set of implications. The engineering-style development of knowledge within a specific technological opportunity implies that the firm can accumulate and have relevant resources and competences internally. The basic scientific style of development of knowledge implies that the firm must continually renew their resources and competencies, to remain at the forefront of the research frontier. This second dimension of a narrower or wider application of technologies to future goods and services also has implications. Again, we can see different trajectories of development for the knowledge-intensive entrepreneurship firm. If the opportunities to use are narrow, then the knowledge-intensive entrepreneurship firm, by definition, has a technological knowledge asset which is narrow and specialized in terms of applications to products and services. These assets can therefore mainly be used to make incremental improvements to existing products and services, and within the same industrial sector. If the opportunities to use are broad, then the firm has a technological knowledge asset which can be applied to many different products and services. These assets can therefore be combined and integrated in such a way that the firm can develop radically new products and services, even within other industrial and service sectors.

These differences in knowledge-intensive entrepreneurship opportunities, in relation to firm competencies and development, help us analyse the case studies presented in Section 4. Let us focus on the idea of opportunities for use, defined in terms of broad or specialized applications to products and services.

In two cases – 'Bioprocess' and 'Diagnostics', these firms could identify, act upon, and realize broad technological opportunities to recombine and to re-use knowledge. In other words, new technological opportunities kept being developed, in ways which the firms could identify and act upon them, based upon the existing knowledge assets of the firm. The R&D activities of 'Bioprocess' and

Table 10.4 Summary of empirical data over the four case firms

Company	*Patent* activity*	*Scientific publication activity*	*Growth in size (employees)*	*Founding year*
'Diagnostic'	1/1/5	53	Slow/steady	1984
'Pharma'	(1**)/7/20	58	Rapid/bankruptcy	1998
'Test'	–/–/–	–	Rapid/turbulent	1990
'Bioprocess'	2/2/4	7	Slow/steady	1960s

Notes

* Patents are measured USPTO granted/USPTO applications/esp@cenet 'worldwide'.

** The patents of 'Pharma' no longer belong to 'Pharma' but have been transferred to another firm.

'Diagnostics' could thereby continuously reinforce and open up new business opportunities of relevance to their existing knowledge. They did so in different ways, however. 'Bioprocess' was able to do so through gradual changes and learning in the understanding of the technologies, which in turn led to an ever increasing refinement and an increasingly sophisticated means of biological production. 'Diagnostics' could instead pursue a sequence of opportunities, by recombining and integrating the original technology with an additional, DNA-based technology. This combination radically opened up for additional product developments, to address new markets.

In the other two cases – 'Pharma' and 'Test' – these firms act upon narrow opportunities for technology re-use. Here, the existing knowledge assets of the knowledge-intensive entrepreneurship firm did not help them generate and renew innovative opportunities. 'Pharma' pursued a lead project, based on receptors, and the accumulated knowledge within 'Pharma' did not reinforce its ability to make money nor did the internal knowledge development open up any significant additional opportunities, such as a product for a different market or disease. 'Pharma' thus had a significant 'decay' or 'depreciation' of the strategic value of its knowledge asset and was thereby different from the two cases discussed above. However, in some ways it did try that strategy. 'Pharma' also tried to leverage its knowledge base by diversifying its research product pipeline. On several occasions, they did so by trying to open up new opportunities mainly based upon 'the same scientific research' but applied to new products. However, they were forced to give up these efforts, due to financial stress and were finally forced to focus entirely upon the initial 'lead' project before the final collapse of the firm.

The final company, 'Test', also faced narrow opportunities for use, and they were not able to use their knowledge assets in the same way to compete, despite their more engineering style of development. They did make efforts to internally recombine knowledge and tried to diversify within their technological area, but these altogether failed. 'Test' also tried to develop new 'gadgets' and new routines for upgrading in their business but these did not reach a broad market. Hence, the range of new opportunities, as emergent out of the accumulated knowledge could thus be said to be rather low, both for 'Pharma' and 'Test' as expected. This is so despite significant differences in the styles of scientific, respective engineering, development.

From the perspective of 'broad' and 'specialized' opportunities to use knowledge to develop products and services, the differences in terms of knowledge networks are also clear in the four cases. The two cases representing 'broad' technologies – 'Bioprocess' and 'Diagnostics' – maintain a diverse and broad set of knowledge network relations. This can be seen both in terms of patents and in terms of scientific relationships as measured by the co-authoring of publications. We interpret this to indicate that the knowledge-intensive entrepreneurship firm has a need for accessing complementary external knowledge resources. For the 'specialized' technologies, such distribution and variance of partners with technological competencies and scientific competencies is not visible in our empirical results. This is either because of overall low activity, such as in the 'Test' case study, or because of the existence of a few but strong ties and well focused patenting activities in core technological areas, as in the 'Pharma' case study. Thus, the two specialized firms primarily used internally accumulated knowledge. They do so either to further strengthen their specialization and thereby act towards increased efficiency and scale as in the case of 'Test' or to build upon existing science and technology for further knowledge generation as in the case of 'Pharma'.

These differences have implications for the growth of the firms, as well as networks. The analysis of our empirical findings has strengthened the theoretical propositions that the main strategic value of the firm with specialized science knowledge is to be constantly opening up (or maintaining) the new opportunity in terms of continued internal knowledge development rather than opening up a broad array of different opportunities for knowledge (re-)combinations for products and services. The strategic value of knowledge for 'broad' firms instead benefits to a greater extent from knowledge recombination in relation to externally generated knowledge that is accessed through networks.

Finally we will briefly discuss and analyse the relationship between the knowledge as an evolving asset and the growth dynamics of the firms. Our explorative analysis of these four ideal types and the specific case studies illustrates quite clear differences along the growth dynamics in terms of employees. Along the dimension capturing a 'broad' knowledge base, two firms have shown slow but steadily growth in number of employees – namely 'Bioprocess' and 'Diagnostics'. For the two firms representing rather 'specialized' technological regimes, 'Pharma' and 'Test', firm growth has been dynamic and rapid, but also turbulent. These latter two firms have been forced to increase as well as decrease the number of employees.[5] We have, however, no direct theoretical reason to infer expected patterns upon the growth dynamics of these ideal types of knowledge-intensive entrepreneurship.

The analysis of the four explorative case studies suggests that the taxonomy may be usefully developed, in order to describe technological knowledge as a dynamic and evolving asset of the knowledge-intensive entrepreneurship firm. An important aspect is the degree to which accumulated technological knowledge can be used in order to solve new sets of problems and in order to develop new products and services. Although the biotechnology sector as a whole may be opportunity-rich, firms active within it may still struggle to survive and to use their knowledge assets to compete according to the specific dynamics of its underlying technologies.

6 Conclusions

This chapter has introduced the idea of diversity of knowledge-intensive entrepreneurship firms, due to the nature of underlying technological opportunities within a high-tech sector. Over time, the impact of sequences of opportunities also impacts the subsequent development of the firm in terms of knowledge networks and growth. The claims that we wish to pursue are hence twofold: first, the innovative opportunities which specific firms identify, act upon and realize may differ – even within a high-technology, opportunity-rich and industrial sector such as biotechnology. Second, by treating knowledge as a dynamic and evolving asset of the firm, which is intrinsically linked to the strategic success and growth of firms, variations of the dynamics within a high-tech industry can at least partially be better understood.

From a theoretical perspective on innovation and knowledge from evolutionary economics, we started by assuming that economic competition is driven by the ability of actors – especially different firms – to develop competencies and to innovate, and to appropriate the potential returns thereof. While that may be the general story, these four case studies demonstrated the difficulties and differences of actually benefiting from knowledge as a dynamic and evolving asset. Thus, we have argued, the nature of the specific technology and ability to use it within products and services explains the evolution of firms. Two aspects of the knowledge-intensive activities of firms have therefore been analysed:

1 the extent to which knowledge development and innovative activities can follow technological or a scientific regime of knowledge development; and
2 the extent to which broad knowledge bases and integration, rather than specialized and focused knowledge bases, gradually open up new opportunities.

Our results may thus lead, in the future, to interesting re-interpretations of existing literature. According to our empirical findings, the growth of knowledge-intensive entrepreneurship firms could hence potentially be explained by taking the contingencies of knowledge as a dynamic and evolving asset into consideration. The dominant view within studies on knowledge-based entrepreneurship, and in high-tech entrepreneurship in particular, usually argues in a normative favour for radical, scientific-type knowledge development and rapid firm growth, and problematizes the absence of such growth.

In contrast, according to our tentative findings, growth of firms within a high-tech sector such as biotechnology must be carefully analysed in relation to the sustainability of such firm growth. For the sustainability of firms, the importance of local learning and the re-use of knowledge for 'opportunity generation' seems more crucial. This is visible in the success of 'Bioprocess', as compared to 'Pharma'. A scientific breakthrough does not automatically translate into a rapidly growing firm, as seen in 'Diagnostics'. On the other hand, for rapid firm growth to take place the gradual unfolding of additional opportunities seems to be of less importance. This can happen quickly, in turbulent jumps. There, short-term market opportunities and temporary exploitation of radical discoveries generate motivations for a rapid build-up of a firm in order to pursue and appropriate upon the initial opportunity at hand.

This chapter thus leads to insights about how knowledge-intensive entrepreneurship affects the firms. Thus our main claim is that the main value and dynamics of entrepreneurial firms is within their potential for technological development for products and services. This suggests that a key competency of the firm must be on innovation and knowledge integration to deliver products and services, and not the technological competencies and scientific competencies per se.

This may help us question the dominant business model in biotechnology, of basic science firms. These cases show that firms which draw benefits from knowledge integration and local learning rather than radical innovations, such as the engineering firm 'Bioprocess', can prosper and grow. They do so, and appropriate profits, even without being especially (radically) innovative and without being especially science-intensive. Scientific and technological knowledge are only strategic assets at particular times and for particular reasons.

This suggests the need for understanding how the dynamics of technological opportunities affect specific firms. In certain industries, such as biotechnology, the developed capabilities and technologies are inherently followed by creative destruction caused by demand instability and/or the imitability of competitors. There is a rapid development of technological competencies and scientific competencies, as well as new results, within the sectoral innovation system as a whole. Hence, the knowledge-intensive entrepreneurial firm must devote resources to the internal application of capabilities and replication of knowledge as well as to the new possibilities opened up by further understandings and development of new science and technologies. Their ability to reconfigure assets and competencies of the firms is therefore crucial.

Quite possibly, we can speculate that biotechnology firms which rely upon basic science play an important role in testing the potential of new ideas, even if they have the worst track record of making money. Possibly, the greatest benefits from actually taking a new technology to market are the verification that the underlying principles and competencies are actually working. For such competencies the main value may lie in the potential for further development. This implies that a successful innovation must be seen in relation to the current knowledge and the integration of gained competencies and understanding with already established ones inside the firm. This is particularly true for the science-specialized type of knowledge-intensive entrepreneurship firm.

From a policy perspective, our analysis and empirical findings imply that the 'bioscience' industry cannot be treated as one heterogeneous industry even if the emergent technologies have opened up a range of technological opportunities and renewal across incumbent industries. The individual bioscience firm is highly dependent upon the idiosyncrasies of its specific technology and competencies, creating great intra-industry variety. Overly focusing upon radical innovations, protection only by patenting, publications and rapid growth of new firms is not the whole story of the bioscience industry. To innovate per se is important, but that is a small part of making a company based upon technological opportunities in biotechnology. A more important function of these biotech companies lies in their ability to integrate knowledge into coherent understanding for the further development of not only more science and technology but especially into use as products and services.

Notes

1 This research has been performed in the context of the VINNOVA-sponsored, RIDE research network, Phase I, administered by IMIT, the Institute for Management of Innovation and Technology, www.imit.se.
2 Analyses and reviews developed in this book demonstrate the variety of ways in which knowledge and entrepreneurship lead to new organizational forms in society and also impact economic growth. Related concepts thereby include R&D intensive firms, technology-based entrepreneurship, and university entrepreneurship.
3 This can be exemplified. Some chapters in this book use a specific quantitative definition such as 'all new firms started within an industrial sector defined by the OECD as high tech', to analyse complex phenomena. However, for our purposes, that high-tech definition focuses on the input to developing knowledge, but it does not allow us to explore how knowledge is actually turned into value. The conceptual definition proposed here therefore includes how knowledge is turned into value (and competencies) within the firms. As such, we can also assume that firms may have difficulties – or are 'struggling' – to realize the opportunities around them.
4 Sales are rather misleading in bioscience industry since the return of licensing and co-development agreements can fluctuate over the years. Measures such as employees, R&D and non-R&D spending are much more stable.
5 Growth (as increase in size) in discovery (science) based firms as opposed to growth in design (technology) based firms has been shown not to imply increased technological diversification, rather these firms maintain technological specialization through the process of growth (Saemundsson 2005).

References

Barney, J. (1991) 'Firm resources and sustained competitive advantage', *Journal of Management*, vol. 17 (1): 99–120.

Brink, J. and McKelvey, M. (2006) 'On the relationship between the evolution of technological firms and their knowledge development regimes', *Journal of Electronic Science and Technology of China*, vol. 4 (4): 1–11.

Dosi, G. (1982) 'Technological paradigms and technological trajectories: a suggested interpretation of the determinants and directions of technical change', *Research Policy*, vol. 11: 147–62.

Eisenhardt, K.M. (1989) 'Building theories from case study research', *Academy of Management Review*, vol. 14 (4): 532–50.

Fagerberg, J., Mowery, D.C. and Nelson, R.R. (2005) *The Oxford Handbook of Innovation*, Oxford: Oxford University Press.

Granstrand, O., Patel, P. and Pavitt, K. (1997) 'Multi-technology corporations: why they have "distributed" rather than "distinctive core" competencies', *California Management Review*, vol. 39 (4): 8–25.

Grant, R. (1996) 'Prospering in dynamically-competitive environments: organizational capability as knowledge integration', *Organization Science*, vol. 7 (4): 375–87.

Holmén, M., Magnusson, M. and McKelvey, M. (2007) 'What are innovative opportunities?', *Industry and Innovation*, vol. 14 (1): 27–45.

Klein, S. and Rosenberg, N. (1986) 'An overview of innovation', in Landau, R. and Rosenberg, N. (eds) *The Positive Sum Strategy: Harnessing Technology for Economic Growth*, Washington, DC: National Academy Press.

Klevorick, A.K., Levin, R.C., Nelson, R.R. and Winter, S.G. (1995) 'On the sources and significance of interindustry differences in technological opportunities', *Research Policy*, vol. 24 (2): 185–205.

McKelvey, M.D. (1996) *Evolutionary Innovations*, Oxford: Oxford University Press.

McKelvey, M. and Orsenigo, L. (eds) (2006) *The Economics of Biotechnology. A Two Volume Reference Collection*, Cheltenham: Edward Elgar.

McKelvey, M., Alm, H. and Riccaboni, M. (2003) 'Does co-location matter for formal knowledge collaboration in the Swedish biotechnology–pharmaceutical sector?', *Research Policy*, vol. 32 (3): 483–501.

Malerba, F. (forthcoming) *KEINS: Knowledge Based Entrepreneurship in Networks*, final report from an EU Project.

Murphy, G., Trailer, J. and Hill, R. (1996) 'Measuring performance in entrepreneurship research', *Journal of Business Research*, vol. 36 (1): 15–23.

Nelson, R.R. and Winter, S.G. (1982) *An Evolutionary Theory of Economic Change*, Cambridge, MA: Harvard University Press.

Nightingale, P. (1998) 'A cognitive model of innovation', *Research Policy*, vol. 27 (7): 689–709.

Prencipe, A. (2000) 'Breath and depth of technological capabilities in CoPS: the case of the aircraft engine control system', *Research Policy*, vol. 29 (7–8): 895–911.

Rosenberg, N. and Birdzell, L.E. (1986) *How the West Grew Rich: the Transformation of the Industrial World*, London: Tauris.

Saemundsson, R. (2005) 'On the interaction between the growth process and the development of technical knowledge in young and growing technology-based firms', *Technovation*, vol. 25 (3): 223–35.

Sanchez, R. and Mahoney, J. (1996) 'Modularity, flexibility and knowledge management in product and organization design', *Strategic Management Journal*, vol. 17 (Winter Special Issue): 63–76.

Scherer (1965) 'Firm size, market structure, opportunity, and the output of patented inventions', *American Economic Review*, vol. 55 (5): 1097–125.

Schumpeter, J.A. (1942) *Capitalism, Socialism and Democracy*, New York: Harper and Brothers.

Shane, S. and Venkataraman, S. (2000) 'The promise of entrepreneurship as a field of research', *Academy of Management Review*, vol. 25 (1): 217–26.

Stankiewicz, R. (2000) 'The concept of "design space"', in Ziman, J. (ed.) *Technological Innovation as an Evolutionary Process*, Cambridge: Cambridge University Press.

Teece, D.J., Pisano, G. and Shuen, A. (1997) 'Dynamic capabilities and strategic management', *Strategic Management Journal*, vol. 18 (7): 509–33.

Vincenti, W.G. (1990) 'Data for design: the air-propeller tests of W.F. Durand and E.P. Lesley. 1916–1926', in *What Engineers Know and How They Know It: Analytical Studies from Aeronautical History*, Baltimore: Johns Hopkins University Press.

Yin, R.K. (1989) *Case Study Research*, Newbury Park: SAGE Publications.

11 Knowledge-based entrepreneurship and international technology transfer in the German laser industry

Guido Buenstorf[1]

1 Introduction

In 1960, US researchers presented the first workable laser to the public. In 1963, there were 32 producers of commercial laser sources worldwide. Only one of them came from Germany (Sleeper 1998). And not only had the German laser industry had a slow start. Its US and UK competitors were already serving the German market when the domestic industry began to develop. This chapter uses the example of the German laser industry to investigate from what sources firms in a new national industry draw their capabilities, and how they develop in the presence of powerful international competition.

To successfully compete in the marketplace, entrants to knowledge-based industries such as lasers depend on distinctive organizational capabilities. In recent years a sizeable literature has developed showing that the initial capability base of entrants, which derives from prior organizational or individual experience, shapes their subsequent strategies and performance (cf. Helfat and Lieberman 2002 for a discussion). We know less, however, about how international knowledge transfer affects the capabilities of entrants to national markets that develop far from the original research underlying the creation of a new industry. The international dimension of industry evolution was already discussed by Vernon (1966). However, he emphasized the role of foreign direct investment, which is more relevant for the offshoring and outsourcing of activities to lower-income countries than for the technology transfer between the leading industrialized economies. The present analysis is intended to help fill this gap in our knowledge on industry evolution.

Given its clear-cut beginnings and international scope, as well as substantial prior research (Albrecht 1997; Sleeper 1998; Klepper and Sleeper 2005; Buenstorf 2007), the laser industry is an ideal object for studying this issue. The present analysis is based on original data covering all German producers of laser sources over the industry's first four decades. We find that entrants acquired knowledge from quite different sources. This heterogeneity is reflected in the pre-entry backgrounds of the early entrants to the industry. Some entrants came out of domestic public research, suggesting that German universities and non-university research organizations were important mediators in the international transfer of laser-related knowledge to the German industry. At the same time, a large number of entrants 'bypassed' German public research. They based their entry to the laser industry on their own R&D efforts that directly took off from research being done in other countries, notably the United States and the UK. Yet other entrants primarily drew on the international transfer of organizational capabilities through spin-off entrepreneurship and sales activities.

The chapter is organized as follows. Section 2 presents a conceptual framework distinguishing alternative sources of organizational capabilities for new entrants. The practical importance of knowledge transfers from these different sources is illustrated in Section 3, outlining five exemplary cases of entry into the German laser industry. Section 4 provides a more detailed characterization of the population of entrants over the industry's first 40 years. Effects of pre-entry background on firm performance are discussed in Section 5, and empirical findings on the longevity of German laser entrants with different backgrounds are presented in Section 6. Section 7 concludes.

2 Channels of entry to new knowledge-based industries

Successful entrants to knowledge-based industries require organizational capabilities enabling them to compete with incumbent firms and other entrants. In this section, alternative sources of knowledge underlying organizational capabilities are discussed for both pre-existing *de alio* entrants and entrepreneurial *de novo* entrants. We then demonstrate that all these sources of knowledge played important roles in the emergence of the German laser industry.

Table 11.1 distinguishes three sources of knowledge on which entrants can draw: (in-house) private research, public research, and the transfer of organizational knowledge from other firms. All three sources are relevant for both *de alio* and *de novo* entry.

A straightforward way for entrants to acquire technological knowledge is through their own private research efforts. This source of knowledge is utilized

Table 11.1 Sources of knowledge underlying firm capabilities

	Source of technological knowledge at entry		
	Private research	*Public research*	*Transfer of organizational knowledge*
De alio entry *(examples)*	Diversification based on in-house R&D *(Siemens; Impulsphysik)*	Diversification based on R&D collaboration/ contract research *(Trumpf)*	Diversification based on acquisition/vertical integration *(Jenoptik; Rofin Sinar)*
De novo entry *(examples)*	Start-up *(Chromatron)*	Academic spin-off *(Lambda Physik)*	Corporate spin-off *(Laser-Optronic)*

Source: own compilation.

by diversifiers that enter to a new industry based on in-house R&D, a pervasive form of entry in the German laser industry. In principle, entrepreneurial *de novo* entrants may likewise base their entry on capabilities built through their own private research efforts. For example, this is done by the tinkerers and garage inventors that account for substantial shares of entry in some industries. However, in the German laser industry there were only a small number of entrepreneurial entrants that did not have close ties to public research or existing laser firms.

Public research is a second source of knowledge available to entrants. In contrast to the private R&D efforts of competing firms, the results of public research are comparatively easy to access since disclosure of findings is an essential feature of the institutional set-up of 'open science' (Dasgupta and David 1994). Access to published scientific findings may not be sufficient as a basis of entry because important parts of the required knowledge and skills tend not to be codified. However, universities and public research organizations are often willing partners of R&D collaborations or contract research. These kinds of interaction can enable pre-existing firms to enter new markets. In the case of academic spin-offs started by university scientists and faculty, *de novo* entry is likewise based on technological knowledge generated through public research. For academic spin-offs, the access to non-codified knowledge is assured by the migration of scientists from public research to the new firm.

Finally, entry into new markets may also be based on the transfer of organizational knowledge from other firms already experienced in these markets. *De alio* entrants frequently base their diversifications on the acquisition of firms that are active in the respective industry. Corporate spin-offs started by employees of industry incumbents are *de novo* entrants that transfer knowledge between organizations. Relevant inter-firm knowledge transfers may relate to technological knowledge, but also to other dimensions of organizational capabilities such as knowledge about markets and customer needs, or general management skills.

The emergence of the German laser industry depended on the international transfer of laser-related knowledge from the United States, where Theodore Maiman presented the first workable laser in 1960. There is no evidence that systematic research into laser technology was conducted in Germany before that date.[2] However, when learning about the advances made in the United States, German university and industry researchers quickly grasped the significance of the new technology. A number of them made substantial efforts toward constructing lasers themselves (cf. Albrecht 1997 for a more extensive discussion). In this process of catching up with the United States, all three of the above sources of knowledge were important in enabling entrants to compete in the emerging industry.

Lasers are light sources based on stimulated photon emission. The distinctive properties of laser light, most importantly its temporal and spatial coherence, are useful in many practical applications. After 1960, lasers became a 'hot' topic among academic physicists. One of Germany's laser pioneers (Wolfgang Kaiser of the Technical University Munich; cited in Albrecht 1997) suggests that almost all German universities engaged in some form of laser research at that time. Going beyond efforts merely to replicate prior US achievements, German laser researchers also began to contribute to laser theory and initiated the development of new laser types. In this context, the dye laser is noteworthy, which was simultaneously and independently developed in 1966 by Fritz Peter Schäfer and his co-workers at the University of Marburg (Germany) as well as by Peter Sorokin and other US researchers working for IBM. Anecdotal evidence suggests that scientists returning from the United States played an important role in the early adoption of laser research at German universities.[3] Thus, German public research was actively engaged in the process of international knowledge transfer from the United States. Through collaborative research with industrial partners as well as academic spin-off activities, laser knowledge generated in German public research was subsequently diffused to the private sector.

Similar to university scientists, researchers working for large and small German firms were quick to embrace laser technology. The available evidence shows that constructing a workable laser was feasible for skilled industry physicists and engineers. Existing German firms with various kinds of backgrounds were thus able to diversify into the laser industry based on in-house technological developments. This established a direct channel of technology transfer from US laser research to the emerging German laser industry, which was not mediated by public research in Germany.[4]

University and industry researchers in Germany based their own laser activities on the prior state of the art in international laser research. Their research efforts enabled the development of marketable products that allowed German firms and entrepreneurs to enter the laser industry. In addition to these research-based channels of international knowledge transfer, further entrants to the German laser industry primarily derived their capabilities from foreign, particularly US and UK, producers of commercial lasers. Several entrepreneurial *de novo* entrants to the German laser industry were started by former employees of foreign laser producers. Accordingly, the spin-off process, which has previously been identified as an important source of entrant capabilities in the US laser industry (Klepper and Sleeper 2005), also had an international dimension. In addition, a substantial number of entrants started out as importers and distributors of foreign lasers before successfully integrating into the manufacturing stage of the industry.

The presence of foreign competitors thus gave rise to international knowledge transfers between firms, providing a further avenue of entry for new German laser firms. This private-sector international knowledge transfer did not operate through direct foreign investment, as was suggested by Vernon (1966), but through spin-off entrepreneurship and also vertical integration from laser imports and sales into manufacturing. These activities have received little prior attention as vehicles of international knowledge transfer.

3 Laser entrants in Germany: five exemplary cases

As argued in the previous section, there were diverse channels of knowledge transfer helping early entrants to the German laser industry acquire the relevant capabilities. In this section, five examples of successful entrants are described to illustrate the alternative sources of knowledge.

3.1 Lambda Physik: laser entrepreneurship out of public research[5]

Göttingen-based Lambda Physik is the most important German laser producer whose origins can be traced back directly to public research. The firm was founded in 1971 by Dirk Basting and Bernhard Steyer, two young researchers working on their chemistry PhDs in Fritz Peter Schäfer's lab. Basting and Steyer initially constructed a high-performance nitrogen laser to be used for their own experiments. Since this laser performed well, they subsequently sold it to other researchers. Initially, they still pursued their PhDs at the Max Planck Institute of Biophysical Chemistry besides running their small-scale firm. Lambda Physik's technology was patented by the optical company Carl Zeiss, a cooperation partner and early customer of the firm.

In the following years, Lambda Physik developed increasingly powerful ultraviolet nitrogen lasers. In 1975, the company acquired the dye laser activities of Zeiss. These had, since 1969, been directed by Werner Schmidt, also a former PhD student of Schäfer's. Schmidt had played a crucial role in the discovery of the dye laser (Köhler 1983). Lambda Physik presented the world's first workable commercial excimer laser in 1977, and it remained a world market leader in this type of laser. The range of applications for excimer lasers was expanded over time, as medical and industrial uses (e.g. in semiconductor production) became increasingly viable. The leading US laser producer, Coherent, acquired the majority of Lambda Physik, and Steyer left the company in 1981. Steyer subsequently started other laser firms. Lambda Physik became a full subsidiary of Coherent when Basting retired in 2003.

The significance of public research as a channel of knowledge transfer enabling entry is evident in the case of Lambda Physik. Being students of one of Germany's foremost laser scientists, Basting and Steyer, the firm's founders, were working at the cutting edge of global laser research. Their advisor, Schäfer, supported their entrepreneurial activities, and through his former student Schmidt he was well connected to the Zeiss company. This connection provided Lambda Physik with a powerful cooperation partner and with access to dye laser technology, allowing the firm to complement in-house developments in its initial years.

3.2 Siemens: diversifying entry based on large-scale industrial R&D

Siemens, Germany's leading electrical engineering and electronics company, immediately reacted to the presentation of the first US lasers (cf. Albrecht 1997, for the following). The company realized that the new technology might strongly affect its product markets, even though the concrete commercial applicability of the laser was still uncertain. Its focus in laser development was on communication technologies, one of the company's main areas of activity.

Siemens' initial laser research was conducted simultaneously with the emerging public laser research in Germany. To a large extent, the company itself became active in the international transfer of laser-related knowledge from the United States, without relying on the intermediation of German public research. Apparently because of the perceived importance of lasers for its future business, Siemens initiated competing laser research activities in two of its research laboratories: the central communication technology laboratory and the central research laboratory. By 1961, both labs had presented workable solid-state lasers. The communication technology laboratory had a close collaboration with Bell Labs; it sourced the rubies for its lasers directly from US suppliers. Research on gas lasers was taken up at Siemens in 1961. A first workable gas laser was presented in 1962.

Since these initial laser activities, Siemens has had a long track record of producing and marketing a broad range of lasers. The company started selling commercial gas lasers in the 1960s. Its helium–neon lasers attained global market leadership. In addition to marketing stand-alone lasers, Siemens also engaged in the large-scale production of laser cavities as well as laser-based machinery (for example, laser marking systems were introduced in the 1970s) and medical equipment. Laser diodes were added to the company's product line in 1981. In addition, KWU, Siemens' power plant technology subsidiary, temporarily acquired laser producer Rofin–Sinar (1987–1995, see below) and also engaged in the development and sale of excimer lasers. Initially intended for use in producing nuclear fuels, they were subsequently developed further for materials processing applications. Siemens' gas laser activities were sold to Zeiss in 1993. Its excimer laser activities became part of Rofin when the firm was spun off again in 1996. Today, through its Osram subsidiary, Siemens is one of two German makers of semiconductors for diode lasers.

The Siemens case is an example of an established company that made an early and large-scale entry into the new field of laser research. Siemens' entry into the laser industry was based on its in-house R&D efforts and targeted the company's established product markets. It had direct contact to US laser researchers. Accordingly, in this case the international knowledge transfer primarily operated through industrial R&D activities, largely bypassing public research in Germany.

3.3 Impulsphysik: diversifying entry based on small-scale industry R&D[6]

Impulsphysik was founded in Hamburg in 1949 by Frank Früngel. This company produced impulse welding and tempering machines, high-speed cameras and stroboscopes, as well as meteorological measurement equipment. Impulsphysik constructed its first laser in 1962; it was first listed as a commercial laser producer in 1964. The firm was the earliest German producer of solid-state lasers for practical application in research and measurement (both for civilian and military use). It also marketed its lasers for applications in materials processing. In addition to Impulsphysik's commercial activities, Früngel in 1967 established a non-profit research institute, 'Impulstechnology Ass. e.V.'. According to his obituary, he authored more than 200 scientific publications and held over 200 patents. Früngel sold his firm in 1987. Afterwards, Impulsphysik's laser department was acquired and divested again by several communication and laser firms (Hagenuk, Jenoptik, Vaisala).

The case of Impulsphysik shows that even smaller companies were able to enter the emerging laser industry on the strength of their in-house R&D and to integrate laser technology into their existing products (Albrecht 1997). A decisive factor underlying the company's ability to transfer laser-related knowledge to Germany seems to have been that Früngel, besides running his firm, remained involved in academic research, thus establishing the firm's absorptive capacities (Cohen and Levinthal 1990) to engage in laser development.

3.4 Laser–Optronic: a spin-off from a foreign laser producer[7]

In the 1970s, Laser–Optronic was Germany's largest producer of solid-state laser systems. The Munich-based firm was organized in 1971 by Falk Strascheg, an engineer who, after leaving the German NCR subsidiary, had built up the German sales division of British laser producer Laser Associates. Laser–Optronic started out both as a distributor of imported laser products and as a manufacturer of solid-state lasers. Its manufacturing activities focused on materials processing applications built around its own laser sources, in particular resistor trimming and marking systems. Laser–Optronic was sold to the US competitor Coherent in 1981 and integrated into Rofin–Sinar in 1991. Strascheg later became a prominent venture capitalist, financing, among others, young German laser firms.

It is noteworthy that Laser–Optronic's founder had an industrial rather than a public research background. Strascheg entered the laser industry as an employee with managerial experience. However, after a brief tenure with his British employer, he was able to start a successful laser firm of his own, also drawing support from public research through R&D cooperations. The founder's background appears to be reflected in the orientation of the firm. Laser–Optronic's manufacturing activities focused on developing practical applications for low-power, solid-state lasers, primarily in the electronics and precision mechanics industries. At the same time, Laser–Optronic broadened the scope of its activities by distributing a variety of lasers and accessories by other producers.

The case of Laser–Optronic shows how prior employment in a laser firm enabled entrepreneurs to identify and occupy viable market niches in the emerging laser industry even if they had no background in laser physics. The importance of spin-off activities in industry evolution has been highlighted before, including studies of the US laser industry (Sleeper 1998; Klepper and Sleeper 2005).

Laser–Optronic suggests that the spin-off process also has an international dimension.

3.5 Rofin–Sinar: entry through vertical integration from laser distribution[8]

Hamburg-based Rofin–Sinar was originally an engineering firm and laser distributor without manufacturing activities of its own. It was co-founded in 1975 by engineer Samuel Simonsson and served as the German distributor of the British firms GTE Sylvania and Rofin, makers of lasers and accessories.

Based on its experience in selling and installing high-power industrial lasers, Rofin–Sinar began developing its own CO_2 lasers in 1979 after GTE Sylvania was acquired by US laser producer Spectra Physics. These lasers were licensed to Spectra Physics upon their introduction on the market. In turn, Siemens–KWU acquired both Rofin–Sinar (in 1987) and the industrial laser division of Spectra Physics (in 1988), which turned the firm into the world market leader in industrial CO_2 lasers. One year later, Rofin–Sinar acquired the laser marking activities of Laser–Optronic from Coherent. In 1996, Siemens withdrew from its industrial laser activities and divested Rofin–Sinar, which regained its independence through an IPO at NASDAQ. It subsequently acquired other German laser producers. Today, the Rofin–Sinar Group is one of the two big manufacturers of industrial lasers in Germany.

Rofin–Sinar could draw on four years' experience in laser imports and application engineering before it started laser manufacturing. Its emergence exhibits some parallels to that of Laser–Optronic. Both firms have an industrial rather than a public research background. In both cases, the ability to integrate laser technology into viable industrial applications was a key capability, which had largely been acquired in the prior activities of the firm or, in the case of Laser–Optronic, its founder. At the same time, the two firms differ not only in the laser types and market segments they concentrated on, but also in the temporal pattern of development. Laser–Optronic combined manufacturing and distribution activities from the beginning, while Rofin–Sinar began making its own lasers only when its main supplier had been acquired.

3.6 Implication

In line with the discussion in Section 2, the five examples indicate the variety of ways in which early German entrants tapped into the existing laser-relevant knowledge, which had been mainly developed abroad. The founders of Lambda Physik acquired their laser knowledge while working in public research. In contrast, researchers at Siemens and Impulsphysik apparently had sufficient absorptive capacities from earlier activities to directly base their laser efforts on prior US research without having to rely on public researchers in Germany. Finally, international knowledge transfer also took place within the private sector. The founders of Laser–Optronic and Rofin–Sinar exemplify how entrepreneurs developed substantial knowledge about laser technology and markets through working for foreign laser firms or selling their products in Germany.

The five examples fall into four out of the six cells of Table 11.1. As regards the two remaining cells, it has already been noted there was little *de novo* entry based on entrants' own private research efforts in the German laser industry. In contrast, at least one prominent diversifier primarily drew on knowledge from public research (Trumpf). However, this firm entered only in the mid-1980s, considerably later than the five firms discussed above.

4 Patterns of entry into the German laser industry, 1964–2003

To learn more about the full population of German producers of laser sources, a new dataset was constructed based on historical information. As an initial step, this required the identification of laser suppliers, their times of entry into and exit from the industry, as well as the kind of lasers they offered in individual years. This information was derived from various sources, most importantly buyer's guides for the German laser market. Germany's first laser-buyer's guide was issued in 1987 by the trade journal *Optronics*. While this publication only appeared once, annual buyer's guides have been produced from 1990 onward by *Laser*, another German trade journal. To obtain information for the industry's earlier history, we primarily used the listings of exhibitors at the biannual *Laser* trade fair in Munich, first held in 1973, as well as the classified listings of laboratory needs suppliers in *Physikalische Blätter*, which began to list laser suppliers in 1963.

The above sources helped identify laser producers as well as their active years in the industry. However, further steps were required to make the data usable for further analysis. Most crucially, systems producers and distribution firms that did not manufacture laser sources had to be excluded from the data. In addition, the pre-entry backgrounds of diversifying firms as well as the founders of *de novo* entrants needed to be identified. We took this information mostly from the trade press. In particular, we made extensive use of all published volumes of *Laser und Optoelektronik* (which, under different names, has been published continuously since 1969) and the more recent *EuroLaser*, utilizing both their edited content and the advertisements. Additional information on potential laser firms was obtained from the *Hoppenstedt* register of German firms, from patent and publication searches, as well as from web resources. Finally, where the published material was insufficient to reliably classify a firm, we contacted the latter by email, telephone, or personally during the 2005 *Laser* trade fair, or consulted industry experts. In the end, sufficient information was obtained to classify all but 16 firms identified in the primary sources. The final dataset contains 143 confirmed German producers of laser sources between 1964 and 2003.[9]

The German laser industry had modest beginnings. Foreign laser firms initially dominated the German market. Newly entering German firms frequently imported foreign lasers rather than (or, as in the case of Rofin–Sinar, before) producing their own. To characterize the early German laser industry, it is instructive to look at the backgrounds of the first entrants (Table 11.2). A total of 12 laser manufacturers were listed in the classified listings of equipment suppliers in *Physikalische Blätter* and/or participated in the Munich *Laser* trade fair before 1975. Seven of them were pre-existing firms that diversified into laser technology. Among these diversifiers, five were major industrial companies (Siemens, two optics firms and two defence contractors); two were SMEs (Impulsphysik and a crystal producer whose founder had previously been a laser researcher at the University of Munich). Four of the five earliest *de novo* entrants had a direct background in academic research. The remaining one was Laser–Optronic, the international spin-off discussed in Section 3.

From 1975 to 1984 the total number of entrants remained small. Diversifiers still accounted for six out of 14 entrants in this decade. Three entrants were earlier laser importers that now also entered laser manufacturing. The *de novo* entrants became more heterogeneous in terms of their backgrounds. Only a single academic spin-off entered in this decade. Two entrants were corporate spin-offs, one of which had an international background (Laser–Licht, which was co-founded by a German capitalist–entrepreneur and a team of former employees of US laser producer University Labs), while the other was an early spin-off from Lambda Physik. Finally, one start-up entrant was organized by an ex-employee of a German firm involved in military laser development, while the background of the remaining firm could not be reconstructed.

From the mid-1980s, entry picked up speed and German laser producers became more prominent not only in the domestic market but also abroad. Industrial applications in materials processing emerged as the stronghold of the German laser industry. While US producers were struggling with slow sales in those years, German producers of industrial lasers grew quickly, and laser exports increased almost fourfold between 1980 and 1987 (Basting 1989). In 1988, German firms accounted for one-third of all industrial lasers produced worldwide, and their share subsequently increased further to around 40 per cent (Heilmann 2003). Moreover, in the 1990s new German laser firms entered the fields of measurement, marking systems and medical applications.

Table 11.2 Entrants to the German laser industry, 1960–2003

	Total	*1960–74*	*1975–84*	*1985–94*	*1995–2003*
Diversifiers	45	7	6	17	15
Integrating distributors	11	0	3	1	7
Start-ups	7	0	1	1	5
Academic spin-offs	28	4	1	9	14
Corporate spin-offs	48	1	2	16	29
Unknown background	4	0	1	3	0
Total	143	12	14	47	70

Source: own compilation.

As can be seen in Table 11.2, there was a considerable shift in entrants' backgrounds after 1985. The relative importance of diversifiers declined as the numbers of both academic and corporate spin-offs strongly increased. Overall, 23 academic spin-offs and 45 spin-offs from existing laser firms entered between 1985 and 2003. This compares to five academic and three corporate spin-offs entering before 1985, indicating that German public research and existing laser producers became increasingly important as sources of capabilities for new entrants. We also observe a growing number of distribution firms integrating into laser manufacturing. While almost all of the later corporate spin-offs had domestic parents, these integrating distribution firms had mostly been active importers before. Imports thus remained a viable channel of international knowledge transfer enabling entry into the German laser industry.

Even though laser producers are found throughout the country, some geographic centres of agglomeration can be identified. Germany's foremost laser location is Munich, which has been a centre of the industry from its inception. A number of early entrants there came from academic backgrounds. In addition to its universities, Munich was also the location of the non-university Institut für Plasmaphysik, which had substantial laser activities and was later integrated into the Max Planck Society. However, some notable entrants in Munich had commercial rather than academic backgrounds, such as Laser–Optronic and Baasel. The latter set out as a manufacturer of laser systems for materials processing, but quickly started making its own laser sources.

Since the mid-1980s, the role of Munich as Germany's centre of laser activities has been challenged by Berlin. In terms of the number of producers present there, Berlin has recently caught up with Munich. Before German reunification in 1990, both West and East Berlin were centres of public laser research, which centred on the Technical University in the West and the Akademie der Wissenschaften in the East. After 1990, new research institutes were established at the Adlershof science park in the former East. A number of entrants have their roots in these research activities, but Berlin entrants also include several non-academic entrants, including corporate spin-offs. Substantial numbers of laser firms also entered in the eastern parts of Germany outside Berlin. This reflects that laser technology was a field where successful research was conducted during socialism (Albrecht 1997 and 2001).

Over its entire history through 2003, 48 of the 143 entrants (34 per cent) were corporate spin-offs organized by former employees of other producers of laser sources (including serial entrepreneurs who had previously founded a laser firm). Diversifiers accounted for 45 entrants (31 per cent), and 28 firms (20 per cent) were founded directly out of an academic background. Eleven entrants (8 per cent) were laser distributors/importers integrating into laser production.

How do these numbers compare to the US laser industry? Total US entry from 1960 to 1993 amounted to 486 firms (Klepper and Sleeper 2005). Accordingly, the German industry had some 30 per cent of the number of entrants observed in the United States (albeit over a shorter period of time), which is roughly in line with the relative size of the two economies. As regards the backgrounds of entrants, the largest difference is in diversifiers. They accounted for 60 per cent of all entrants to the US industry, but only for 31 per cent of the German entrants. In contrast, spin-off activities out of existing laser firms were more widespread in Germany than in the United States (34 per cent versus 16 per cent of all entrants). Only a minority of the corporate spin-offs were strategically organized by the management of incumbent firms. Even if these are disregarded and the most restrictive measure of genuinely 'entrepreneurial' spin-offs (Helfat and Lieberman 2002) is adopted, the number of corporate spin-offs in Germany still exceeds that of the US laser industry.

The large number of corporate spin-offs observable in the German laser industry is suggestive of a workable spin-off process. For the US laser industry, spin-off activities out of existing firms have been studied in detail by Klepper and Sleeper (2005). Their key conclusion is that spin-offs were the result of employee learning in incumbent firms. Spin-offs tended to produce similar lasers to their parent firms, and more successful incumbents had higher spin-off rates than less successful ones. These findings resonate with those obtained for other industries. For example, in the historical US auto and tyre industries, better-performing firms had more spin-offs than their inferior competitors, and on average these spin-offs were also more successful themselves (Klepper 2002b and 2007; Buenstorf and Klepper 2009). High spin-off rates thus seem to indicate the strength rather than weakness of incumbent firms; they are apparently caused by the superior learning environment that these firms provide to potential spin-off founders.

Deliberately adopting a comparative perspective and using a similar methodology to Klepper and Sleeper (2005), Buenstorf (2007) studied the determinants of the spin-off likelihood of German laser firms. The analysis points to striking parallels between the spin-off processes in both countries. Annual spin-off rates were higher in more successful firms, where performance was proxied by the total number of years a firm was active in the laser industry or in a specific submarket (defined by types of lasers; see below for more detail). Similar to the US industry, this suggests that employees acquired knowledge on the job, enhancing their ability to start firms of their own, with more successful firms being superior 'training grounds'.

Alternatively, when including the current laser experience of firms as an explanatory variable, it was found that spin-off rates initially increased with firms' accumulated industry experience. They reached their peak at about 14 years of presence in the laser industry. Again, this is consistent with employee learning driving the spin-off process, as well as with prior US results. Presumably, the laser-related capabilities of incumbent firms initially grew with their industry experience, enhancing the learning potential for employees.

Both the effects of firm performance and accumulated experience were more pronounced within specific submarkets (defined by types of lasers) than for the aggregate industry. This suggests that much of the knowledge favoring spin-off activities was rather specific in nature. In contrast, no evidence was found that the pre-entry background of incumbent firms affected their own propensity for spin-offs.

As regards exit from the German laser industry, the role of acquisitions is notable. Qualitative evidence suggests that the acquired firms included some successful producers taken over by competitors interested in their technology base. Early on, such acquisitions tended to be made by foreign firms (for example, as was noted above, both Laser–Optronic and Lambda Physik were acquired by the US competitor Coherent). More recently, German laser producers have likewise been active in acquiring competitors.

5 Industry evolution and the pre-entry experience of producers

Early research on industry evolution (Gort and Klepper 1982; Klepper and Graddy 1990; Klepper 1996) showed that most (narrowly defined) industries evolve according to a life-cycle pattern. The industry life-cycle is characterized by a pronounced shakeout in the number of active firms while the demand for the industry's product is still increasing. The growing numbers of active firms in the German laser industry (Figure 11.1) as well as its US counterpart (Klepper and Thompson 2006) show that the laser industry diverged from the conventional pattern of industry evolution.

This divergence can be explained by the tendency of laser producers to specialize along submarkets (Klepper 1997; Klepper and Thompson 2006). Laser light can be generated using a variety of media, including some gases (most importantly CO_2 and helium–neon), solid crystals doted with small concentrations of specific molecules such as neodymium, and also semiconductors. A crucial feature of laser technology is that general-purpose lasers do not exist. Lasers using the different media are fundamentally different in terms of wavelengths,

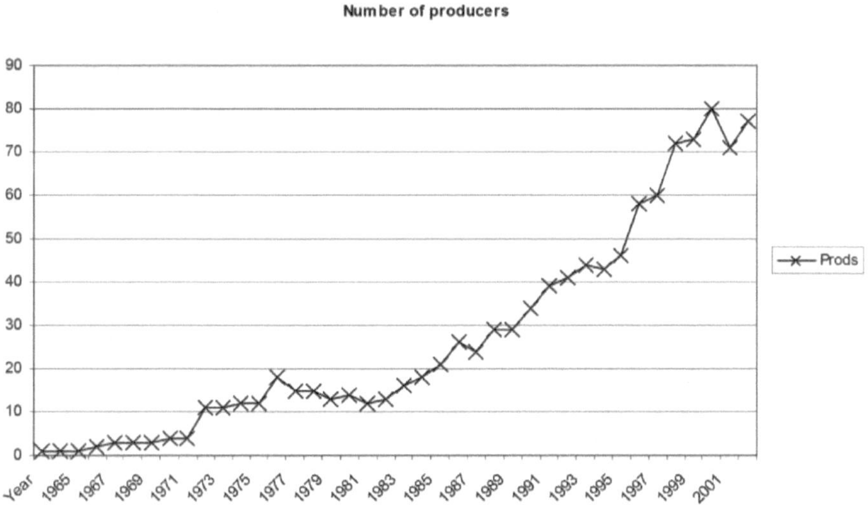

Figure 11.1 Active firms, entrants and exiters in the German laser industry.

power ranges and other characteristics. As a consequence, different types of lasers are used in different applications. In turn, the successful marketing of new laser types has critically hinged on finding economically viable applications. Entrants able to identify such applications and develop suitable lasers have tended to be competitive vis-à-vis the more established laser producers because economies of scope across laser types were of little importance. The specialization of producers has been pronounced, with the average producer being active in 1.6 types of lasers.

In the 40 years of evolution in the German laser industry, innovative new laser types have become competitors of more established lasers in individual submarkets. Specifically, semiconductor lasers have often been substituted for helium–neon lasers, while diode-pumped solid-state lasers (DPSS) have increasingly challenged CO_2 lasers in the lower parts of their power range. However, as is indicated by Table 11.3, these partial replacements of laser types have not caused shakeout dynamics at the submarket level of the industry, but the numbers of active producers have grown over time for all laser types.

In addition to regularities in how firm populations in industries evolve over time, recent studies have also found the prospects of entrants to differ systematically according to their pre-entry experience. Given the heterogeneity of entrants to the German laser industry, these findings are directly relevant for the present study.

As regards diversifying entry, Klepper and Simons (2000) found that the US television receiver industry was dominated by firms with pre-entry experience in radio production. Although radio producers represented a minority of all entrants, no firm without a background in radios ever captured a large market share in TV receivers. Diversifying radio producers entered earlier than the typical entrant, and their rate of exit hazard was lower. Strong performance effects of related diversification were also found in various other industries, including

Table 11.3 German laser producers by submarket, 1965–2003 (distribution activities by laser producers are excluded)

	Total number	*Mean number; 1965–74*	*Mean number; 1975–84*	*Mean number; 1985–94*	*Mean number; 1995–2003*
Solid	89	2.1	5.7	13.8	38.2
Semiconductor	77	0	2.8	12.1	33.2
Dye	15	0.2	1.2	2.0	3.4
CO_2	35	0	2.9	9.7	12.3
Helium–neon	27	1.7	3.7	7.2	9.4
Ion	15	0	0.3	2.4	4.9
Excimer	14	0	0.2	2.2	4.1
Others	9	0.5	1.6	1.0	1.0

Source: own compilation.

Note
The distinction of laser types follows Klepper and Sleeper (2005) except for the omission of helium–cadmium lasers, which played no important role in the German laser industry.

the historical US shipbuilding (Thompson 2005) and farm tractor (Buenstorf 2008) industries. In the context of the US laser industry, Sleeper (1998: Chapter 8) reports detailed results on the performance of diversifiers. These indicate that the closer the experiences of pre-existing firms were related to the laser industry, the better the firms tended to perform in this industry.

Several industry studies found that spin-offs started by ex-employees of industry incumbents were significantly more successful than other *de novo* firms (whose founders had no prior experience in the respective industry). An oft-cited example for the prominence of spin-offs is semiconductors, where spin-offs – including Fairchild Semiconductors and Intel – played a most crucial role (Moore and Davis 2004). Likewise, spin-offs were highly successful in the disk drive industry (Agarwal *et al.* 2004), the automobile industry (Klepper 2002b) and the tyre industry (Buenstorf and Klepper 2009).

6 Pre-entry experience and the longevity of entrants in the German laser industry

Above we have shown that entrants to the German laser industry were heterogeneous in terms of their pre-entry backgrounds. Prior evidence has moreover been discussed indicating that firms differing in background also differ in performance, which suggests that capabilities based on different sources of knowledge vary in how useful they are for entrants.

To what extent does their pre-entry background explain the performance of firms in the German laser industry? We approach this issue by adopting the number of a firm's active years in the laser industry as a measure of firm performance and conduct a hazard rate analysis of exit from the industry. Using survival in the market as a performance criterion is dictated by the lack of financial data covering the complete firm population over a period of 40 years. At the same time, longevity is a suitable proxy for performance for various reasons. Theoretically, it is supported by a straightforward opportunity-cost argument. If firms leave a market that they once entered, they may either be forced to (i.e. they are bankrupt), or the owners may consider better uses for their capital. Given irreversibility of at least some of their prior investments, the decision for voluntarily leaving the market will generally indicate substandard performance or the expectation thereof.[10]

Alternatively, firms may disappear from the dataset because other laser firms have acquired them (or their laser activities). In these cases, we cannot tell whether a particularly promising firm was swallowed by a competitor with deeper pockets, or whether a marginal producer sold out to prevent an impending bankruptcy. In other words, the fact of being acquired does not contain systematic information on firm performance. Exit by acquisition is accordingly not taken into account in our performance measurement; only the fact that the firm survived up to the time of its acquisition enters the analysis. In more technical terms, acquisitions by other laser firms are treated as censored exits in the hazard rate analysis.[11]

Generally speaking, hazard rate analysis models the hazard of experiencing some event at time t conditional on a vector of explanatory variables. There are various specifications of hazard rate models. Here, the parametric Gompertz specification is used.[12] It assumes that the baseline hazard $h_0(t)$ and the systematic hazard conditional on the vector x_j of independent variables (for subject j) are multiplicatively related:

$$h(t|x_j) = h_0(t)\exp(x_j\beta_x) \tag{11.1}$$

where the baseline hazard $h_0(t)$ is specified as

$$h_0(t) = \exp(\beta_0)\exp(\gamma\, t) \tag{11.2}$$

The baseline hazard is thus assumed to increase or decrease monotonically over time. For firm survival, one would expect to find $\gamma<0$, i.e. a falling baseline hazard as the lifetime of a firm increases (reflecting the 'liability of newness' of firms). We further assume that the effect of independent variables on the hazard is proportional, i.e. the age dependence of the hazard does not vary between firms that differ in their x_j vectors.[13]

The size of the dataset severely limits the feasible scope of quantitative analyses. In particular, the small number of only 49 non-censored exits in the dataset reduces the power of the hazard rate analysis. It has nonetheless been feasible to arrive at some distinctive findings.

The first model (Model 1 in Table 11.4) focuses on the longevity of academic spin-offs, an important group of entrants that have also been subject to much attention in recent policy debates in Germany. A dummy variable is included in this model that assumes the value one for all 28 firms started out of universities and non-university public research organizations. The coefficient estimate obtained for this variable is positive and significant (at the 0.01 level). It implies

Table 11.4 Hazard rate analysis of German laser producers, 1964–2003 (Gompertz specification)

	Model 1	*Model 2*	*Model 3*
Academic spin-off	0.847*** (0.330)		
Diversifier		−0.721** (0.345)	
Corporate spin-off		−0.884** (0.383)	
Distributor		−1.370* (0.750)	
Munich region			−0.225 (0.370)
Berlin region			0.558 (0.368)
Constant	−3.341*** (0.239)	−2.569*** (0.265)	−3.192*** (0.240)
Age	0.014 (0.024)	0.014 (0.025)	0.011 (0.024)
No. of obs. (failures)	143 (49)	143 (49)	143 (49)
Log-likelihood (chi^2)	−124.173 (5.75**)	−123.027 (8.04**)	−125.539 (3.02)

Note
Standard errors in parentheses; *** $p \le 0.01$; ** $p \le 0.05$; * $p \le 0.10$.

that the hazard of exit was twice as large for academic spin-offs as for the average non-academic entrant. This suggests that even in a high-tech industry like lasers, where technology is largely based on recent scientific research, entrepreneurs coming from public research may be disadvantaged compared to other entrants.

The next model looks at the effects of various kinds of non-academic pre-entry experience. In this specification (Model 2 in Table 11.4), the academic spin-offs, as well as the seven start-ups without industry experience and the four entrants with unknown backgrounds (which on average lasted longer than the academic spin-offs), make up the control group. Three dummy variables for firm backgrounds are included, which indicate, respectively, diversifying firms, corporate spin-offs from laser firms, and prior distribution firms that integrated into laser manufacturing.

The coefficient estimates indicate that all three groups of entrants had exit hazards ranging between 25 and 50 per cent of the hazard of the control group. They are significant at the 0.05 level for the diversifiers and the corporate spin-offs. The coefficient estimate for the integrating distributors suggests that these firms may have been the group with the lowest hazard, but it is significant only at the 0.10 level. In any case, the results of Model 2 point to powerful performance effects of pre-entry experience acquired outside public research.[14]

Finally, in Model 3 we probe into the effect of geography on firm survival, concentrating on the two major centres of laser production, the Munich and Berlin regions (postal codes 80000–89999 and 10000–15999, respectively). Model 3 (Table 11.4) includes two dummy variables that identify firms entering in one of the two regions. The coefficient estimates for the two variables suggest that, on average, Munich firms had a 20 per cent lower hazard than the control group of entrants located outside the two centres, while the hazard of entrants in Berlin was 75 per cent above average. However, both coefficient estimates are not significant at conventional levels, and the overall fit of the model is poor. Accordingly, there is no evidence that being located in the two major industry clusters had a positive effect on firm performance.

7 Conclusions

International knowledge transfer enabling entry into a new national industry operates through different channels. The diversity of transfer mechanisms has been exemplified by five early entrants to the German laser industry. Subsequently, the effects of pre-entry knowledge on firm performance have been studied for the full population of German laser producers over the laser industry's life span from the 1960s onward.

Both the backgrounds of key early entrants and the quantitative findings on firm performance indicate the importance of knowledge transfer channels that are not directly based on public research. The majority of entrants did not come from academia. Non-academic backgrounds apparently provided entrants with useful capabilities, as is indicated by the finding that academic spin-offs had the shortest survival times of all entrants to the German laser industry. This is an intriguing finding given that academic spin-offs have received substantial public attention and benefit from a number of public policy measures.

In contrast, the large number and longevity of diversifying entrants is evidence that existing firms with sufficient absorptive capacities could successfully enter the new industry. Some of these diversifiers were large firms with sizeable and well-staffed R&D departments, but successful entrants also included SMEs with backgrounds in other high-tech activities. The examples of Siemens and Impulsphysik show that diversification did not require close collaboration with German public research, but could be based on in-house R&D.

The study has also found a large number of corporate spin-offs that were started out of existing laser firms. Corporate spin-offs fared well in the laser industry, which suggests that spin-off founders acquired valuable knowledge in their prior employment and were able to transfer this knowledge to their new ventures. Both the performance of spin-offs and earlier findings on the characteristics of laser firms that affected their spin-off rates (Buenstorf 2007) identify the spin-off process as an important source of organizational capabilities, similar to what has been found before for several US industries (e.g. Agarwal *et al.* 2004; Klepper and Sleeper 2005). The existence of spin-offs from parent firms located outside Germany shows moreover that the spin-off process also serves as a vehicle of international knowledge transfers.

Further *de novo* entrants based their entry into laser production on prior experience in the import and distribution of lasers. Similar to the spin-offs of foreign firms, their pre-entry experience can be traced back directly to the presence of foreign competitors to the German laser industry. Their performance indicates that this experience helped them acquire organizational capabilities, again suggesting substantial international knowledge spillovers between firms.

On a more theoretical note, the success of corporate spin-offs and integrating distributors raises questions as to what exactly their founders might have learned during their previous tenure – i.e. what specific kind of knowledge was at the root of their knowledge-based entrepreneurship. One interpretation of the above findings is that the success of corporate spin-offs was not only due to technological know-how 'inherited' from their parent firms. In an R&D-intensive industry such as lasers, one would expect that firms from academic backgrounds were similarly well-endowed with this kind of knowledge; yet they performed markedly worse than the corporate spin-offs.

It would also seem that the integrating distributors, being remote from the R&D and the production process of the firm whose products they sold, should have been less able than the spin-offs to absorb technological knowledge (even though most of these firms were run by academically educated physicists or engineers, and their business did not merely consist in selling lasers 'off the shelf', but they performed application engineering and consulting functions as well). Similarly, since the integrating distributors were not part of the organization of the manufacturing firms they represented, it is unlikely that they learned as much as the spin-offs in terms of organizational routines.

The most straightforward explanation for the performance of the integrating distributors thus seems to be that, similar to the corporate spin-offs, they were well positioned to learn about market opportunities prior to actually entering the market. Qualitative evidence from various industries indicates that spin-offs tended to make products that were related to those of their parent firm but focused on a different market segment. And it is easy to imagine how employees of incumbent firms learned about market opportunities that (for whatever reason) the firm did not pursue itself. Similarly, distributor firms were in close contact with customers, learning about their needs and the corresponding opportunities.

Overall, the present analysis finds that Germany's laser industry developed in ways rather similar to its US analogue. This runs counter to many assertions on the German economy in the period under investigation, which suggest both a widespread reluctance to enter into new, innovative activities and a general lack of entrepreneurship. It would seem worthwhile exploring whether the laser industry was favoured by special characteristics and circumstances, or whether perhaps conventional wisdom does not fully do justice to the German economy of the 1980s and 1990s.

Notes

1 Max Planck Institute of Economics, Evolutionary Economics Group, Kahlaische Strasse 10, 07745 Jena (Germany) and University of Kassel. Phone: (+49) 3641 686821; fax: (+49) 3641 686868; email: buenstorf@uni-kassel.de.

2 Even in the United States, research into lasers, starting from the related maser technology developed after 1950, did not really take off before 1957. In 1959/1960, however, a race to construct the first laser began in which six US universities and industry laboratories were involved. In close succession, ruby (1960), helium–neon (1960), glass (1961) and semiconductor lasers (1962) were presented by these laboratories (Sleeper 1998: Chapter 3).

3 For example, the laser theorist Hermann Haken in 1960 migrated back to Germany from Bell Labs, where he had been involved in the discussion of solid-state lasers. He was a friend of Wolfgang Kaiser's, who came to Munich from Schawlow's group, also at Bell Labs, and who had alerted Haken to the work of that group (Albrecht 1997).

4 This is not to deny the more indirect but still important role that public research plays in the education of industry scientists (cf. also Zellner 2003).

5 This example draws primarily on an interview with Dirk Basting in the trade journal *Laser und Optoelektronik*, no. 2, 1982: 10–12; the article '25 Jahre Lambda Physik', *Laser und Optoelektronik*, no. 1, 1997: 2; Basting (1998); as well miscellaneous pieces of information from trade publications, from 1971 to present.

6 This example draws primarily on a biographical sketch in *Laser und Optoelektronik* (1986); an obituary of Frank Früngel in Physikalische Blätter (1995: 687); as well as Albrecht (1997).

7 The primary sources for this example are an article on Laser-Optronic in *Laser und Optoelektronik* (no. 2, 1977: 24–5) as well as various biographical articles on Falk Strascheg.

8 This example draws on 'Rofin-Sinar Laser: Siemens' adrette Adoptivtochter greift über den großen Teich', *Laser und Optoelektronik*, no. 3, 1988: 22–3, as well as other material from trade publications.

9 The 16 unidentified firms are disregarded in the following. In most cases, they were listed only very briefly, and no other evidence on their laser production could be obtained in the written material and the expert interviews, which raises doubts regarding their status as actual laser producers and/or market relevance. Eliminating these short-lived firms appeared to be the most conservative option with regard to the survival analysis. Furthermore, two firms (Rofin–Sinar and Zeiss) entered twice. Their spells in the laser industry were treated as separate entries.

10 It has to be noted that survival in the present context relates only to the firm's activities in the laser industry. Exit does not imply the end of the firm's existence, but only indicates its withdrawal from the laser market. Given the sources used to construct the dataset, leaving the market may consist simply in ceasing to make any effort to sell to customers that would be affected by trade fair participation or listings in buyer's guides.

11 Acquisitions by non-laser firms are disregarded as long as the firm remained active as a laser producer. We also ran stratified competing risks Cox regressions to study the effect of acquisitions on the results of the hazard rate models in more detail (cf. footnote 14).

12 Parametric survival models assume a specific functional form both for the effect of the independent variables and for the effect of time on the baseline hazard.

13 The validity of this assumption was tested in several ways. First, all model specifications reported below were re-estimated as fully specified Gompertz models that included both age-invariant and age-dependent effects for all covariates (cf. Klepper 2002a). All age-dependent effects were found to be insignificant, and the model fit was not significantly increased relative to the corresponding proportional specifications. Second, the models were re-estimated as semi-parametric Cox regressions. Doing so did not yield qualitatively different results. The proportionality assumption underlying the Cox specification was neither rejected for the model specification nor for the individual covariates (Grambsch and Therneau 1994).

14 Stratified competing risks Cox regressions (Lunn and McNeill 1995) making the same distinctions between entrants' backgrounds as Models 1 and 2 were run to further analyse how background affected the likelihood of being acquired. The results indicate that academic and corporate spin-offs had very similar chances of being acquired. Smaller coefficient estimates were obtained for the acquisition hazard of diversifiers and integrating distributors, but these differences were not statistically significant.

References

Agarwal, R., Echambadi, R., Franco, A.M. and Sarkar, M.B. (2004) 'Knowledge transfer through inheritance: spin-out generation, development and survival', *Academy of Management Journal*, vol. 47 (4): 501–22.

Albrecht, H. (1997) 'Laserforschung in Deutschland 1960–1970', Habilitation Thesis, University of Stuttgart, Stuttgart.

Albrecht, H. (2001) 'Die innovation des lasers in Deutschland – Forschungen an der TU Berlin und der Universität Jena im Vergleich', in Abele, J., Berkleit, G. and Hänseroth, T. (eds) *Innovationskulturen und Fortschrittserwartungen im geteilten Deutschland*, Cologne: Böhlau, pp. 263–76.

Basting, D. (1989) 'Lasertechnologie – Made in Germany', *Laser und Optoelektronik*, no. 21: 32–4.

Basting, D. (1998) 'Technologietransfer durch Ausgründung', in Max–Planck–Gesellschaft (ed.) *Wirtschaft und Wissenschaft – eine Allianz mit Zukunft in Deutschland?*, Ringberg–Symposium in München, October, pp. 245–57.

Buenstorf, G. (2007) 'Evolution on the shoulders of giants: entrepreneurship and firm survival in the German laser industry', *Review of Industrial Organization*, vol. 30 (3): 179–202.

Buenstorf, G. (2008) 'Comparative industrial evolution and the quest for an evolutionary theory of market dynamics', in Elsner, W. and Hanappi, H. (eds) *Advances in Evolutionary Institutional Economics*, Cheltenham: Edward Elgar, pp. 59–78.

Buenstorf, G. and Klepper, S. (2009) 'Heritage and agglomeration: the Akron tyre cluster revisited', *Economic Journal*, vol. 119 (537): 705–33.

Cohen, W.M. and Levinthal, D.A. (1990) 'Absorptive capacity: a new perspective on learning and innovation', *Administrative Science Quarterly*, vol. 35 (1): 128–52.

Dasgupta, P. and David, P.A. (1994) 'Toward a new economics of science', *Research Policy*, vol. 23 (5): 487–521.

Gort, M. and Klepper, S. (1982) 'Time paths in the diffusion of product innovations', *Economic Journal*, vol. 92 (367): 630–53.

Grambsch, P.M. and Therneau, T.M. (1994) 'Proportional hazards tests and diagnostics based on weighted residuals', *Biometrika*, vol. 81 (3): 515–26.

Heilmann, R. (2003) 'Eine kurze Geschichte der Lasertechnik in Deutschland', *Laser Magazin*, no. 5–6: 8–14.

Helfat, C.E. and Lieberman, M.B. (2002) 'The birth of capabilities: market entry and the importance of pre-history', *Industrial and Corporate Change*, vol. 11 (4): 725–60.

Klepper, S. (1996) 'Entry, exit and growth, and innovation over the product life cycle', *American Economic Review*, vol. 86 (3): 562–83.

Klepper, S. (1997) 'Industry life cycles', *Industrial and Corporate Change*, vol. 6 (1): 145–81.

Klepper, S. (2002a) 'Firm survival and the evolution of oligopoly', *RAND Journal of Economics*, vol. 33 (1): 37–61.

Klepper, S. (2002b) 'The capabilities of new firms and the evolution of the US automobile industry', *Industrial and Corporate Change*, vol. 11 (4): 645–66.

Klepper, S. (2007) 'Disagreements, spinoffs and the evolution of Detroit as the capital of the U.S. automobile industry', *Management Science*, vol. 53 (4): 616–31.

Klepper, S. and Graddy, E. (1990) 'The evolution of new industries and the determinants of market structure', *RAND Journal of Economics*, vol. 21 (1): 27–44.

Klepper, S. and Simons, K.L. (2000) 'Dominance by birthright: entry of prior radio producers and competitive ramifications in the U.S. television receiver industry', *Strategic Management Journal*, vol. 21 (10/11): 997–1016.

Klepper, S. and Sleeper, S.D. (2005) 'Entry by spinoffs', *Management Science*, vol. 51 (8): 1291–306.

Klepper, S. and Thompson, P. (2006) 'Submarkets and the evolution of market structure', *RAND Journal of Economics*, vol. 37 (4): 861–86.

Köhler, H. (1983) *30 Jahre Forschung und Entwicklung im Zeiss–Werk Oberkochen. Band 1*, Oberkochen: Carl Zeiss.

Laser und Optoelektronik, various volumes.

Lunn, M. and McNeill, D. (1995) 'Applying Cox regression to competing risks', *Biometrics*, vol. 51 (2): 524–32.

Moore, G. and Davis, K. (2004) 'Learning the Silicon Valley way', in Bresnahan, T. and Gambardella, A. (eds) *Building High-Tech Clusters: Silicon Valley and Beyond*, Cambridge: Cambridge University Press, pp. 7–39.

Sleeper, S.D. (1998) 'The role of firm capabilities in the evolution of the laser industry: the making of a high-tech market', PhD Dissertation, Pittsburgh: Carnegie Mellon University.

Thompson, P. (2005) 'Selection and firm survival: evidence from the shipbuilding industry, 1825–1914', *Review of Economics and Statistics*, vol. 87 (1): 26–36.
Vernon, R. (1966) 'International investment and international trade in the product life cycle', *Quarterly Journal of Economics*, vol. 80 (2): 190–207.
Zellner, C. (2003) 'The economic effects of basic research: evidence for embodied knowledge transfer via scientists' migration', *Research Policy*, vol. 32 (10): 1881–95.

12 Are the dynamics of knowledge-based industries any different?

Ricardo Mamede,[1,2] *Daniel Mota*[3] *and Manuel Mira Godinho*[4]

1 Introduction

The concept of 'knowledge-based industries' has been widely used both in academic circles and in policy-making over the last decade. The OECD (1999) has defined such industries as ones which are relatively intensive in their inputs of technology and/or human capital. This definition includes not only those manufacturing industries that are found to be highly dependent on R&D inputs, but also a number of service activities that are intensive users of high technology and/or have the relatively highly skilled workforce required to fully benefit from technological innovation (OECD 1999).

Knowledge-based industries (henceforth KBIs), thus defined, have been shown to play an increasing role in contemporary advanced economies. Between the mid-1980s and the mid-1990s, the contribution of these industries to the business value-added of the advanced economies grew from around 45 per cent to more than 50 per cent (OECD 1999). After a downturn at the beginning of the millennium (e.g. the share of high- and medium–high-technology manufacturing fell to about 7.5 per cent of total OECD value-added in 2002, compared to about 8.5 per cent in 2000), the relative importance of KBIs in value-added, international trade, and GDP growth has since shown clear signs of recovery (OECD 2005).

Such a secular rise in the relevance of KBIs has drawn the attention of an increasing number of scholars to the study of such industries. Qualitative and quantitative studies on activities such as ICT-related industries, pharmaceuticals, biotechnology and business services, among others, have proliferated in recent decades. While the detailed study of specific KBIs has certainly contributed to a better understanding of each industry – and has often provided crucial insights that could be generalized to different contexts – it is, however, not clear to what extent the 'knowledge-based' category is actually useful if one is concerned with understanding the evolution of firms and industries. Put differently, while there are strong reasons to draw the attention of academics and policy-makers to those industries which are heavily based on knowledge inputs, it is sensible to ask whether the dynamics of KBIs significantly differ from the dynamics of other industries.

There might be at least three different reasons for expecting KBIs to have a different behavioural pattern than the one found in other industries: demand directed to KBIs grows faster; these industries tend to be younger in terms of their industry and technology life-cycles; and they may face different sets of entry and exit barriers, particularly with regard to the need to invest in specialized knowledge assets.

As is well known, the empirical evidence drawn from several studies on industry dynamics has enabled the identification of some statistical regularities, which are often taken as stylized facts/stylized results on how firms and industries evolve over time (for surveys see Caves 1998; Dosi *et al.* 1997; Geroski 1995). Such statistical regularities include the following: the entry and exit of firms are both frequent phenomena, implying high levels of firm turnover in most industries; a high number of new firms leave the market shortly after entry; firm entry and exit are very often correlated, frequently being associated with similar determinants; entering and exiting firms are typically smaller than the average incumbent (which means that the impact of entry and exit on worker turnover tends to be lower than the impact on firm turnover); both the size and the age of firms have a positive impact on their survival chances, and usually a negative impact on growth.

In this chapter, we intend to discuss the extent to which KBIs differ from other industries in relation to some of the stylized facts and regularities of industry dynamics that have just been mentioned. In particular, we analyse the patterns and the determinants of firm entry and post-entry performance (measured in terms of the survival of new firms), comparing groups of KBIs with the remaining industries and using data for the Portuguese economy in the second half of the 1990s.

The chapter is organized as follows: in the following section, we present the data sources and propose our own classification for KBIs (an alternative to the one used by the OECD); Section 3 includes some relevant statistics on the entry and survival of new firms; in Section 4, the models used in the chapter are described; Section 5 presents results for industry entry rates and Section 6 deals with the determinants of new-firm survival; finally, Section 7 concludes.

2 Data and the classification of KBIs

The study presented in this chapter uses two main sources of data. The first one is the 'Quadros de Pessoal' (QP henceforth) database, which compiles information on employers and employees collected on a yearly basis by the Portuguese Ministry of Employment and Social Solidarity (MTSS). The survey that supports this database was first carried out in 1982 (at present, the data available for research covers the period 1985–2005) and annual reporting to this survey is compulsory for all firms employing paid labour in Portugal. The survey includes questions related to the characteristics of both firms (e.g. total employment, industry classification, location, legal status, ownership, number of plants, etc.) and their employees (gender, date of birth, educational background, professional category, type of contract, etc.).

Both firms and workers are identified by their social security numbers, which in principle should make it possible to follow them over time. This feature of the QP database makes it an extremely valuable source of data for conducting analysis of both industry dynamics (see, for example, Cabral and Mata 2003) and labour market dynamics (e.g. Portugal and Cardoso 2006). In practice, however, some problems may arise in the use of these data for the purposes of longitudinal analysis. On the one hand, data on workers' characteristics were not collected either in 1990 or in 2001, restricting the availability of continuous series to the periods 1985–1989, 1991–2000 and 2002–2005. On the other hand, there was a change in the industry classification system from 1994 to 1995, which means that one cannot build continuous series of data on industries without making some rather arbitrary assumptions. Therefore, in this chapter, we will restrict our use of data from the QP database to the period 1995–2000.

In order to enrich the information available at the industry level (namely, by adding information on revenues and costs, which is absent from the QP dataset), we also draw on the 'Inquérito às Empresas Harmonizado' (IEH), an annual survey conducted by the Portuguese National Statistical Office (INE). The IEH dataset includes information on all firms with 100 or more employees and on a representative sample of firms with less than 100 employees belonging to all industries, with the following exceptions: financial intermediation (ISIC classes 65–7); public administration and defence, and compulsory social security (ISIC class 75); private households with employed persons (ISIC class 95); and extra-territorial organizations and bodies (ISIC class 99).

A second group of industries was further excluded from our study, due either to data quality problems in the QP dataset or to our decision to focus our analysis on essentially market-oriented activities. This second group includes: agriculture, hunting and forestry (ISIC classes 1 and 2); fishing (ISIC class 5); mining and quarrying (ISIC classes 10–14); electricity, gas and water supply (ISIC class 40); education services (ISIC class 80), except driving schools and professional training; health services (ISIC class 80), except dental practices, clinical tests and veterinary activities; sewage and refuse disposal, sanitation and similar activities (ISIC class 90); activities of membership organizations (ISIC class 91); recreational, cultural and sporting activities (ISIC class 92), except motion picture and video production, distribution, and projection; other service activities (ISIC class 93), except washing and (dry-)cleaning of textile and fur products, hairdressing and other beauty treatment, and funeral and related activities.

The list of industries that were retained includes 337 classes of an ISIC rev.3-equivalent classification system disaggregated to five digits, comprising 386,510 firms (166,604 of which entered the market between 1995 and 2000).

Since our final aim is to compare the dynamics of KBIs and those of other industries, we need to establish some criteria for what we understand as 'knowledge-based'. Besides considering the OECD's criteria for identifying KBIs (which includes both high-technology and medium–high-technology manufacturing industries, together with knowledge-based services), we have put forward an alternative classification of KBIs (KBI–MMG, which we will compare with KBI–OECD), taking advantage of the information available in the QP database on the employees' educational background. Thus, according to the KBI–MMG classification, an industry is considered to be a KBI if it is among the 10 per cent of industries with the highest average proportion of employees holding a university degree (in the period 1995–2000).

The use of this alternative criterion for the classification of KBIs is beneficial in at least two senses: first, while the criteria used by the OECD to classify the knowledge intensity of manufacturing and service industries is based on information drawn from several countries over different periods of time, our classification was built on information about the firms which are actually under analysis, during the same period of observation; second, since the concept of 'knowledge-basedness' is far from being an established one, the use of two different classification systems allows us to assess the extent to which the (dis) similarities between the dynamics of KBIs and those of other industries depend on the way in which we define the former.

In Table 12.1 we can see that the two ways of classifying KBIs – the one used by the OECD and our KBI–MMG – overlap only partially.

3 Descriptive statistics on entry, growth and survival

A simple inspection of some basic descriptive statistics on firms and industries allows us to detect a number of similarities and differences between the dynamics of KBIs and other industries.

The values displayed in Table 12.2 are the industry annual averages of the corresponding variable for the period 1995–2000. The variables are defined as follows:

- The *entry rate* in year *t* is the ratio of the number of firms that were created at *t* to the total number of incumbents in the same year.
- The *exit rate* is the ratio of the number of firms that ceased to exist after *t* to the total number of incumbents in that year.
- The *growth rate* is measured in terms of the relative change in the total number of employees in an industry since the previous year.[1]
- The *Hirschman–Herfindhal concentration index* is, as usual, the sum of squares of the market shares of every firm in each industry, with the market share of firm *i* being measured here as the proportion of employees working for *i* in relation to the total number of employees in the industry.

Table 12.1 Number of industries (at the five-digit level) in different categories

OECD classification	*MMG classification*		*Total*
	KBIs	*Other industries*	
KBIs	15	29	44
Other industries	19	274	293
Total	34	303	337

Table 12.2 Annual average industry statistics (1995–2000)

		All	*KBI-OECD*	*KBI-MMG*
Entry rate	Mean	0.11	0.11	0.16
	Coeff. Var.	0.55	0.64	0.44
Exit rate	Mean	0.08	0.07	0.08
	Coeff. Var.	0.50	0.57	0.38
Growth rate	Mean	0.05	0.08	0.11
	Coeff. Var.	2.60	1.63	2.27
Hirschman-Herfindhal Index	Mean	468	714	762
	Coeff. Var.	1.60	0.99	1.68
Median firm size	Mean	6.96	9.68	6.24
	Coeff. Var.	1.16	1.15	1.72
Price–cost margin	Mean	0.12	0.16	0.16
	Coeff. Var.	1.00	0.81	0.94
Advertising intensity	Mean	0.01	0.01	0.02
	Coeff. Var.	2.00	2.00	1.50
Proportion of graduates	Mean	0.06	0.14	0.25
	Coeff. Var.	1.33	0.86	0.40

Note
The data of the variables in this table is derived from the *Quadros de Pessoal* database, with the exception of the 'price–cost margin' and 'Advertising intensity' variables, which were calculated with IHE-INE data.

- The *median firm size* refers to the distribution of firm sizes in terms of the number of employees.
- The *price–cost margin* consists of the average industry ratio $(R-C)/C$ (where R represents total revenues and C total costs, in both cases for all firms) weighted by firm size.
- The *advertising intensity* is the industry average of the ratio of each firm's total spending on advertising to its total revenues, weighted by firm size.
- The *proportion of graduates* is measured as the industry average proportion of employees in each firm who hold a university degree, weighted by firm size.

From Table 12.2, it is worth emphasizing the following aspects:

1 According to some dimensions, during the period under analysis, KBIs, regardless of the way in which they are defined, seem to behave in a fashion which is markedly different from the remaining industries – this is particularly so in the case of the growth rate, the price–cost margin, and the proportion of graduates; while the latter is essentially a result of the way industries are classified, the fact that, on average, KBIs grew faster and had higher price–cost margins during this period, confirms the idea that these industries contributed strongly to growth and value added.

2 In some other aspects, the behaviour of the KBI–OECD group is not the same as the behaviour of the KBI–MMG group; in particular, the latter group shows higher entry rates (while the entry rate of KBI–OECD industries is not distinct from the entry rates of all sectors together) and lower median sizes (while the OECD's KBIs seem to be composed of firms which are bigger than the average).

3 Finally, as far as the coefficients of variation are concerned, the information displayed suggests that there may be non-negligible differences between industries belonging to the same group of KBIs; this is particularly true in the case of the industries' growth rate, median firm size and advertising intensity (suggesting that industries which are considered as KBIs according to certain criteria may be quite distinct in many respects).

Beyond these particularities, Table 12.2 confirms the idea that entry and exit are fairly pervasive phenomena in all sorts of industries. Furthermore, the inspection of correlations between entry rates and exit rates (Table 12.3) confirms that these two phenomena are not only pervasive, but also positively correlated – confirming what many studies of industry dynamics have already shown.

In keeping with the stylized facts of industry dynamics, it can be seen that across all the groups of industries that we are analysing firms tend to enter with relatively small sizes and then to grow after entry (if they survive), typically at decreasing rates.

Table 12.4 shows, for different entry cohorts in each of the groups of industries under consideration, how the average size of new firms changes as they grow older.

Table 12.4 suggests that firms in knowledge-based industries tend to enter the market with higher sizes than firms in other industries. It also suggests that the individual firms in the KBI–MMG group that entered the respective industries grew more slowly than the rest of the firms during the period under analysis, particularly those in the KBI–OECD group.

Finally, further differences between KBIs and other industries can be detected by analysing the patterns of new-firm survival. Table 12.5 shows the percentage of firms per number of years of duration, again for different entry cohorts in each of the groups of industries under analysis.

Table 12.5 clearly shows that firms in KBIs tend to survive longer than the average: while 31 per cent of all firms that entered the market between 1995 and 2000 exited after the first two years, less than 25 per cent of the firms in either of

Table 12.3 Correlation between industry entry and exit rates

	All	*KBI-OECD*	*KBI-MMG*
Pearson correlation	0.632**	0.738**	0.485**

Notes
** Correlation is significant at the 0.01 level (2-tailed).

Table 12.4 Average dimension of firms per age for each entry cohort

	Year of entry	*Firms' age*						*Total no. of firms*
		1	2	3	*4*	*5*	*6*	
All industries	1995	4.0	4.7	5.3	6.0	6.7	7.5	21,410
	1996	4.0	4.8	5.3	5.9	6.4	–	21,582
	1997	3.9	4.8	5.7	6.4	–	–	24,892
	1998	3.9	4.8	5.6	–	–	–	27,990
	1999	3.7	4.6	–	–	–	–	29,573
	2000	3.9	–	–	–	–	–	41,157
	Total	**3.9**	**4.7**	**5.5**	**6.1**	**6.6**	**7.5**	**166,604**
KBI-OECD	1995	3.7	4.4	6.0	7.8	9.9	14.1	1,750
	1996	3.9	4.9	5.7	6.0	6.7	–	1,743
	1997	4.4	5.9	6.7	7.4	–	–	2,055
	1998	5.0	7.2	8.6	–	–	–	2,299
	1999	4.7	6.0	–	–	–	–	2,389
	2000	6.2	–	–	–	–	–	3,508
	Total	**4.9**	**5.8**	**6.9**	**7.1**	**8.3**	**14.1**	**13,744**
KBI-MMG	1995	3.0	3.6	4.2	4.9	5.3	5.7	1,964
	1996	3.3	4.0	4.4	5.0	5.5	–	2,066
	1997	3.8	4.9	5.4	6.5	–	–	2,413
	1998	4.6	5.6	6.9	–	–	–	2,713
	1999	3.9	4.9	–	–	–	–	3,008
	2000	6.3	–	–	–	–	–	4,617
	Total	**4.5**	**4.7**	**5.4**	**5.5**	**5.4**	**5.7**	**16,781**

Table 12.5 Percentage of firms per number of years of duration for each entry cohort

	Year of entry	*Firms' duration (no. of years)*						*Censored**	*Total*
		1	*2*	*3*	*4*	*5*	*6*		
All industries	1995	18.4	9.7	7.7	6.2	5.5	6.1	46.4	100.0
	1996	17.3	9.4	8.1	7.1	7.7	–	50.4	100.0
	1997	17.4	9.8	8.8	9.5	–	–	54.5	100.0
	1998	16.6	11.0	11.8	–	–	–	60.6	100.0
	1999	18.7	14.4	–	–	–	–	66.9	100.0
	2000	22.6	–	–	–	–	–	77.4	100.0
	Total	**18.9**	**12.1**	**7.6**	**4.9**	**3.0**	**1.7**	51.8	**100.0**
KBI-OECD	1995	14.3	8.4	6.7	5.6	6.0	5.3	53.7	100.0
	1996	13.0	7.9	6.7	6.5	7.0	–	58.9	100.0
	1997	13.7	8.1	7.2	6.7	–	–	64.3	100.0
	1998	10.7	8.5	9.4	–	–	–	71.4	100.0
	1999	15.0	9.4	–	–	–	–	75.6	100.0
	2000	18.4	–	–	–	–	–	81.6	100.0
	Total	**14.6**	**9.8**	**6.6**	**4.3**	**3.0**	**1.5**	60.2	**100.0**
KBI-MMG	1995	14.6	8.0	6.5	5.9	5.4	5.9	53.7	100.0
	1996	13.3	7.8	7.2	6.9	7.1	–	57.7	100.0
	1997	14.6	7.6	7.1	6.8	–	–	63.9	100.0
	1998	11.4	8.3	9.5	–	–	–	70.8	100.0
	1999	14.9	9.5	–	–	–	–	75.6	100.0
	2000	17.8	–	–	–	–	–	82.2	100.0
	Total	**14.9**	**9.9**	**6.5**	**4.2**	**2.7**	**1.5**	60.3	**100.0**

Notes

* The observations relating to the duration of firms are considered as censored when a firm was still active after the end of the period under consideration (i.e. 2000). We use the files related to 2001 and 2002 to check for censoring.

the two KBI groups did so; similarly, almost two-thirds of KBI firms survived after the fourth year, while the survival rate for the whole set of firms is about 56.5 per cent.

To summarize, the descriptive statistics discussed above suggest that the dynamic behaviour of Portuguese knowledge-based firms and industries presented some distinctive features during the second half of the 1990s. In particular, it was shown that KBIs have grown faster, their price–cost margin has been higher, and firms within these industries have survived longer than the average. On the other hand, there seem to be some differences between the two KBI groups, namely in terms of entry rates.

In the following sections, we seek to understand to what extent such patterns are caused by the different general conditions and characteristics of the firms/industries that are included in each group or, alternatively, by differences in the factors that determine firm entry and firm survival in each industry type.

4 Theoretical considerations on panel data and survival models

The available data are organized in two dimensions – a time dimension and a cross-sectional dimension – which is the typical framework of panel data models. These two sources of variation will allow us to disentangle more accurately the effects of selected variables on industry entry. Consider the following benchmark model:

$$y_{it} = \mathbf{X}_{it}\beta + \varepsilon_{it}$$

where

$$\varepsilon_{it} = \alpha_i + \gamma_t + \eta_{it} \qquad (12.1)$$

The subscripts refer to the individual – i – and to time – t. The dependent variable will be either the industry entry rate or the firm growth rate. $\mathbf{X}$ stands for a $(1 \times k)$ vector of explanatory variables and ε is the disturbance term. This model is also known as an error components model due to its underlying error structure – it comprises three distinct variables, all of which are independent and identically distributed with mean zero, but with different variances, namely σ^2_α σ^2_γ and σ^2_η. We assume that the error term is uncorrelated with the regressors and that each of the first two components will model unobserved effects that are either specific to individuals or time. The final component is a purely random error. This specification neatly accommodates the possible presence of individual autocorrelation.

The model for the determinants of entry rates will be estimated by maximum likelihood. Our preference for a random effects model over a fixed effects one depends mainly on a twofold argument. First, we want to consider time-invariant variables as regressors and only a random effects specification allows us to estimate parameters associated with those variables. Second, our dataset is rather lengthy in cross-sectional terms but short in terms of its time dimension. Estimating a fixed effects model under such circumstances would drastically decrease the degrees of freedom. Nevertheless, our specification will be tested against other standard models.

There is still one important issue to be discussed concerning the data panel models applied to the analysis of entry determinants. In these models, the entry rate will be the dependent variable. By definition, the entry rate is constrained to lie between zero and one. This fact alone implies inherent heteroskedasticity of the model. However, the severity of this problem is not very great. For the sake of simplicity a linear model is preferred although the model may predict entry rates outside the (0, 1) interval.

While the study of entry is based here on data panel models, the determinants of new-firm survival are investigated using a piecewise-constant exponential hazard model, a semi-parametric type of approach to the statistical analysis of duration data (Jenkins 2004; Lancaster 1990).

A central concept in this type of analysis is the *hazard function.* In the present context, the hazard function corresponds to the instantaneous probability of a firm exiting the industry at time t, given that it actually stayed in the market until t. Formally,

$$h(t) = \lim_{\Delta t \to 0} \frac{P(t \leq T < t + \Delta t | T \geq t)}{\Delta t} = \frac{f(t)}{1 - F(t)} = \frac{f(t)}{S(t)} \tag{12.2}$$

where $f(t)$ is the probability density function, $F(t)$ is the distribution function, $S(t)$ is the survival function (i.e. the probability that a firm will survive after t).

One model that has often been used in this context is the Proportional Hazards (PH) model. Such a framework is characterized by the fact that it satisfies a separability assumption:

$$h(t, X) = h_0(t)e^{X'\beta} \tag{12.3}$$

where $h_0(t)$ is a 'baseline hazard' function which depends on t and $e^{X'\beta}$ is an individual specific non-negative function of covariates **X**, which does not depend on t. This property greatly simplifies the estimation of the model; it implies that:

1 the pattern of 'duration dependence' is monotonic and common to all firms (i.e. the probability of survival is monotonically increasing or monotonically decreasing); and
2 the role of firms' characteristics and other covariates (such as industry characteristics or macroeconomic conditions) is to scale up or down the survival–duration profile.

The basic idea underlying the piecewise-constant hazard model, which we will be using below, is the following. The time axis is partitioned into a number of intervals using cut-points (which are chosen by the researcher – in the present case, each interval corresponds to one year), and it is assumed that the baseline hazard is constant within each interval, although it may differ between intervals. The hazard function therefore becomes:

$$h(t;X)=\begin{cases} h_1e^{X_1'\beta} & 0<t\leq c_1 \\ h_2e^{X_2'\beta} & c_1<t\leq c_2 \\ \dots & \\ h_Ke^{X_K'\beta} & c_{K-1}<t \end{cases} \tag{12.4}$$

where the time axis is divided into K intervals by points c_1, c_2 ..., c_{K-1}. Besides the already mentioned flexibility concerning the shape of the hazard function (note that there will be one baseline hazard for each interval), this specification provides a relatively simple way of incorporating time-varying covariates – a relevant feature in the context of the present chapter, where we have data for the same variables for successive years.

Regarding the likelihood function, it is worth noting that we are dealing with annual data. This means that we do not know the exact time t at which firms exit the market; we only know the year interval in which exit (or censoring) occurs. Let $\{c_k\}$ represent, as before, the end points of the K intervals into which the data is grouped (with $c_0=0$ and $c_K=\infty$). Thus, the individual contribution to the likelihood function will be:

$$L_i=[f(c_k)^{\delta_i}S(c_{k-1})^{1-\delta_i}]^{d_{ik}} \tag{12.5}$$

with

$$\begin{cases} d_{ik}=1 & \text{if firm's i's exit or censoring happen in the interval }]c_{k-1}, c_k] \\ d_{ik}=0 & \text{otherwise} \end{cases}$$

and

$$\begin{cases} \delta_i=1 & \text{if firm i exits in the interval }]c_{k-1}, c_k] \\ \delta_i=0 & \text{if firm i is censored in the interval }]c_{k-1}, c_k] \end{cases}$$

Noting that $f(c_i)$, the probability of firm i exiting during the interval, corresponds to

$$f(c_k)=\Pr(c_{k-1}\leq T<c_k)=S(c_{k-1})-S(c_k) \tag{12.6}$$

each firm's contribution to the likelihood function becomes

$$L_i=\{[S(c_{k_i-1})-S(c_{k_i})]^{\delta_i}S(c_{k_i-1})^{1-\delta_i}\}^{d_i}=\left\{\left[1-S\frac{(c_{k_i})}{S(c_{k_i-1})}\right]^{\delta_i}-S(c_{k_i-1})\right\}^{d_i} \tag{12.7}$$

It is possible to show (see Lancaster 1990: 176–81) that the elements of equation (12.7) above can be expressed in terms of the hazard function presented in (12.6) as follows:

$$\frac{S(c_{k_i})}{S(c_{k_i-1})} = \begin{cases} \exp\{-h_k e^{X_k^i \beta}(c_k - c_{k-1})\} & \text{if} \quad k = 1, 2, \ldots K-1 \\ 0 & \text{if} \quad k = K. \end{cases}$$

$$S(c_{k_i-1}) = \exp\left\{\sum_{k-1}^{l=1} h_l e^{X_l^i \beta}(c_l - c_{l-1})\right\}, \quad k = 2, \ldots K$$

5 Determinants of industry entry rates

In order to analyse the determinants of entry, we use a model inspired by the seminal paper by Orr (1974). In general terms, we assume that the entry of new firms into a market is influenced by three groups of factors: incentives to entry, entry barriers and the behaviour adopted by incumbents in order to prevent entry. Both datasets available for the purpose of this study allow us to build approximate indicators for each of those groups of factors. Thus, the following explanatory variables were used in the regressions:

- *Price–cost margin*: this is considered as an incentive to entry, being used as a proxy for expected profitability. We expect it to have a positive effect (if any) on entry rates.
- *Industry growth*: this is used as another proxy for expected profitability. We expect it to have a positive effect on entry rates.
- *Median firm size*: this is used as a proxy for scale economies. It is expected to have a negative effect on entry rates, since firms must enter with larger sizes if they are to avoid incurring higher average costs. The barrier to entry stems from the need to raise more capital for start-up purposes, given imperfections in the capital markets.
- *Advertising intensity*: this is used as a proxy for product differentiation. It is expected to have a negative effect on entry rates, since new firms must incur additional advertising costs or lower revenues when entering the market.
- *Hirschman–Herfindhal (HH) index*: higher concentration levels are expected to facilitate collusion among the incumbents. This will make the entry of new firms harder.

The dependent variable in the regressions is the entry rate at the five-digit industry level, computed as the proportion of incumbent firms in each industry at period t which have entered the market in that same period. Table 12.6 shows the bivariate correlations between the variables used in our regressions.

Table 12.6 shows that the price–cost margin, the industry growth rate, the median size of firms per industry and the HH index are all significantly correlated with the industry entry rate, with the expected signs; on the contrary, the bivariate correlation between entry rates and advertising intensity is non-significant.

Table 12.6 Pearson indexes of the bivariate correlations between the determinants of industry entry rates

	Price-cost margin	*Industry growth*	*Median firm size*	*Advertising intensity*	*HH index*
Entry rate	0.056(*)	0.437(**)	−0.367(**)	0.007	−0.090(**)
Price–cost margin		0.097(**)	0.037	−0.060(*)	0.050(*)
Industry growth			−0.157(**)	0.044	−0.014
Median firm size				0.002	0.118(**)
Advertising intensity					0.149(**)

Notes
* Correlation is significant at the 0.05 level. ** Correlation is significant at the 0.01 level.

Although the PCM is admittedly a rather crude proxy for expected profitability, we can see that it is significantly and positively related with the rate of industry growth and the HH index (as is to be expected), and that it is significantly and negatively related to advertising intensity, possibly suggesting that intensive advertising is associated here with tougher competition (and, accordingly, with higher costs and/or lower revenues).

While many of the explanatory variables are significantly correlated between one another, the Pearson correlation indexes (in absolute values) are not sufficiently high to raise problems of multicollinearity in a multivariate regression context.

Table 12.7 shows the results of the regressions for the complete dataset of industries.

In column 1, we can see that all explanatory variables have the expected impact on entry rates, although the parameter relating to the intensity of advertisement is not significant. It is worth noting that the individual (i.e. industry)

Table 12.7 Regression results for the determinants of industry entry rates (all industries included)

	1 (All industries)	*2 (All industries)*	*3 (All industries)*
Price–cost margin	0.020882**	0.020480**	0.018573**
Industry growth	0.096306**	0.095980**	0.095321**
Median firm size	−0.002548**	−0.002569**	−0.002509**
Advertising intensity	−0.024738	−0.025961	−0.055734
Hirschman-Herfindhal index	−0.000009**	−0.000009**	−0.000009**
KBI-OECD		0.004708	
KBI-MMG			0.037608**
Intercept	0.128008**	0.127717**	0.124947**
$Sigma^2$	0.002740**	0.002740**	0.002624**
Schwarz B.I.C.	−3,004.16	−3,000.66	−3,011.23
Log likelihood	3,033.83	3,034.04	3,044.61
Number of observations	1,685	1,685	1,685

Notes
** significant at 5%.

effect is quite relevant, suggesting that, on average, there are particular features of each industry not captured by the explanatory variables used in the regression that play a relevant role in explaining the entry rates; this is evident when one compares the SBIC statistics of the chosen models with the statistics of models estimated by the plain ordinary least squares method or models estimated by the within-group estimator even after accounting for the yearly effects (usually associated with fixed effects models – see the results in the Appendix).

In order to assess whether being a KBI can influence the level of the entry rate in each industry – after taking into account the five explanatory variables listed above – we include in regressions 2 and 3 a dummy variable for KBIs (in regression 2 we use the OECD classification, and in regression 3 our own classification). The results shown in Table 12.7 indicate that, unlike the KBI–OECD group, the KBI–MMG group is positively and significantly associated with higher entry rates at the industry level (even when several determinants of entry are taken into account).

We are also interested in understanding the extent to which the determinants of entry differ for distinct groups of industries. More specifically, we want to analyse whether the factors that affect entry in KBIs are different from the ones that affect entry in general. For this purpose, we ran separate regressions for each group of KBIs using the same explanatory variables as before. The results of these regressions are shown in Table 12.8.

Regressions 4 and 5 show us that, contrary to the general case, the impact of the PCM was not significant as a determinant of entry rates within knowledge-based industries in Portugal during the period 1995–2000. On the other hand, the growth rate of industries and the median size of firms both affect entry rates in all types of industries, irrespective of their knowledge intensity. Finally, while the level of market concentration had only a rather small, although significant, effect on entry in the general case, its impact is non-significant in the case of the KBI–MMG group.

We will leave the discussion of these results for the concluding section of the chapter. Before that, however, we will discuss the determinants of firm survival.

Table 12.8 Regression results for the determinants of industry entry rates (separate results for different groups of industries)

	1 (All industries)	*4 (KBI-OECD)*	*5 (KBI-MMG)*
Price–cost margin	0.020882**	0.004539	−0.003443
Industry growth	0.096306**	0.065881**	0.058102**
Median firm size	−0.002548**	−0.001762**	−0.000195**
Advertising intensity	−0.024738	0.070825	−0.223831
Hirschman–Herfindhal index	−0.000009**	−0.000021**	−0.000005
Intercept	0.128008**	0.137150**	0.166514**
Sigma^2	0.002740**	0.003798**	0.004201**
Log likelihood	3,033.83	385.383	268.016
Number of observations	1,685	220	170

Notes
** significant at 5%.

6 Determinants of new-firm survival

In the last two decades, there has been a proliferation of studies on the determinants of new-firm survival. Firm survival has been found to be robustly related with firm-specific variables such as size and age – which typically have a negative impact on survival chances (e.g. Audretsch and Mahmood 1995; Dunne *et al.* 1989; Mata and Portugal 1994; Wagner 1999) – and less robustly related with other industry-specific and macroeconomic variables.

The specific relationship between survival and the age of firms which is identified in the empirical exercises can be very sensitive to the models used for estimations. As was pointed out in Section 4 above, the Proportional Hazards (PH) model is often used in studies on the determinants of new-firm survival, largely due to the simplicity of its estimation. However, this model implicitly assumes that the pattern of 'duration dependence' is monotonic, i.e. that the probability of survival monotonically increases or monotonically decreases with a firm's age – which need not be the case.

In fact, while there are reasons to believe that age makes firms less prone to failure – e.g. it may take time for firms to discover the true worth of their competencies (as suggested by Jovanovic 1982) or to develop specific knowledge, trust and appropriate routines (as put forward by organizational ecologists; see Carroll and Hannan 2000) – it is not obvious that this relationship has to be monotonic. In fact, it can be argued that new firms will stay in the market until such time as they have exhausted the initial resources that they gathered for the new venture. This may imply that the probability of survival will decrease in the years immediately after entry, and increase only after new firms have either succeeded in stabilising their cash-flows or, alternatively, exhausted their start-up financial resources and therefore exited the market. The model we use here, the piecewise-constant hazard model, does not impose any specific form on the relationship between survival and age, therefore allowing for the identification of possible non-monotonic shapes for such a relationship.

In addition to age and size, we have included as an explanatory variable the proportion of graduates (as a proxy for human capital). Furthermore, we have controlled for the annual growth rate of the industry that each firm belongs to, as well as for the annual GDP growth (in order to account for the influences of the general economic climate upon firms' survival chances).

Tables 12.9 and 12.10 present the results of the regressions, with the hazard rate being used as the dependent variable (see Section 4).

Table 12.9 presents the regression results for the determinants of new-firm hazards, considering all firms that entered the market between 1995 and 2000. The results of regressions 6 to 8 reveal that the relationship between survival and the age of firms is in fact non-monotonic: the hazard rates of firms increase from the first to the fourth year, and then decrease with age thereafter. In regression 6, all other determinants of firm survival have the expected impact – a firm's survival is positively (or the hazard rate is negatively) related with its size, its

Table 12.9 Regression results for the determinants of new firm hazards (all firms that entered between 1995 and 2000 are included)

	6 (All new firms)	*7 (All new firms)*	*8 (All new firms)*
Firm age = 6 years	1.10065**	0.544396**	0.446487**
Firm age = 5 years	1.59327**	0.969400**	0.855364**
Firm age = 4 years	1.76050**	1.08398**	0.961047**
Firm age = 3 years	1.70149**	1.03472**	0.913830**
Firm age = 2 years	1.61168**	0.964947**	0.848016**
Firm age = 1 year	−0.646441**	−0.614049**	−0.613525**
Current size	−0.002843**	−0.002847**	−0.002809**
Proportion of graduates	−0.165898**	−0.020851	0.027766
Industry growth	−0.090384**	−0.090328**	−0.088400**
GDP growth	−0.698987**	−0.532830**	−0.501292**
KBI-OECD		−0.092620**	
KBI-MMG			−0.099781**
Schwarz B.I.C.	163,939	164,697	165,106
Log likelihood	−163,879	−164,631	−165,039
Number of observations	166,604	166,604	166,604

Notes
** significant at 5%.

Table 12.10 Regression results for the determinants of new firm hazards (separate results for firms belonging to different groups of industries)

	6 (All new firms)	*9 (New firms in KBI-OECD)*	*10 (New firms in KBI-MMG)*
Firm age = 6 years	1.10065**	3.42088*	1.09060**
Firm age = 5 years	1.59327**	4.71319*	1.71049**
Firm age = 4 years	1.76050**	4.87920*	1.84687**
Firm age = 3 years	1.70149**	4.78600*	1.77675**
Firm age = 2 years	1.61168**	4.51827*	1.64271**
Firm age = 1 year	−0.646441**	−0.841141*	−0.820166**
Current size	−0.002843**	−0.004981*	−0.002819**
Proportion of graduates	−0.165898**	−0.126927*	−0.086692**
Industry growth	−0.090384**	−0.187366*	−0.133616**
GDP growth	−0.698987**	−1.65182*	−0.809479**
Schwarz B.I.C.	163,939	11,148.0	13,745.1
Log likelihood	−163,879	−11,100.3	−13,696.5
Number of observations	166,604	13,744	16,781

Notes
** significant at 5%.

human capital endowment, the growth of the industry that it belongs to and GDP growth – with all coefficients being statistically significant at a 5 per cent level.

Introducing dummy variables in regressions 7 and 8 for the two KBI groups renders the human capital variable insignificant (while the coefficients of such dummies are both significant and have the expected sign). This fact can have two alternative explanations: either the proportion of graduates is in fact not important per se, only appearing to be significant in regression 6 due to the exclusion of what would be the 'true' determinant of firm survival (i.e. being or not a part of a KBI); or, alternatively, the fact that the inclusion of the KBI dummies renders the human capital coefficient statistically insignificant may just be a result of the non-negligible correlation between the latter variables and each of the KBI dummies.

One way to sort out this issue is to run separate regressions for different groups of firms, according to the type of industry that they are part of. Table 12.10 displays the results of such regressions.

Table 12.10 shows that the main results discussed for the general case (regression 6) remain essentially unchanged, regardless of the industry group under analysis (regressions 9 and 10): the hazard rates are non-monotonically related with age (increasing until the fourth year and decreasing thereafter) and negatively related with the size of firms, the proportion of graduates among the labour force and the growth rates of both the specific industry and the economy as a whole.

7 Conclusions

This chapter has compared the dynamics of knowledge-based industries (KBIs) with the universe of all market-oriented industries in Portugal between 1995 and 2000. Our main goal was to discuss the extent to which KBIs differ from other industries in relation to some basic stylized facts and regularities (concerning firm entry, and post-entry growth and survival) which have been identified in a number of previous studies.

In our comparative exercise, we took two classifications of knowledge-based industries, the one put forward and used by the OECD (KBI–OECD) (OECD 1999) and another one, suggested by us (KBI–MMG), based on the proportion of graduate workers in each industry. The data analysed stems mainly from the Portuguese social security database ('Quadros de Pessoal') based on an annual survey of all firms with paid workers, and was complemented with data on costs and revenues at the industry level from the INE, the national statistical office.

The descriptive analyses of the data have basically confirmed a number of well-known statistical regularities concerning the evolution of firms and industries, namely: entry and exit are quite pervasive in all sorts of industries; the firms' average size grows as they grow older; and the proportion of firms leaving the market in the first few years after entry is quite high.

In the same vein, the regression analyses have confirmed some stylized results, in particular:

- The entry of new firms responds positively to incentives (in particular to industry growth) and negatively to barriers to entry (namely, the presence of economies of scale measured in terms of the median firm size in each industry).
- The survival of new firms is positively related with size, even after controlling for human capital endowment, industry growth, and GDP growth.

As far as the impact of age on firm survival is concerned, we found a non-monotonic relationship, with the hazard rates of firms increasing up to the fourth year after entry and declining thereafter. While this result deviates from the frequently found positive relationship between the age and survival of firms, this deviation may well be explained by the fact that many previous studies on firm survival did not allow for a non-monotonic relationship between the two variables.

More relevant to the aim of this chapter is the discussion of the specificities of KBIs in the domains just mentioned. In fact, we found that KBIs (regardless of how they are classified) and the firms within them show some signs of distinctiveness in their dynamics in comparison with the general case. In particular, on average:

- KBIs have higher growth rates.
- Entry within the KBI groups is less responsive to incentives (contrary to the general case, the impact of the PCM on entry is not significant; and the impact of industry growth, though significant, is less expressive).
- KBI firms have higher survival chances.
- The survival of firms belonging to KBIs is notably more dependent on industry growth than in the general case.

These results do much more than simply confirm the idea that knowledge-based industries were fast-growing industries in the second half of the 1990s. They seem to suggest that the creation of firms in these industries responds to incentives in a different way than in other industries, a result with potentially relevant policy implications. In particular, with the variables accounting for profitability expectations and entry rates not showing the same pattern as in the general case, incentive-based policies (e.g. tax incentives for new technology-based firms) might be less effective in promoting entrepreneurship in these industries than in other cases (in other words, the determinants of new-firm creation are probably more closely related with the availability of relevant skills and competences than with market incentives).

These results also seem to suggest that the survival of firms in knowledge-based industries is more susceptible to downturns in the business cycle, highlighting the need to pay greater attention to the sustainability of new knowledge-based enterprises when devising mechanisms for the support of such activities.

Appendix

Table 12.A1 Pooled regression results (estimation by OLS)

	1 (All industries)	*2 (All industries)*	*3 (All industries)*
Price–cost margin	0.017412	0.016900	0.007924
Industry growth	0.133204**	0.132938**	0.123093**
Median firm size	−0.002440**	−0.002449**	−0.002401**
Advertising intensity	0.014832	0.012862	−0.104619
Hirschman–Herfindhal index	−0.000005	−0.000005	−0.000006
KBI-OECD		0.001557	
KBI-MMG			0.039507**
Intercept	0.123856**	0.123832**	0.122829**
Schwarz B.I.C.	−2,527.40	−2,523.76	−2,563.69
Log likelihood	2,549.69	2,549.76	2,589.70
Number of observations	1,685	1,685	1,685

Notes
** significant at 5%.
* significant at 10%.

Table 12.A2 Effects regression results (estimation by OLS)

	1 (All industries)	*2 (All industries)*	*3 (All industries)*
Price–cost margin	0.027740	Estimation not possible due to variables being constant in time	Estimation not possible due to variables being constant in time
Industry growth	0.025084		
Median firm size	−0.003702**		
Advertising intensity	−0.137820		
Hirschman–Herfindhal index	−0.000010		
KBI-OECD			
KBI-MMG			
Intercept	0.123856**		
Schwarz B.I.C.	−2,284.25		
Log likelihood	3,554.70		
Number of observations	1,685		
F-test of individual intercepts	9.179**[1]		

Notes
** significant at 5%.
* significant at 10%.
1 the null is accepted at a 5% level, meaning that the individual intercepts are statistically significant.

Notes

1 Corresponding author.
2 Dinâmia and ISCTE Lisbon University Institute, Avenida das Forças Armadas, 1649–026 Lisboa, Portugal (ricardo.mamede@iscte.pt).
3 INE (Instituto Nacional de Estatística) Av. António José de Almeida 1000–043 Lisboa, Portugal (daniel.mota@ine.pt).
4 UECE and ISEG (Instituto Superior de Economia e Gestão, Universidade Técnica de Lisboa) Rua Miguel Lupi, 1200 Lisboa, Portugal (mgodinho@iseg.utl.pt).
5 The available data does not allow for the use of any other proxy for industry growth at the level of sectoral disaggregation that we are working with.

References

Audretsch, D. and Mahmood, T. (1995) 'New firm survival: new results using a hazard function', *Review of Economics and Statistics*, vol. 77 (1): 97–103.

Cabral, L.M.B. and Mata, J. (2003) 'On the evolution of the firm size distribution: facts and theory', *American Economic Review*, vol. 93 (4): 1075–90.

Carroll, G. and Hannan, M. (2000) *The Demography of Corporations and Industries*, Princeton: Princeton University Press.

Caves, R. (1998) 'Industrial organization and new findings on the turnover and mobility of firms', *Journal of Economic Literature*, vol. 36 (4): 1947–82.

Dosi, G., Malerba, F., Marsili, O. and Orsenigo, L. (1997) 'Industrial structures and dynamics: evidence, interpretations and puzzles', *Industrial and Corporate Change*, vol. 6 (1): 3–24.

Dunne, T., Roberts, M.J. and Samuelson, L. (1989) 'The growth and failure of US manufacturing plants', *Quarterly Journal of Economics*, vol. 104 (4): 671–98.

Geroski, P.A. (1995) 'What do we know about entry?', *International Journal of Industrial Organization*, vol. 13 (4): 421–40.

Jenkins, S.P. (2004), *Survival Analysis.* Unpublished manuscript, Institute for Social and Economic Research, University of Essex, Colchester, UK.

Jovanovic, B. (1982) 'Selection and the evolution of industry', *Econometrica*, vol. 50 (3): 649–70.

Lancaster, T. (1990) *The Econometric Analysis of Transition Data*, Cambridge, MA: Cambridge University Press.

Mata, J. and Portugal, P. (1994) 'Life duration of new firms', *Journal of Industrial Economics*, vol. 42 (3): 227–45.

OECD (1999) *OECD Science, Technology and Industry Scoreboard: Benchmarking Knowledge-based Economies 1999*, Paris: Organisation for Economic Co-operation and Development.

OECD (2005) *OECD Science, Technology and Industry Scoreboard 2005*, Paris: Organisation for Economic Co-operation and Development.

Orr, D. (1974) 'The determinants of entry: a study of the Canadian manufacturing industries', *Review of Economics and Statistics*, vol. 56 (1): 58–66.

Portugal, P. and Cardoso, A.R. (2006) 'Disentangling the minimum wage puzzle: an analysis of worker accessions and separations', *Journal of the European Economic Association*, vol. 4 (5): 988–1013.

Wagner, J. (1999) 'The life history of cohorts of exits from German manufacturing', *Small Business Economics*, vol. 13 (1): 71–9.

13 Entrepreneurship and growth

David B. Audretsch and Max Keilbach

1 Introduction

The decade of the 1990s was fraught with economic challenges for Europe. Growth was stagnant and unemployment ratcheted upwards throughout the decade. Thus, it was not surprising when the European Union introduced a new initiative at the turn of the century to reignite growth and sustainable high-quality employment in Europe. However, the focus of that European growth policy may have been less anticipated – knowledge-based entrepreneurship. Romano Prodi (2002: 1), who at the time served as President of the European Commission, proclaimed that the promotion of entrepreneurship was a central cornerstone of European economic growth policy: 'Our lacunae in the field of entrepreneurship needs to be taken seriously because there is mounting evidence that the key to economic growth and productivity improvements lies in the entrepreneurial capacity of an economy.' With the 2000 Lisbon Proclamation, the European Council made a commitment to becoming not just the leader in knowledge but also the entrepreneurship leader in the world in order to ensure prosperity and a high standard of living throughout the continent (Audretsch 2007a and 2007b).

Europe was not alone in focusing on entrepreneurship as a key factor generating economic growth. From the other side of the Atlantic, Mowery (2005) observes:

> During the 1990s, the era of the 'New Economy', numerous observers (including some who less than 10 years earlier had written off the U.S. economy as doomed to economic decline in the face of competition from such economic powerhouses as Japan) hailed the resurgent economy in the United States as an illustration of the power of high-technology entrepreneurship. The new firms that a decade earlier had been criticized by such authorities as the MIT Commission on Industrial Productivity (Dertouzes *et al.* 1989) for their failure to sustain competition against large non-U.S. firms were seen as important sources of economic dynamism and employment growth. Indeed, the transformation in U.S. economic performance between the 1980s and 1990s is only slightly less remarkable than the failure of most experts in academia, government, and industry, to predict it.
>
> (Mowery 2005: 1)

The purpose of this chapter is to explain why public policy across the globe has increasingly been turning to entrepreneurship to provide an engine of economic growth. Entrepreneurship has emerged as a missing link to economic growth policy, because investments in new knowledge may not automatically spillover for commercialization and economic growth. Rather, by providing a conduit for the spillover of knowledge, entrepreneurship enables public policy to accrue a return on its investments in knowledge.

In particular, this chapter provides empirical evidence showing that:

1 entrepreneurship tends to be greater in higher knowledge contexts, and
2 entrepreneurship has a positive impact on economic growth.

While the first provides insight as to why entrepreneurship has become more prevalent as knowledge emerges as a key factor generating competitiveness in the global economy, the second explains why entrepreneurship has become such an important goal for public policy.

2 Entrepreneurship as opportunity recognition and action

The prevalent theoretical framework for analysing entrepreneurship within the economics literature has been the general model of income choice or what has become known as entrepreneurial choice. The model of entrepreneurial choice has its roots in Knight (1921), but was more recently extended and updated by Lucas (1978), Kihlstrom and Laffont (1979), Holmes and Schmidt (1990) and Jovanovic (1994). According to the model of entrepreneurial choice, individuals face a choice of earning their income either from wages earned through employment in an incumbent enterprise or else from profits by becoming an entrepreneur and founding a new enterprise.

The decision-making process of individuals on whether or not to pursue an entrepreneurial opportunity by starting a new firm, E, is characterized by the model of occupational (or entrepreneurial) choice, where E reflects the decision to become an entrepreneur, π^* is the profits expected to be accrue from starting a new firm, and w is the anticipated wage that would be earned from employment in an incumbent enterprise.

$$E = f(\pi^* - w) \tag{13.1}$$

Accordingly, there is a solid research tradition primarily in labour economics, focusing on the decision confronting individuals to start a firm. Theory and empirical evidence provide compelling reasons to conclude that characteristics specific to the individual help shape the cognitive processes guiding the entrepreneurial decision which is characterized by the model of income or entrepreneurial choice.

3 The knowledge filter and entrepreneurial opportunities

The literature on entrepreneurship has generally focused on the cognitive process by which individuals recognize opportunities and then make the decision to become an entrepreneur by starting a new firm. However, relatively little is known about the actual source of such entrepreneurial opportunities. In most models of entrepreneurship, entrepreneurial opportunities are assumed to be exogenous. After reviewing the entrepreneurship literature, Plummer *et al.* (2007) conclude 'A notable lack of research focused on the origins of opportunity', while Companys and McMullen (2007) suggest that:

> Despite the advances and the importance of entrepreneurial opportunities to strategy and entrepreneurship, there have been surprisingly few recent studies that explore the nature of opportunities ... Currently, scholarly understanding of the origins of entrepreneurial opportunities remains limited owning to the fact that most studies have taken opportunities for granted when exploring strategic and entrepreneurial processes.
>
> (Companys and McMullen 2007)

Audretsch and Keilbach (2007a and 2007b) and Audretsch *et al.* (2006) identify one particular source of entrepreneurial opportunities. According to Audretsch and Keilbach (2007a and 2007b) and Audretsch *et al.* (2006), in drawing upon Audretsch's (1995) knowledge spillover theory of entrepreneurship, new knowledge and ideas created in an incumbent firm but not completely and exhaustively commercialized by that incumbent firm, provide a valuable source of entrepreneurial opportunities. However, in order to appropriate the value of the knowledge underlying those opportunities, the individual or team of individuals are not able to do so in an incumbent firm and therefore must enter into entrepreneurship by starting a new firm. By starting a new firm, that is through entrepreneurship, the entrepreneur is both serving as an agent of change and also as a conduit for the spillover of knowledge from the firm in which it was created to a new firm where it is commercialized.

The reason why a new firm is needed for the spillover and commercialization of knowledge is because of the existence of what Audretsch *et al.* (2006) and Acs *et al.* (2004) term 'the knowledge filter'. As Arrow (1962) emphasized, new knowledge is inherently characterized by uncertainty, asymmetries and a high cost of transaction across agents. An implication of Arrow's characterization of new knowledge is that agents across a decision-making hierarchy may disagree about the value of any new idea, resulting in the knowledge filter.

While the knowledge filter impedes the commercialization resulting from investments in knowledge it also creates entrepreneurial opportunities. According to the knowledge spillover theory of entrepreneurship (Audretsch and Keilbach 2007a and 2007b), opportunities for entrepreneurship are the duality of the knowledge filter. The higher is the knowledge filter, the greater are the divergences in the valuation of new ideas across economic agents and the decision-making hierarchies of incumbent firms. It is these divergences in the valuation of ideas across agents and their inability to penetrate the knowledge filter that generates entrepreneurial opportunities. When individuals choose to pursue those opportunities through entrepreneurship, entrepreneurship serves as the conduit for the spillover of knowledge.

4 Linking entrepreneurship to economic growth

The starting point for models of economic growth in the Solow tradition is that the factors of production are exogenous. Adding the factor of knowledge, as in the Romer, Lucas and Jones models,

$$Q_i = h(t)f(C_i, L_i, K_i) \qquad (13.2)$$

where Q represents economic output, C is the stock of capital, L the labour force and K is knowledge capital. As the plethora of studies have shown, the inclusion of the factor of new knowledge in the model, shifts the spatial unit of observation towards a more localized unit of analysis, which is represented by the subscript i. In the Romer, Lucas and Jones models, knowledge automatically spills over and is commercialized,

$$K = Kc \qquad (13.3)$$

reflecting the Arrow observation about the non-excludability and non-exhaustive properties of new knowledge. Thus, investments in R&D and human capital not only automatically impact output, but in a multiplicative manner, because of its external properties, suggesting that new knowledge, K, is tatamont to economic knowledge, K^c.

As discussed above, the emphasis on, or rather assumption about the non-excludability property is better suited for information rather than knowledge. Information has, by its definition, a very low level of uncertainty and its value is not greatly influenced or shaped by asymmetries across economic agents possessing that information. Thus, information can certainly be characterized as being non-excludable and non-exhaustive. By contrast, as Arrow points out, there is a gap between new knowledge and what actually becomes commercialized, or new economic knowledge, $K - K^c > 0$. In fact, the knowledge filter, θ, is defined as the gap existing between investments in knowledge and the commercialization of knowledge, or economic knowledge.

$$\theta = K^c / K \qquad (13.4)$$

It is the existence of the knowledge filter, or knowledge not commercialized by incumbent enterprises, that generates the knowledge spillover entrepreneurial opportunities. As long as the incumbent enterprises cannot exhaust all of the commercialization opportunities arising from their investments in knowledge, opportunities will be generated for potential entrepreneurs to commercialize that knowledge by starting a new firm. By applying knowledge created in one organizational context to create a new organizational context – a new firm – the new entrepreneurial start up serves as a vehicle for the spillover of knowledge, λ, where

$$\lambda = E/K \tag{13.5}$$

and E reflects entrepreneurial start-ups commercializing knowledge spilling over from an incumbent firm, university or other source investing in the production of new knowledge. Nelson and Winter (1982) emphasized that the knowledge conditions will tend to shape the competitive advantage of incumbents versus new entrepreneurial startups in commercializing opportunities emanating from knowledge investments made by incumbents. Winter (1984) distinguished between what he termed as the routinized technological regime from the entrepreneurial technological regime (Breschi *et al.* 2000).

According to Winter (1984), knowledge conditions bestowing the innovative advantage to large incumbent firms are characterized by the routinized regime. Underlying knowledge conditions bestowing the innovative advantage to entrepreneurial small firms and start-ups are characterized by the entrepreneurial regime.

The decision-making process of individuals on whether or not to pursue an entrepreneurial opportunity by starting a new firm, E, was characterized by equation 13.1 in the model of occupational (or entrepreneurial) choice previously explained. As explained above, most of the literature has generally taken the source of entrepreneurial opportunities for granted, or assumed that it is exogenous. However, according to the knowledge spillover theory of entrepreneurship, investments in knowledge by incumbent firms and other organizations, such as universities and research institutes, provide an important source of entrepreneurial opportunities, as long as the knowledge filter exists.

If g represents entrepreneurial opportunities which have nothing to do with new knowledge or ideas, then

$$\bar{E} = (\pi^*(g) - w) \tag{13.6}$$

where $\bar{E}$ represents entrepreneurship from traditional sources of opportunity, that is non-knowledge sources, such as regional growth. The knowledge spillover theory of entrepreneurship suggests adding two additional factors as influencing entrepreneurial opportunities. The first additional source is the amount of new knowledge and ideas being produced in incumbent firms and organizations. The second additional source is the magnitude and extent of the knowledge filter, which will constrain the extent to which new investment in new knowledge can spill over for commercialization. These two additional factors shape the extent of knowledge spillover entrepreneurship – the amount of investment in creating new knowledge, K, and the magnitude of the knowledge filter, θ. Thus, knowledge spillover entrepreneurship, E^* is given by

$$E^* = (\pi^*(K, \theta) - w) \tag{13.7}$$

However, the perception of entrepreneurial opportunities may be necessary but not sufficient to induce knowledge spillover entrepreneurship. A broad range of barriers, that can be legal, institutional and historical in nature, can serve as a barrier to entrepreneurship (Audretsch 2007). Thus, equation 13.7 needs to modified to include the impact of such barriers to entrepreneurship, β,

$$E^* = (1/\beta)(\pi^*(K, \theta) - w) \tag{13.8}$$

where β represents barriers to entrepreneurship. The existence of such barriers, or a greater value of β, explains why economic agents would choose not to enter into entrepreneurship, even when endowed with knowledge that would otherwise generate a potentially profitable opportunity through starting a new firm.

Since $\bar{E} < E$, the total amount of entrepreneurial activity exceeds that generated by knowledge spillovers. Combining entrepreneurial activity from such traditional sources with knowledge spillover entrepreneurship yields the total amount of entrepreneurship,

$$E = \bar{E} + E^* \tag{13.9}$$

or

$$E = (1/\beta)(\pi^*(g, K, \theta) - w) \tag{13.10}$$

Thus, the first hypothesis to emerge from the knowledge spillover theory of entrepreneurship is that $\frac{dE}{dK} \geq 0$, or what we term as the endogenous entrepreneurship hypothesis, which suggests that the degree of entrepreneurship will tend to be positively related to the degree of new knowledge investments in a particular context.

The second hypothesis emerging from the knowledge spillover theory of entrepreneurship is the growth hypothesis, which suggests that, because entrepreneurship serves as a conduit of knowledge spillovers, economic growth will be greater in a context with a greater degree of entrepreneurial activity, *ceteris paribus*.

5 Empirical evidence

In this section empirical evidence is provided for testing both the growth hypothesis and the endogenous entrepreneurship hypothesis. To test these two hypotheses a two-equation model is estimated where the first equation is a Cobb–Douglas function of the form

$$Y_i = K_i^{\alpha} L_i^{\beta} R_i^{\eta} E_i^{\gamma} \qquad (13.11)$$

where Y is economic performance of region i, measured as GDP, K is the region's i endowment with capital, L is labour and E represents the region's entrepreneurship. Hence, this specifies formally that entrepreneurship contributes to the economic output of regions. The variables used to estimate equation 13.11 are:

- *Output* (Y_i) is measured as Gross Value Added corrected for purchases of goods and services, VAT and shipping costs, measured in year 2000. Statistics are published every two years for *Kreise* by the Working Group of the Statistical Offices of the German Länder, under 'Volkswirtschaftliche Gesamtrechnungen der Länder'.
- *Physical Capital* (K_i) is the stock of capital used in the manufacturing sector of the *Kreise* and has been estimated using a perpetual inventory method, which computes the stock of capital as a weighted sum of investments done in the producing sector in the period 1980 to 2000. In the estimates we used a β-distribution with $p=9$ and a mean age of $q=14$. For a detailed description of this procedure see, for example, Audretsch and Keilbach (2003).
- *Labour* (L_i) is measured using data on from the Federal Labour Office, Nürnberg, that report the number of employees liable to social insurance by *Kreise*.
- Knowedge is measured by *R&D* (R_i) and is the *number of employees engaged in R&D* in the public (1992) and in the private sector (1991) as a share of total employment. The data are from the *Stifterverband für die Wissenschaft* under obligation of secrecy. A measure of *spatial lags* for R&D is also included to control for the impact of spatial knowledge spillovers from adjoining regions. This variable is constructed using a binary spatial weight matrix **W** that by pre-multiplying to the vector of R_i yields $\widetilde{R}_i$, the sum of R&D activities of neighbouring regions for each region i. In the regressions, the Cobb–Douglas Function is augmented by this variable $\widetilde{R}_i$.
- *Entrepreneurship* (E_i) is measured using new-firm start-up rates. The ZEW in Mannheim has provided the start-up rate data, which are provided biannually by *Creditreform*, the largest German credit-rating agency. In addition, we also use two additional measures of entrepreneurship. The first measures only start-up activity in high-technology manufacturing industries (with an R&D-intensity exceeding 2.5 per cent). The second measure includes only start-ups in ICT industries.

There are reasons to suspect that entrepreneurship may be endogenous and must be estimated endogenously. In particular, studies have found that entrepreneurship is not influenced by a number of regional-specific characteristics, such population density (a positive impact on start-up rates), population growth (positive impact on start-up rates), skill and human capital levels of the labour force (positive impact), and mean establishment size (negative impact on start-up rates). In order to endogenize the entrepreneurship rate, a second equation is specified:

$$E_i = f(y_i, x_i) \tag{13.12}$$

where y_i is a set of variables describing economic growth and x_i is a vector of other regional-specific characteristics influencing the level of entrepreneurship.

The model is estimated for all 440 German counties ('Kreise'). All of the variables measures are for the year 2000 unless stated otherwise. Measures of the variables used to estimate the entrepreneurship rate are based on 'Statistik Regional', the CD-Rom that is published by the German Regional Statistical Offices.

GDP growth measures general economic opportunities that would also include (non-knowledge) entrepreneurial opportunities. GDP growth is measured as $g_Y = \ln(Y_{t1} - Y_{t0})/(t_1 - t_0)$, with $t_0 = 1992$ and $t_1 = 2000$, measuring the regions' average growth rates between years 1992 and 2000. Rather than the stock measure of GDP, GDP growth is a measure of the regions' past economic performance.

Knowledge generated in a region is measured by *R&D intensity*, which is measured as the number of researchers in the public and private sectors, relative to the number of all employees. *Population density* is measured as the number of inhabitants in each county relative to its surface.

Governments can influence entrepreneurship in a region through programmes promoting start-ups. The impact of such policies is reflected by using a measure of *subsidies per person*, which is the sum of all subsidies in a region divided by the population of the region. The *unemployment rate* for the region is included to control for the impact of unemployment on entrepreneurship.

To control for the impact of taxes on entrepreneurship, a measure of taxes is included. While taxes are influenced more by federal rather than state tax policy, *regional business tax level* reflects the regional component of business taxes.

There are two types of diversity included as influencing entrepreneurship in a region. The first is the extent of social diversity, which may influence the degree of entrepreneurship in a region. The measure used to capture social diversity is based on regional voting behaviour for the 1998 German Parliament vote. The measure is an entropy index that takes into account all major political parties but also smaller ones (six in total). The variable is transformed to [0, 1], where zero indicates no variety and the maximum variety is indicated by one.

The second type of diversity reflects the extent of diversity in the economic activity of the region, *industrial diversity* which is measured by a Herfindhal index. The degree to which a region is hospitable and offers amenities is reflected by a measure of *locational attractiveness*. While it is not easy to capture the extent to which a region is attractive, this measure is the number of a region's hotel beds divided by the size of the region, which is a proxy measure for the attractiveness of the region.

Entrepreneurship in a given region may be influenced by the degree of entrepreneurship in adjoining regions. To control for the spatial structure of entrepreneurship across regions, spatial lags of entrepreneurship capital are included. The spatial variables are calculated by pre-multiplying a spatial weight matrix **W** with the different measures of E_i.

The two regression equations are estimated simultaneously using the method of three-stage least squares. This estimation procedure enables a correction for the endogeneity bias in the estimation occurring

The empirical results from estimating the three-stage regression results are presented in Table 13.1.

The empirical results generally support the growth hypothesis. Not only do labour, physical capital and R&D influence economic performance, but the degree of entrepreneurial activity does as well. Those regions exhibiting a higher degree of entrepreneurial activity also have a higher level of growth. This is consistent with the hypothesis that entrepreneurship is positively linked to economic growth.

Similarly, the empirical results are consistent with the knowledge spillover theory of entrepreneurship. After controlling for regional-specific factors, both high-tech entrepreneurship and ICT entrepreneurship are greater in those regions exhibiting higher amounts of knowledge, as measured by R&D. This is consistent with the hypothesis that higher knowledge contexts have a greater extent of entrepreneurial activity than do low knowledge contexts.

6 Conclusions

According to the knowledge spillover theory of entrepreneurship, entrepreneurship serves as a key conduit of knowledge spillovers and therefore promotes economic growth. It should be recognized that there are many other important mechanisms which also transmit knowledge spillovers, such as the mobility of scientists and workers, and informal networks, linkages and interactions. Similarly, there are certainly new firms started that have nothing to do with the spillover of knowledge. Still, the spillover theory of entrepreneurship suggests that there will be additional entrepreneurial activity as a rationale and cognitive response to the creation of new knowledge. Those contexts with greater investment in knowledge should also experience a higher degree of entrepreneurship, *ceteris paribus*.

By endogenously facilitating the spillover of knowledge created in a different organization and perhaps for a different application, entrepreneurship may provide what Audretsch *et al.* (2006) and Acs *et al.* (2004) term as the missing link to economic growth. Confronted with a formidable knowledge filter, public policy instruments emerging from the new growth theory, such as investments in human capital, R&D and university research, may not adequately result in satisfactory economic growth. One interpretation of the European Paradox, where

Table 13.1 Estimating the impact of entrepreneurship on economic growth

	Dependent variable: GDP			
Constant	0.1965 (0.95)	0.3117 (1.38)	−0.3803 (−1.51)	0.0677 (0.33)
Capital	0.1037*** (4.65)	0.1297*** (5.79)	0.1345*** (5.63)	0.0930*** (4.13)
Labour	0.7976*** (31.87)	0.7419*** (29.15)	0.7425*** (26.33)	0.8168*** (31.96)
R&D	0.0335*** (3.98)	0.0290*** (3.35)	0.0440*** (5.08)	0.0371*** (4.43)
Spatial lag of R&D	0.0018 (0.24)	−0.0019 (−0.25)	0.0015 (0.19)	0.0012 (0.17)
General entrepreneurship	0.3054*** (8.18)			
High-tech entrepreneurship		0.1762*** (7.80)		
ICT entrepreneurship			0.0789*** (3.56)	
Low-tech entrepreneurship				0.2843*** (7.60)
'R-sq'	0.9408	0.9402	0.9370	0.9405
	All	*High-tech*	*ICT*	*Low-tech*
Economic output	−0.0132 (−0.67)	0.0324 (1.12)	−0.0014 (−0.05)	−0.0198 (−0.97)
Economic growth	0.2275 (2.34)	0.4520*** (3.19)	0.0746 (0.56)	0.1988 (1.99)
R&D intensity	1.2913 (1.43)	6.1519*** (4.64)	5.3232*** (4.29)	0.2528 (0.27)
Population density	1.0649*** (3.30)	2.0077*** (4.24)	1.8958*** (4.26)	0.8883* (2.68)
Capital investment	−0.0003 (−0.63)	−0.0006 (−0.8)	0.0005 (0.63)	−0.0003 (−0.53)
Subsidies	0.0037 (1.26)	0.0009 (0.21)	−0.0117** (−2.79)	0.0052 (1.71)
Unemployment rate	0.0009 (0.26)	−0.0406*** (−8.03)	−0.0525*** (−10.81)	0.0085 (2.51)
Business tax	−0.0635 (−1.67)	−0.0222 (−0.4)	−0.0732 (−1.40)	−0.0548 (−1.41)
Social diversity	0.0857 (0.68)	1.0858*** (5.75)	0.2474 (1.43)	−0.0237 (−0.18)
Industrial diversity	0.9795*** (6.60)	1.1909*** (5.46)	1.2875*** (6.29)	0.9579*** (6.28)
Locational attractiveness	0.2672 (2.31)	0.0569 (0.34)	0.2275 (1.45)	0.2927 (2.46)
Spatial lag of 'all'	0.0661*** (2.90)			
Spatial lag of 'high-tech'		0.0707 (2.30)		
Spatial lag of 'ICT'			−0.0099	

Table 13.1 continued

	Dependent variable: GDP			
			(−0.34)	
Spatial lag of 'low-tech'				0.0801***
				(3.42)
Constant	−5.8698***	−9.9603***	−8.7212	−5.8799
	(−19.84)	(−21.36)	(−20.51)	(−19.21)
Pseudo R-sq	0.2777***	0.5058***	0.6369***	0.2588***
Number of observations	433	433	433	433

such investments in new knowledge have certainly been vigorous and sustained, is that the presence of such an imposing knowledge filter chokes off the commercialization of those new investments, resulting in diminished innovative activity and ultimately stagnant growth.

By serving as a conduit for knowledge spillovers, entrepreneurship is the missing link between investments in new knowledge and economic growth. Thus, the spillover theory of knowledge entrepreneurship provides not just an explanation of why entrepreneurship has become more prevalent as the factor of knowledge has emerged as a crucial source for comparative advantage, but also why entrepreneurship plays a vital role in generating economic growth. Entrepreneurship is an important mechanism permeating the knowledge filter to facilitate the spillover of knowledge and ultimately generate economic growth.

References

Acs, Z.J., Audretsch, D.B., Braunerhjelm, P. and Carlsson, B. (2004) 'The missing link: the knowledge filter and entrepreneurship in endogenous growth', Centre for Economic Policy Research (CEPR) Discussion Paper.

Arrow, K.J. (1962) 'Economic welfare and the allocation of resources for invention', in Nelson, R.R. (ed.) *The Rate and Direction of Inventive Activity*, Princeton: Princeton University Press, pp. 609–26.

Audretsch, D.B. (1995) *Innovation and Industry Evolution*, Cambridge, MA: MIT Press.

Audretsch, D.B. (2007a) 'Entrepreneurship capital and economic growth', *Oxford Review of Economic Policy*, vol. 23: 63–78.

Audretsch, D.B. (2007b) *The Entrepreneurial Society*, New York: Oxford University Press.

Audretsch, D.B. and Keilbach, M. (2003) 'Entrepreneurship capital and economic performance', CEPR Discussion Paper no. 3678.

Audretsch, D.B. and Keilbach, M. (2007a) 'The theory of knowledge spillover entrepreneurship', *Journal of Management Studies*, vol. 44 (7): 1242–54.

Audretsch, D.B. and Keilbach, M. (2007b) 'The localization of entrepreneurship capital: evidence from Germany', *Papers in Regional Science*, vol. 86 (3): 351–65.

Audretsch, D.B., Keilbach, M. and Lehmann, E. (2006) *Entrepreneurship and Economic Growth*, New York: Oxford University Press.

Breschi, S., Malerba, F. and Orsenigo, L. (2000) 'Technological regimes and sectoral patterns of innovation', *Economic Journal*, vol. 110 (463): 388–410.

Companys, Y.E. and McMullen J.S. (2007) 'Strategic entrepreneurs at work: the nature, discovery, and exploitation of entrepreneurial opportunities', *Small Business Economics*, vol. 29 (4): 301–22.

Dertouzos, M., Lester, R. and Solow, R. (1989) *Made in America: Regaining the Productive Edge*, Cambridge, MA: MIT Press.

Holmes, T.J. and Schmitz Jr., J.A. (1990) 'A theory of entrepreneurship and its application to the study of business transfers', *Journal of Political Economy*, vol. 98 (2): 265–94.

Jovanovic, B. (1994) 'Entrepreneurial choice when people differ in their management and labor skills', *Small Business Economics*, vol. 6 (3): 185–92.

Kihlstrom, R.E. and Laffont, J.J. (1979) 'A general equilibrium entrepreneurial theory of firm formation based on risk aversion', *Journal of Political Economy*, vol. 87 (4): 719–48.

Knight, F. (1921) *Risk, Uncertainty and Profit*, Chicago: University of Chicago Press.

Lucas Jr., R.E. (1978) 'On the size distribution of business firms', *Bell Journal of Economics*, vol. 9 (2): 508–23.

Mowery, D. (2005) 'The Bayh-Dole Act and high-technology entrepreneurship in U.S. universities: chicken, egg, or something else?', paper presented at the Eller Centre Conference on Entrepreneurship Education and Technology Transfer, University of Arizona, 21–22 January 2006.

Nelson, R.R. and Winter, S.G. (1982) *An Evolutionary Theory of Economic Change*, Cambridge, MA: Harvard University Press.

Plummer, L.A., Hayene, J.M. and Godesiabois, J. (2007) 'An essay on the origins of entrepreneurial opportunity', *Small Business Economics*, vol. 29 (4): 363–79.

Prodi, R. (2002) 'For a new European entrepreneurship', public speech, Instituto de Empresa in Madrid.

Winter, S.G. (1984) 'Schumpeterian competition in alternative technological regimes', *Journal of Economic Behavior and Organization*, vol. 5 (3): 287–320.

Part IV

Policy conclusions

14 Knowledge-intensive entrepreneurship in Europe

Some policy conclusions

Franco Malerba and Nicholas Vonortas

1 Introduction

Several policy conclusions and recommendations emerge from this book. They add to the traditional policies that foster entrepreneurship such as the creation of conditions favourable to the entry of new firms, financial support and venture capital, the support of later stages of firms' growth related to the transformation of those entrants that have been successful.

In contrast to the tradition in entrepreneurship literature, this book has gone beyond the individual-centric approach to emphasize the major role of knowledge, innovation systems, networks and institutions in fostering knowledge-intensive entrepreneurship. This book has taken a system view of knowledge-intensive entrepreneurship. It has viewed entrepreneurship as an activity related to generation, use, coordination and integration of knowledge. Firms have been perceived as knowledge processors, integrating scientific and technological knowledge with knowledge about applications and markets (Cohendet and Llerena 2006). They coordinate their domains of knowledge, their organizational structure and their product range (Brusoni and Prencipe 2006). Knowledge-intensive entrepreneurship has been looked at in its broader form of start-up entrepreneurship, corporate entrepreneurship and academic entrepreneurship. In addition, knowledge has been perceived to be distributed among agents, thus redirecting attention from the individual person or organization to the group/network. Finally, this book has been based on the working hypothesis that knowledge-intensive entrepreneurship is highly affected by the sectoral context, as well as by the national and regional/local innovation systems. That is to say, for instance, that the specific learning environment will affect the type and intensity of knowledge-intensive entrepreneurship through the creation of specific opportunity conditions, the cumulativeness of the advancements, and the intellectual property regime underlying innovations.

This book benefited from prior analyses that had already pointed out the naivety of many entrepreneurship studies that promoted the simplistic view of the rugged individualist entrepreneur who single-handedly built great companies. It thus went a long way from the five common 'myths' regarding entrepreneurship in the United States as these myths were captured by the National Commission on

Entrepreneurship (NCOE) (2001) and Blithe (2000): *the risk-taking myth* (most successful entrepreneurs take wild, uncalculated risks in starting their companies); *the high-tech invention myth* (most successful entrepreneurs start their companies with a breakthrough invention – usually technological in nature); *the expert myth* (most successful entrepreneurs have strong track records and years of experience in their industries); *the strategic vision myth* (most successful entrepreneurs have a well-considered business plan and have researched and developed their ideas before taking action); and the *venture capital myth* (most successful entrepreneurs start their companies with millions in venture capital to develop their idea, buy supplies and hire employees). Instead, this book took into consideration right from the start NCOE's (2002) guidance that the issues ranking at the top of the agenda of new and growing businesses include finding and retaining quality people, accessing capital, intensive networking, and infrastructure and institutional support.

2 Policy recommendations

We divide the policy recommendations of this book into general and more specific recommendations. General recommendations relate to the targets, levels of intervention and context conditions of knowledge-intensive entrepreneurship policy. Specific policy recommendations address issues such as the creation of new knowledge and the establishment of appropriate knowledge infrastructure, the build-up of networks for distributed knowledge, the sectoral context and the differences between sectors that imply differences in the practice of knowledge-intensive entrepreneurship, the institutional setting affecting knowledge-intensive entrepreneurship, and the factors influencing the link between knowledge-intensive entrepreneurship and regional growth.

2.1 General policy recommendations

1 The *broad policy targets* to promote knowledge-intensive entrepreneurship should enable the creation and commercialization of knowledge in various ways and forms (Figure 14.1):

- Start-ups of new and small independent firms.
- Spin-offs and spin-outs from universities and larger established corporations.
- Knowledge-intensive entrepreneurship within larger corporations and technological diversification.
- Public research laboratories.
- Knowledge-intensive entrepreneurship in networks.

2 In order to act on knowledge-intensive entrepreneurship, various *levels of intervention* could be envisaged:

- The individual.
- The firm.

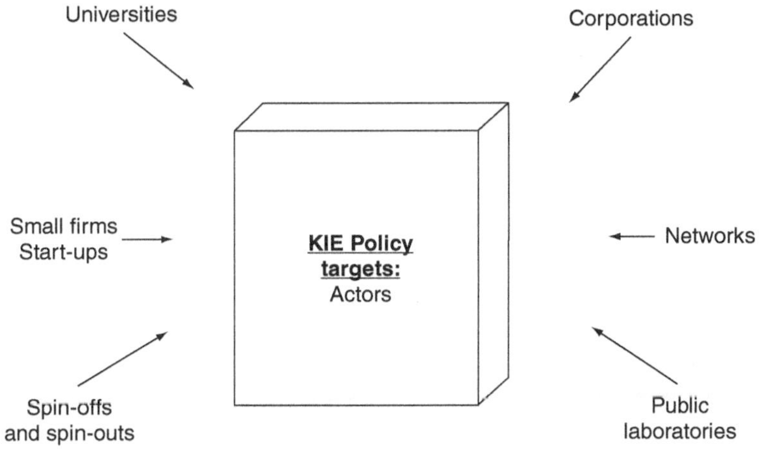

Figure 14.1 Policy targets for Knowledge Intensive Entrepreneurship (KIE).

- Other organizations (universities, banks/VCs).
- Clusters and networks.

3 Policies should be cognizant of the different *context conditions* in which knowledge-intensive entrepreneurship takes place. These include:

- Innovation systems.
- Sectors and sectoral clusters.
- Geographical clusters.

The idiosyncrasies of firm technology and competencies also create significant intra-industry variety.

Given the multidimensionality and context dependence of knowledge-intensive entrepreneurship, then, various policy instruments can be used either singularly or jointly, from R&D support to seed capital or venture capital, to training and human resource formation, to regulation and standardization (Figure 14.2). Moreover, combinations of central policy actors (ministries) and decentralized policy actors (regions, municipalities, local government) must be used extensively.

2.2 Specific policy recommendations

4 *New knowledge* and *knowledge infrastructure* are prerequisites of knowledge-intensive entrepreneurship:

- Basic research.
- Migration from science to commercial sector and back.
- Quantity and quality of both scientific and managerial education.

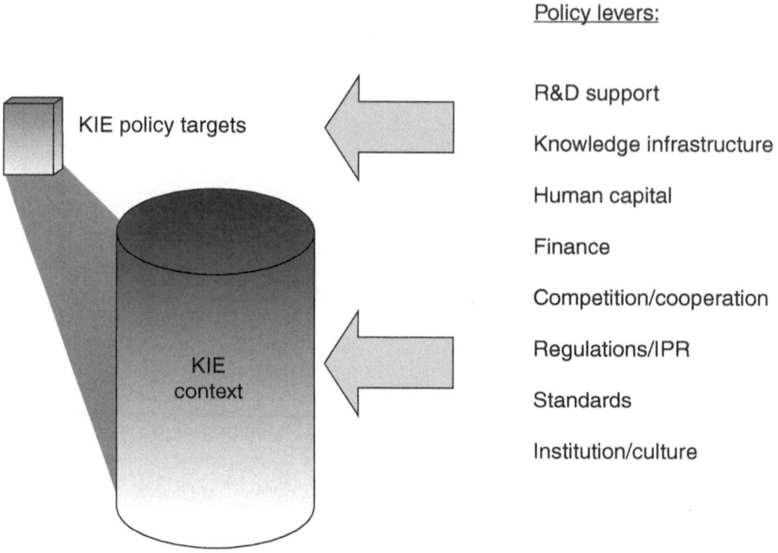

Figure 14.2 Policy instruments for Knowledge Intensive Entrepreneurship (KIE).

- Governance of higher education institutions to promote academic entrepreneurship.
- Effectiveness of IPR retention by universities.

Policies in favour of knowledge-intensive entrepreneurship should support the creation of new knowledge and the building of knowledge infrastructure that will enable the creation and commercialization of knowledge in various ways and forms.

The support of basic research will directly or indirectly foster knowledge-intensive entrepreneurship. There is a certain preference on the part of policy-makers for more applied research. The reason for this preference may be that the returns on public money spent on applied research can more easily be identified. Results of applied research in the form of new technologies and products often turn out to be commercially successful on a rather short term and, thus, to contribute to new employment and economic growth. This book indicates, however, that there is a very substantial commercially valuable contribution also made by basic research which is usually overlooked or mixed up with the educational services of the university system. This contribution is the training on the job of scientists who later migrate into the commercial sphere. Basic research in the sciences and engineering, e.g. in laser technology, material sciences, nano-technology, biochemistry, and many more, is almost always connected with very sophisticated technicalities of constructing and handling equipment, running experiments, testing materials, developing high-end instruments, etc. These technicalities are to a large extent based on non-codified knowledge or knowledge that becomes codified only much later. For this reason, the sophisticated technicalities developed in basic research need to be acquired by participating in basic research, i.e. on the job, usually at a postgraduate qualification level. To develop these technicalities on the one hand and to individually acquire them on the other are usually not profitable activities. However, once these technicalities have been individually acquired on the job, their command is a commercially most valuable resource.

Policies should not make difficult the migration from scientific activities to the commercial sector. Moreover, because of also largely tacit know-how, new scientific knowledge can only be transferred into the commercial sector by the migration of those who carry the knowledge, i.e. by attracting scientists with the most recent technical know-how into the commercial sector. Thus, migration must be understood as complementary to other mechanisms of transferring scientific knowledge that can resort to codified knowledge. Scientists who leave basic research institutions (usually after acquiring a doctoral degree or after a few years of post-doc researcher experience) and who migrate into the commercial sector to a large extent engage in R&D. In fact, commercial R&D very heavily relies on the influx of such researchers. With the innovative potential they carry they can initiate and coordinate the firms' R&D activities in which commercial applications of the latest brand of technicalities are developed.

The prerequisite of the transfer of embodied knowledge is that properly funded, internationally competitive basic research is run in domestic research institutions. The enormous commercial value of the capabilities acquired on the job in doing basic research – that is often not sufficiently recognized by innovation policy – can be assessed by the firms' willingness to pay for the realization of the potential that migrating scientists bear. The upshot of the policy related insight to be derived is then that the creation of a national (or, for that matter, EU-wide) innovative capacity crucially depends on investment in high-quality scientific problem-solving potential and on the indirect support by means of funds going to efficient basic research institutions. This may be an alternative to direct subsidies for specific commercial R&D projects and programmes. This is the more true, the more the generation of innovations rather than their exploitation is the focus of innovation policy. Support of basic research can in the longer term promote the expansion of technological boundaries and simultaneously provide the human input in the form of migrating scientists required by the commercial sector to take advantage of the expanding boundaries. While the former effect of basic research seems well understood, the latter is less so and deserves more attention in deciding on innovation policies.

Policies should aim at increasing the quantity and quality of both scientific as well as managerial education. The main point here is that only scientific education is not enough, but has to be complemented by managerial skills and training. The survival of firms having a university background is below that of all other entrants. It indicates that relative to other entrants, academic entrepreneurs suffer from a lack of market- and customer-related capabilities. This is not to suggest that academic entrepreneurship is a bad thing to have. Rather, one possible implication is that policies should not only focus on the quantity of academic entrepreneurship, but also on its quality, i.e. they should help academics to acquire knowledge on markets and customers, or to find suitable partners.

In order to foster academic entrepreneurship (and academic patenting) policies should pay attention to the governance of higher education institutions. The information one can draw from the chapters in this book regarding academic patenting shows that the percentage of academic patents (i.e. patents whose inventors include at least one university professor) out of the total number of patents in each of the European countries considered (France, Italy and Sweden) is roughly the same figure one can calculate for the United States, but with an important difference: while well over 60 per cent of US academic patents are owned by the universities that employed the academic inventor at the time of the patent, European universities own at most 10 per cent of their scientists' patented inventions. Conversely, well over 60 per cent of European academic patents are owned by business companies, as opposed to around 25 per cent in the United States. These data suggest that common perceptions of European universities as 'good in science, bad in technology transfer' (which are strictly related to the presumed existence of a 'European Paradox') may be flawed. Rather, European universities seem to be as good at technology transfer as their US counterparts (at least to the extent that patents are a good indicator of technology transfer), but they are not as good as them in retaining the Intellectual Property Rights (IPRs) over the transferred technology. This line of reasoning brings us to a change of perspective. The traditional question 'why do European universities not patent as much as the US ones' should be replaced with two different ones: 'why do European universities retain the property of patents over their scientists' invention to a much lesser extent than the US universities?' and 'are there any reasons to encourage European universities to retain more the property of patents over their employees' inventions?' The most immediate answer to the first question points to the legal framework of European universities, which lack the financial autonomy of their US counterparts, especially the private ones. In addition, European university professors are, or regard themselves as, civil servants, whose careers do not depend much on their ability to bring financial resources to the institutes they work for. Finally, countries such as Sweden have retained until recently the so-called 'professor's privilege', which leaves all IPRs over academic research results to the employee (the scientist) instead of the employer (the university). All of these features have a common origin in the institutional profile of higher education and research in most European countries, which sees central governments as the key actors regulating the professors' career and economic incentives, and leaving little margins to universities to act as full-fledged employers, which can set up incentive schemes and also profit from their employees' research. This suggests that, in the absence of more radical and comprehensive policies on the governance of higher education institutions, the current policy trend of targeting university–industry technology transfer and IPR issues per se may have only limited results.

High IPR retention by university is not necessarily desirable. Is a higher rate of 'IPR retention' by universities desirable? If yes, why? Do we expect that patents owned by universities would be better marketed than those owned by business companies? Or that patents owned by universities could become a relevant source for self-financing? Answering these questions will require more research, starting from the chapters in this book that have discussed university patenting. In the future it will be necessary to test whether university-owned patents are better exploited, for commercial purpose, than business-owned ones; and whether they are a meaningful source of revenues for the universities. A more aggressive stance of universities towards 'IPR retention' can create problems in the relationship between firms and universities, as well as among academic scientists. In the first case, tension may arise between firms and universities over the share of IPRs over cooperative research results to be retained by each part. In the second case, conflicts may exist between researchers over inventorship attribution. A research project may result in both a number of publications and one or more patents. However, especially in the life sciences and pharmaceuticals, some authors of the publications are not recognized as inventors of the patent(s), that is, they are given no economic rights over the inventions. The most natural explanation for the mismatch is the existence of 'gift authorship': technicians and academic colleagues who have contributed only marginally or not in a creative way to the research are rewarded with the insertion of their names in the list of authors (a common practice in academia, despite requests in the contrary directions by the editors of academic journals), but are left out of the patent in order to avoid legal problems of mis-attribution. However, junior scientists, who have contributed creatively to the research, are more at risk of not being credited with inventorship than senior scientists whose contribution to the research may be described not so much as 'creative' but rather as 'entrepreneurial'. If confirmed, the phenomenon could lead to the recommendation of setting up, within TTOs or other university offices, an ethical committee devoted to check inventorship attributions.

5 *Networks* are key for the creation and survival of knowledge-intensive entrepreneurship – competencies, resources likely to be widely distributed:

- Facilitate the access by knowledge-intensive entrepreneurship to knowledge resources and stimulate interfaces with the market.
- Support not just individual entrepreneurs but networks (implications for ETPs).

Public policy must recognize that new firms stemming out from knowledge-intensive entrepreneurship cannot be treated as homogenous. The individual firm is highly dependent upon the idiosyncrasies of its technology and competencies, creating great intra-industry variety in the evolution of firms. Policy decision-makers must be aware of these heterogeneities.

Policies should facilitate the access by knowledge-intensive entrepreneurship to knowledge resources and stimulate interfaces between knowledge-intensive entrepreneurship and the market. Complex processes such as R&D are likely to be widely distributed and difficult to access. Second, the very newness and complexity that form the base of the knowledge-intensive entrepreneurship can increase the uncertainty of the venture. This can in turn make it difficult to persuade both suppliers and customers to invest in the knowledge-intensive entrepreneurship's products or services since firms are not willing to pay for the risks unless there are considerable potential gains to be made. Both problems imply that it becomes all the more important for the founders to make use of their personal networks. This both in order to access resources and to try to associate the knowledge-intensive entrepreneurship with well-known firms or individuals that can increase the legitimacy of the firm and thus reduce uncertainty in the eyes of both customers and suppliers. The importance of this has also been observed empirically in start-ups where an intense period of networking has preceded the build-up of acceptance for new technologies to be able to attract customers at all. Second, the relations formed may influence the knowledge-intensive entrepreneurship's perceptions of the opportunity and how to pursue it. At the same time as joint partnerships and collaborative agreements for R&D etc. provide the knowledge-intensive entrepreneurship with resources, the knowledge-intensive entrepreneurship may run the risk of becoming over-embedded in these relations. That is, the close cooperation likely required for the accommodation of, for example, R&D can shelter or blindfold the knowledge-intensive entrepreneurship to new information from other relevant upstream, horizontal or downstream actors.

Two implications follow from this. While the upstream ties are important for the supply of R&D, it is important that these relations do not come to define the business. Rather it is the customers and what customers value that ought to define the business. It is thus important for a knowledge-intensive entrepreneur to cultivate and explore ties with downstream actors from the start. This not least because new firms seldom start out with more than a hypothesis of how to act upon their opportunity. For that reason it is moreover important to try to entertain a set of relations and to keep a balance between exploration and exploitation in both supply and customer oriented interactions. Moreover, it is important not only to form ties but also to be able to cut them. Relations may not only blindfold knowledge-intensive entrepreneurship to new information, but the social exchange developed in strong ties can also come to supersede economic imperatives. As seen in the case of the Xerox spin-off companies, knowledge-intensive entrepreneurship may thus at times be better off at a farther distance from its collaborative partner. However, in distancing itself from one actor, knowledge-intensive entrepreneurship may also run the risk of distancing itself from the larger community of which that actor is a part. In order to facilitate such a shift and not become too reliant on only one actor, it can thus be important to try cultivating a set of strong ties even though this might require much time and resources. These issues of shifting relations are furthermore important also in knowledge-intensive entrepreneurship's progress from emergence to early growth since with the growth and development of knowledge-intensive entrepreneurship, the resource base is likely to shift and with that also knowledge-intensive entrepreneurship's requirements. Finally, it is also important to keep in mind that the growth of knowledge-intensive entrepreneurship is influenced by its organizational legacy. Depending on the origin, the appropriability of the technology, the need for complementary assets etc., different strategic choices have to be made, each of which affects the network formed by knowledge-intensive entrepreneurship to mobilize resources. For knowledge-intensive entrepreneurship it thus becomes important to assess the restrictions and opportunities imposed on it by its specific technology-base and context.

Translating these conclusions into policy implications it is possible to highlight at least two issues. The first issue concerns the influence conveyed in relations. Given the importance of information for matching the technology with the market, it is important for knowledge-intensive entrepreneurship to access both kinds of information. However, depending on the origin, knowledge-intensive entrepreneurship may be at a disadvantage from the outset to access relevant market information. For example, one might assume that a significant portion of founders of knowledge-intensive entrepreneurship are researchers. Since personal networks are historically determined to some extent it is further plausible to assume that the networks of these researchers more often extend to other researchers than to downstream actors such as customers. As a consequence, knowledge-intensive entrepreneurship originating from, for example, academia may be overly influenced by information from upstream actors and may find it difficult to access relevant downstream information. The resulting policy implication is that it should be beneficial to provide such knowledge-intensive entrepreneurship with market interfaces in terms of, for example, forums or technology licensing offices (a function for which TLOs in Sweden at least are poor providers). The other way around, such forums could also allow potential customers to spot technological opportunities.

Public policy should support not just individual entrepreneurs but networks. Since relations not only influence the perceptions and actions of individual firms, but also those of communities at large it may be insufficient to provide help to single knowledge-intensive entrepreneurship. Instead, efforts need to be directed at the networks of firms and individuals that constitute these communities in order to create acceptance or change a prevailing behaviour.

6 Policies must take into account the *sectoral context* of knowledge-intensive entrepreneurship. Sectoral differences may concern:

- Type of knowledge.
- Actors involved in innovation.
- Networks.
- Institutions.

Policy intervention is especially important when knowledge-intensive entrepreneurship innovate in new (emerging) technological areas that do not reinforce, but rather overturn, incumbent firms' asset values.

Policies should take into account the sectoral context in which knowledge-intensive entrepreneurship takes place. In fact sectors differ very much in terms of type of knowledge, actors involved in innovation, networks and institutions. So the conditions affecting knowledge-based entrepreneurship could be radically different from sector to sector, as this book has shown. Being a new firm in biotechnology requires a different set of skills and relationships with the university than being a new firm in software. While the problem of mobilizing resources to identify, act upon and realize opportunities is common for knowledge-intensive entrepreneurship whatever the sector, different industries likely provide different environments to form networks. The reason is that a knowledge-intensive entrepreneurship's technology-base opens up for action along some dimensions, and restricts it along others. From a policy point of view it thus becomes important to direct measures that take the different environments into consideration.

In relation to the previous point, policy may be especially important when knowledge-intensive entrepreneurship innovates into new (emerging) technological areas that do not reinforce, but rather overturn, incumbent firms' asset values. In such cases, the ability to form networks with incumbent firms at least is likely to be particularly difficult and the need for policy support thus greater.

7 The importance of the *institutional setting* cannot be overemphasized.

- New member countries need support.
- Open standards may be important in complex environments (decreasing uncertainties).
- Policies that obstruct the spin-off process, such as non-compete clauses, should be reconsidered.

New member countries are in higher need of policies that support knowledge-intensive entrepreneurship, as several chapters in this book have shown.

Policies should support open standards. Governments and firms acting in committees can reduce technological uncertainty and stimulate innovation diffusion. In particular, the greater the openness in the standard, the greater the number of firms that can act as agents of diffusion in convincing third-party governments to adopt the standard.

Policies that obstruct the spin-off process, such as non-compete clauses, should be reconsidered. Spin-offs should be seen as legitimate targets of policies supporting entrepreneurial activities. Spin-offs from industry incumbents are widespread and tend to be successful. This book has shown that the determinants of spin-off frequency in Europe are similar to the US laser industry. They suggest that spin-offs are fundamentally based on employee learning and knowledge transfer. In this or other industries, they have been shown to increase product variety, to help overcome organizational inertia, to re-utilize capabilities of failed firms (including renewed employment opportunities), and to give rise to regional clustering. In the least, policy should not block spin-offs.

8 Both supply-side and demand-side factors are important to linking knowledge-intensive entrepreneurship to regional growth. Regions need both sets of factors to achieve high levels of economic growth.

- Supply side: contexts rich in knowledge will tend to generate more entrepreneurial activities (knowledge spillover theory of entrepreneurship).
- Demand side: a lot of factors, institutions, laws, traditions and cultures shape the amount of region-specific entrepreneurship capital.

Policy can affect both supply and demand factors.

The entrepreneurship capital of a society has been defined as the regional milieu of agents and institutions that is conducive to entrepreneurial behaviour, i.e. being ready to bear risks of developing uncertain new knowledge to new products or technologies. Entrepreneurship capital reflects a number of different legal, institutional and social factors. As this book has shown, regions with a high degree of entrepreneurship capital facilitate the start up of new knowledge-intensive firms. Moreover, entrepreneurship can influence economic performance through the increase in competition and the increased diversity in a region. All three processes – knowledge spillovers, increased competition and increased diversity in a region – contribute to economic growth.

While a region with a high endowment of entrepreneurship capital can be expected to show higher levels of economic performance, the inverse relationship can also be expected to hold. A strong economic performance will provide a larger market for all products and, thus, it will provide a higher level of entrepreneurial opportunities in general. That is, it will increase the region's entrepreneurship capital. In addition, by providing opportunities, a high level of knowledge creation endogenously increases a region's knowledge-based entrepreneurship capital. Whereas economic performance can be considered as a general driver of entrepreneurship capital, new knowledge is a specific driver of knowledge-based entrepreneurship.

3 By the way of a conclusion

The topic of knowledge-intensive entrepreneurship cannot be exhausted in one book. This book has taken the broadest approach possible, looking at knowledge-intensive entrepreneurship in the form of new company creation, incumbent corporate entrepreneurship, and academic entrepreneurship paying attention to sectoral systems, regional/national systems, and institutions. While the policy implications are, in our view, compelling, the project has just scratched the surface and left several aspects for further investigation. For example, we did not deal extensively at all with the very important topic of knowledge-intensive entrepreneurship finance. Similarly we did not discuss in detail the regulatory frameworks, legislation, tax policies and social customs (Figure 14.3). Such issues will be left for future research.

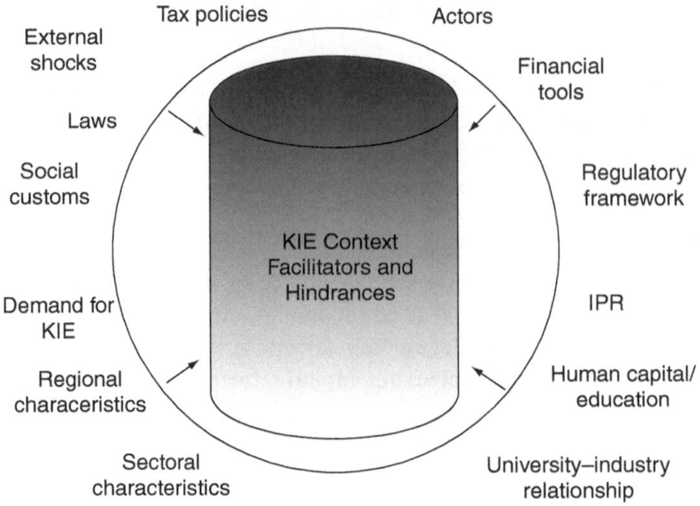

Figure 14.3 Factors affecting Knowledge Intensive Entrepreneurship (KIE).

This book, however, has produced important overall messages which should guide future analyses of knowledge-intensive entrepreneurship and the required policies. They include:

Systems: knowledge-intensive entrepreneurship is an inherently systemic policy issue that must be tackled from various angles simultaneously for long-term results. Single policy fixes will work for a short while but will not change attitudes for the long haul.

Socio-economic incentives: the simplistic view of the rugged individualist entrepreneur who takes uncalculated risks and single-handedly builds great companies is far from reality. Builders of knowledge-intensive enterprises respond to economic and social incentives that can be influenced to a significant extent by policy.

Competence building: the issue at hand – the building of new activities and new structures – invariably depends on achieving 'new combinations' of capabilities and competence building.

Knowledge application: knowledge-intensive entrepreneurship is about the application of knowledge to new activities. Policies need to focus on application rather than on the creation of new knowledge.

References

Blithe, A. (2000) *The Origin and Evolution of New Businesses*, New York: Oxford University Press.

Brusoni, S. and Prencipe, A. (2006) 'Making design rule: a multi-domain perspective', *Organization Science*, vol. 17 (2): 179–89.

Cohendet, P. and Llerena, P. (2006) 'Knowledge-based entrepreneur: the need for a relevant theory of the firm', KEINS, Knowledge base entrepreneurship: institutions, networks and systems, EU project no. CT2–CT–2004–506022, deliverable 6.

National Commission on Entrepreneurship (2001) *Five Myths about Entrepreneurs*, Washington, DC: NCOE.

National Commission on Entrepreneurship (2002) *Entrepreneurship: a Candidate's Guide*, Washington, DC: NCOE.

Index

References to notes are prefixed by *n*. Illustrations not included in the text page range are indicated by *italics*.